DYNAMIC PROGRAMMING

Applications to Agriculture and Natural Resources

DYNAMIC PROGRAMMING
Applications to Agriculture and Natural Resources

JOHN O. S. KENNEDY

School of Economics, La Trobe University
Melbourne, Australia

ELSEVIER APPLIED SCIENCE PUBLISHERS
LONDON and NEW YORK

ELSEVIER APPLIED SCIENCE PUBLISHERS LTD
Crown House, Linton Road, Barking, Essex IG11 8JU, England

Sole Distributor in the USA and Canada
ELSEVIER SCIENCE PUBLISHING CO., INC.
52 Vanderbilt Avenue, New York, NY 10017, USA

WITH 50 TABLES AND 45 ILLUSTRATIONS

British Library Cataloguing in Publication Data

Kennedy, John O.S.
Dynamic programming: applications to agriculture and natural resources
1. Farm management 2. Dynamic programming
I. Title
630′.68 S561

Library of Congress Cataloging in Publication Data

Kennedy, John O.S.
Dynamic programming

Bibliography: p.
1. Agricultural resources—Management—Mathematical models
2. Natural resources—Management—Mathematical models
I. Title
S494.5 M3K46 1986 630′.68 85-31110

ISBN 0-85334-424-8

Printed in Great Britain by Galliard (Printers) Ltd, Great Yarmouth

To

Ann, Jessica and Vanessa

PREFACE

Humans interact with and are part of the mysterious processes of nature. Inevitably they have to discover how to manage the environment for their long-term survival and benefit. To do this successfully means learning something about the dynamics of natural processes, and then using the knowledge to work with the forces of nature for some desired outcome. These are intriguing and challenging tasks. This book describes a technique which has much to offer in attempting to achieve the latter task.

A knowledge of dynamic programming is useful for anyone interested in the optimal management of agricultural and natural resources for two reasons. First, resource management problems are often problems of dynamic optimization. The dynamic programming approach offers insights into the economics of dynamic optimization which can be explained much more simply than can other approaches. Conditions for the optimal management of a resource can be derived using the logic of dynamic programming, taking as a starting point the usual economic definition of the value of a resource which is optimally managed through time. This is set out in Chapter 1 for a general resource problem with the minimum of mathematics. The results are related to the discrete maximum principle of control theory. In subsequent chapters dynamic programming arguments are used to derive optimality conditions for particular resources.

A second reason for learning about dynamic programming is the important practical one that it provides a means of solving dynamic and stochastic resource problems numerically. It is often said that the range of types of dynamic programming problem is so great that it is not possible to develop standard solution routines in the way routines have been developed to solve other types of problem such as linear programming problems. The task of having to devise a computer program, albeit a straightforward one, has perhaps deterred some from applying dynamic programming. However, certain processes are common to the solution of many different types of dynamic programming problem. The use of some general purpose dynamic programming (GPDP) routines written in BASIC is explained at the end of each chapter in Part II after the exposition of methodology. These sections can be skipped without loss of continuity in the development of the methodology. The GPDP routines cannot be used to exploit the special structures of some problems, so there is no claim that they can be used to solve all problems efficiently. However, they can be used to reduce the tedium of solving relatively simple problems numerically.

Since Bellman expounded and popularized dynamic programming in the 1950s, several texts on dynamic programming with management applications have appeared. However, the management applications have been mainly to business and industry. There does not seem to have been a dynamic programming text published with applications to the management of agricultural and natural resources. Books on operations research usually devote a chapter or two to dynamic programming. The treatment is inevitably limited, seldom progressing for example to the solution of infinite-stage problems. Students could even be forgiven for forming the impression that one of the uses of dynamic programming is as a highly inefficient method of solving linear programming problems.

The techniques of dynamic programming are explored in Parts I and II. They are illustrated by applying them to a simple crop-irrigation problem which is developed progressively throughout Part II.

In Parts III and IV the nature of some of the problems in managing particular resources, and how dynamic programming can be and has been applied to aid in their management, are considered. Part III deals with agricultural applications and Part IV with natural resource applications. There is a worked example application in each chapter.

The book should be of interest to those who want to know more about dynamic programming and about how it can be applied to solving agricultural and natural resource problems. It should also be suitable as a student text. A minimum course would consist of Chapters 1 to 3 covering basic methodology, followed by some subset of the application chapters in Parts III and IV.

Mention should be made of three points of presentation. For clarity of exposition, the arguments of functions are enclosed in curly brackets.

Secondly, forward instead of backward numbering of stages is used in formulating the dynamic programming recurrence equations. Whilst this may not be so mathematically concise, it may avoid some mental somersaults. Thirdly, the terms 'uncertainty' and 'risk' are often used interchangeably. There is a convention of reserving uncertainty for outcomes for which there are no known probabilities of their occurrence. However, I accept that decision makers can and should employ subjective probabilities, whether these are based on objective probabilities if they exist, or on well-considered degrees of belief.

Finally, retaining responsibility for remaining shortcomings, I would like to thank all those who have commented on drafts at various stages and who have offered encouragement, including Lee Anderson, John Dillon, John Freebairn, Nick Hastings, Bharat Hazari, Greg O'Brien, John Quilkey, Keshav Vishwakarma and Jim Watkins. Particular thanks go to Oscar Burt for his many useful suggestions, and for sharing some of his philosophy. Typists Sandra Barnes, Helen Cook and Heather Watkins deserve special mention for their good work in typing the original manuscript under difficult circumstances.

JOHN O.S. KENNEDY

CONTENTS

III Dynamic Programming Applications to Agriculture

V Conclusion

Appendices

(For details of ordering these programs and other files on disk, please refer to the order form at the back of the book).

PART I

Introduction

Chapter 1

THE MANAGEMENT OF AGRICULTURAL AND NATURAL RESOURCE SYSTEMS

1.1 The Nature of Agricultural and Natural Resource Problems

Human survival and welfare have always depended on the successful management of agricultural and natural resources. However, over time the management of these resources has become more complex as a consequence of growth of various sorts. Increases in world population and material standards of living circumscribe old notions of natural resources being virtually limitless. As Boulding (1966) has argued in *The Coming of Spaceship Earth*, until recently man could concentrate on maximizing resource flows through the economy, without much concern for the impact on resource stocks. It is increasingly important to consider the extent to which future consumption may be restricted as a result of current usage. In other words, the *user costs* of the current consumption of resources should be evaluated.

Along with population and economic growth there has been an increase in the scale of exploiting agricultural and natural resources. For example, widespread land clearing, fertilizing and fishing have placed strains on the maintenance of the resource base, leading to problems of pollution and damage to natural habitats.

Another type of growth has been the increase in technical knowledge about the exploitation and conservation of resources. For example, research continues to reveal more about the conversion of feed inputs to meat output, and about the migration and reproduction of fish. Because better decisions can usually be made with more information, the management of resources entails the processing of more information than there was previously.

The question of what is successful management is more difficult to answer now that it is widely recognized that the sum of individuals' interests may not be the same as the interest of society. Problems at the individual level are more likely to take account of environmental constraints imposed by society, and problems at the aggregate level are more likely to be formulated with multi-dimensional goals.

Operations research techniques can be used to tackle the increased complexity of resource management. Many resource problems entail decisions which are sequential, risky and irreversible. Dynamic programming is a versatile technique with considerable scope for helping to solve such problems.

The management of immobile and inanimate resources such as minerals involves exploration and extraction decisions which are clearly sequential and subject to risk. The management of living resources is even more complex because they interact with other natural systems which are uncontrollable and imperfectly understood. For example, fish respond to changes in temperature and currents, which in turn are affected by and influence the weather. Decisions are further complicated by the changing state of the biological unit as it passes through phases of birth, growth, reproduction and death.

The relationship between outputs and inputs not only changes through time, but is also uncertain. Uncertainty is increased if inputs such as water and solar energy are determined by the weather, if the biological units are free ranging or migratory, and if the biological units are part of a food chain which maintains a dynamic equilibrium with other living systems.

Another feature of the management of agricultural and natural resource systems is that decisions can have irreversible effects on the natural resource base. In the case of inanimate resources there are examples of decisions which are for all intents and purposes irreversible, such as to exhaust a mine or exploit a wilderness area. The decisions are technically reversible but are economically irreversible because the costs of reversal are unacceptably high. Georgescu-Roegen (1971) has interpreted the second law of thermodynamics as support for the proposition that the consumption of a resource in the economic process irreversibly reduces the usefulness or value of the resource for any further consumption.

Decisions with respect to living systems are often reversible. Living systems are typically well-buffered and are capable of reacting against adverse stimuli. Biomass which is destroyed will often regenerate automatically given time. However, decisions which lead to the extinction of species are technically irreversible, at least at the present stage of genetic engineering.

In summary, current developments in the usage of natural resources are increasingly raising complex and socially important issues. The next section briefly reviews some of the management techniques used to study resource problems, and considers the scope for the greater use of dynamic programming. In Section 1.3 the control variables employed in resource management are broadly classified prior to the formulation of a generalized resource problem in Section 1.4. Optimality conditions for the generalized problem are derived using the reasoning of dynamic programming. The aim is to provide some intuitive insight into the conditions required for intertemporal optimality.

Besides being used as an analytical device, dynamic programming is also used as a numerical solution technique. Numerical solution methods are explained in Chapters 2 and 3. The types of resource problem amenable to numerical solution are introduced briefly in Section 1.5.

1.2 Management Techniques Applied to Resource Problems

Given the variety of types of management problem encountered in agriculture and natural resource industries it is not surprising that all of the common operations research techniques, including dynamic programming, have been applied to problems in these industries. What is surprising is that dynamic programming is not used more frequently than it is. Some of the special problems that occur in agriculture, forestry and fisheries, and the techniques used to solve them, are examined in turn.

1.2.1 Farm management

There is a long tradition of applying operations research techniques to solving farm management problems (Throsby, 1974; Martin, 1977). However, it is probably fair to say that the most popular techniques are linear programming for whole-farm management, and marginal analysis for single-enterprise management.

Linear programming was originally used to solve farm planning problems which were deterministic and static. The linear programming framework has been extended to solve more complex whole-farm problems which may be characterized as dynamic, stochastic, non-linear or discrete. Whilst linear programming is computationally much more efficient than dynamic programming for solving deterministic problems with a linear objective function and linear constraints, dynamic programming may be more suitable for solving more intractable problems.

Examples of the use of marginal analysis for solving static, deterministic single-enterprise problems are widespread in texts dealing with production economics and farm management. The classic application is the decision on the optimal application of fertilizer to a growing crop. To the extent that the fertilizer problem can be typified as a single-period, deterministic, point-input, point-output process, straightforward calculus techniques can be applied.

Marginal analysis has been extended to stochastic, single-period, single-enterprise problems in order to provide further insight into the theory of the firm (Magnusson, 1969) and to aid the efficient management of farm enterprises (Dillon, 1977; Anderson *et al.*, 1977). If the statistical moments of financial outcome as a function of input levels are known and utility is a well-defined function of financial outcome, then stochastic utility maximization problems can be solved using calculus.

The extension of marginal analysis to the solution of multi-period single-enterprise problems has been less straightforward. Dynamic systems are controlled by the timing as well as the level of inputs and outputs. Dillon (1977, p.97) notes that some early work in livestock production functions 'did not really comprehend the problem of profit maximization over time'. He gives the marginal conditions which must hold for the solution of point-input, point-output processes. The necessary marginal conditions for more complex problems are best obtained by formulating the problem in an optimal-control framework and applying the maximum principle. This approach is described in Section 1.4 for the discrete-time case. However, in practice solution is often difficult.

There is scope for the wider adoption of control theory formulations of crop and livestock problems. In general, time in such formulations is best treated as a discrete variable because decisions are made at intervals. There is a range of solution techniques besides dynamic programming for solving problems formulated in this way, such as iterative gradient methods. However, dynamic programming is a technique particularly suited for obtaining numerical solutions to problems which involve functions which are non-linear and stochastic, and state and decision variables which are constrained to a finite range of values. Applications to agriculture are considered in Part III.

1.2.2 Forestry management

Trees take much longer to reach maturity than do crops and livestock. The longer time to maturity makes it all the more desirable to formulate forestry problems as dynamic and stochastic. The opportunity cost of resources locked up in the forest is much more important in determining the optimal length of the replacement cycle for the forest than for the agricultural enterprise. Over the life of the forest sequential inputs such as fertilizer may be applied, and intermediate outputs from thinning may be obtained before final harvesting. The final returns from harvesting are particularly uncertain over the life of the forest, partly because of the physical risks of fire and disease, and partly because of the uncertainty of future timber prices and rates of interest.

Chappelle and Nelson (1964) approached the thinning problem using marginal analysis. Since then dynamic programming, along with other dynamic optimizing techniques, have been applied quite extensively to thinning and rotation problems in forestry, though surprisingly most of the problems have been formulated as deterministic rather than stochastic. Applications are considered in Chapter 9.

1.2.3 Fisheries management

A fish stock is typically managed in a quite different institutional environment to that in which agricultural and forestry resources are managed. The fact that fish are mobile in three-dimensional space, and often migratory, leads to various difficulties in their management. For example, it is difficult, and sometimes impossible, to assign property rights in fish. The fishery is the archetypal example of a common-property resource, and therefore tends to be overexploited. The decision problem is simple for each boat operator if all boat operators work independently. With no property rights in future fish stocks, each fisherman tends to concentrate on returns from harvesting in the current period only. However the decision problem for the industry is complex if the present value of social returns from the fishery is to be maximized. This problem can be decomposed into: (1) the dynamic optimization problem; and (2) the institutional problem of finding how each boat operator can be induced to carry out his part of the optimal plan.

Another dimension of difficulty is uncertainty. The total number of fish is uncertain, as is the breakdown of the total number by age category. There are no inputs such as feed to control the biological growth process except in the case of aquaculture. Knowledge about the growth, migration, recruitment and natural mortality of a fishery is invariably fragmentary.

Perhaps because of the common-property nature of the fishery and the paucity of technical data many models have been developed to answer qualitative rather than quantitative questions. The interest has often centred on whether it is socially optimal to fish to extinction, and the existence and stability of steady-state equilibria. To this end optimal control models based on the maximum principle have been used extensively (Clark, 1976).

Dynamic programming has also been used for examining these questions, and for finding quantitative policies for particular fisheries. Applications are considered in Chapter 10.

1.3 Control Variables in Resource Management

In this section some of the basic types of control required in the management of different resource systems are considered. The discussion paves the way for the consideration of a generalized resource problem in the next section.

All resource systems are managed for the eventual harvesting of some product. However, the pattern of harvesting varies between resource systems. In the case of a renewable biological resource, harvesting may continue indefinitely, whereas in the case of an exhaustible resource harvesting finishes with depletion or with extraction costs high relative to resource price. Resource systems are classified by pattern of harvesting in Table 1.1, which also shows the basic type of control variables often used in the management of different resource systems.

The management of an agricultural or naturally-occurring resource may be typified as a series of decisions on the levels of inputs and outputs. As a generalization, the management of an agricultural resource requires a sequence of both input and output decisions, whereas the management of a natural resource requires a sequence of output decisions only. Types of input, output and replacement decisions are described in turn. The distinctions are important for determining the formulation of control problems.

1.3.1 Input decisions

Two types of input control are acquisition inputs and maintenance inputs. Acquisition inputs are the resources themselves or inputs required to gain access to the resource. In agriculture they may be land, seeds or young stock; they include hunting inputs in the case of the fishery and exploration inputs in the case of a mineral. Maintenance inputs are those inputs which

TABLE 1.1
Input and Output Decisions Required in Resource Management

Enterprise	*Input decisions*		*Output decisions*		
	Acquisition	*Continual maintenance*	*Continual produce*	*Repro-duction*	*Final biomass*
(1) Once-only harvesting or culling					
Forestry (no thinning)	×				×
Horticulture, grains	×	×			×
Poultry, pigs, beef	×	×		×	×
(2) Continual harvesting until exhaustion					
Mining	×		×		
(3) Continual harvesting until producing stock is culled					
Forestry (with thinning)	×		×		×
Viticulture, fruit, tropical sorghum	×	×	×		
Eggs, wool, dairy	×	×	×	×	×
(4) Continual harvesting indefinitely					
Aquaculture	×	×	×		
Ocean fishing	×		×		

must be supplied by man for the survival or growth of the resource. For the management of some natural resources no maintenance inputs may be required, as in the case of the fishery, or relatively few and mainly for protection, as in the case of forest stands and exhaustible resources. However, in agriculture maintenance inputs are typically very important, being a major determinant of the timing, quantity and quality of the eventual produce. In crop production the maintenance inputs may be irrigation, fertilizers and pesticides; in livestock production they may be feed, water, shelter and veterinary supplies.

1.3.2 Output decisions

Three types of output can be distinguished: output continually harvested from the resource stock; young stock resulting from reproduction by the parent stock; and the biomass of an adult stock after harvest or slaughter.

Fruit, eggs, wool and milk are examples of continual agricultural outputs. Exhaustible resources mined and timber from thinning forest stands are examples of continual natural resource outputs. The two further categories of output relate only to living resources. The production of young stock can usually be controlled, although the case of the deep-sea fishery is an exception. Examples of valuable biomass of a once-living stock are vegetables, grains, meat and felled trees.

1.3.3 Timing and replacement decisions

The scope for deciding the timing of harvesting depends on the type of output. In the case of inanimate resources such as minerals the product is always available and timing of harvesting is a decision variable. Some outputs from living resources either appear at a point in time (e.g. eggs and young stock) or mature ready for harvest over a relatively short time period (e.g. wool and milk). In these cases, the output decision is one of whether or not to harvest the product when it becomes available. Availability of product depends on time, weather and season, and on any inputs injected by man. The situation is different in the case of output which is final biomass because it accumulates continuously. There are many opportunities for harvesting the biomass, either continually (e.g. forest thinnings) or finally (e.g. forest clear felling). In this case the timing of harvesting is an important decision variable.

For living-resource categories (1) and (3) in Table 1.1 a decision to cull is accompanied by the decision on whether or not to replace the culled stock with new stock of the same or of a different kind. The final harvesting decision is followed by an acquisition decision. If there is to be replacement with the same kind of stock, the replacement may be obtained either by retaining the output of young stock or from outside the system.

Whatever optimal decisions have to be made, whether they relate to the timing and levels of inputs or outputs, the same principle applies. The current consequences of a decision must be weighed against the future consequences of the decision. The way in which this should be done for continual maintenance and harvesting and decisions is demonstrated in the marginal conditions for intertemporal optimality which are derived in the next section using the logic of dynamic programming. The solution of culling and replacement problems is also considered.

1.4 A Simple Derivation of the Conditions for Intertemporal Optimality

The consequence of an input into a resource system is an immediate loss exchanged for some future gain. For example, feed is fed to livestock in the expectation of marketable weight gain in the future. Conversely, the consequence of extracting an output from a resource system is an immediate gain exchanged for some future loss. Harvesting an additional fish now means that it cannot be harvested later; that the effort required to catch another fish later may be increased because there will be fewer fish per unit volume of sea; and that the growth in the biomass of the fish stock will be reduced if growth is an increasing function of biomass. If optimal decisions are to be made on the timing and level of a sequence of inputs or outputs, and the goal of the decision maker is a function of current and future returns, then both the immediate and future consequences of a current decision have to be taken into account. McInerney (1976, 1978, 1981) gives a useful diagrammatic exposition of how they should be taken into account in a simple two-period model.

The rules which summarize the optimality conditions which hold for control problems in general are referred to as the maximum principle. The maximum principle can be defined for either continuous-time or discrete-time problems. One simple method for deriving the maximum principle for continuous-time problems is via the reasoning of dynamic programming (Sethi and Thompson, 1981). The maximum principle for discrete-time problems can be obtained using calculus and dynamic Lagrange multipliers (Benavie, 1970). However, as shown below, the maximum principle for discrete-time problems can also be derived easily and directly using the reasoning of dynamic programming. A discrete-time formulation is appropriate for most resource problems because in general outputs are obtained (inputs are applied) periodically rather than continuously. The derivation shows how dynamic programming can be used as an analytical device. For comparison, the same rules are obtained in Appendix 1.A by working with a dynamic Lagrangian expression.

1.4.1 The general resource problem without replacement

For the generalized problem let the level of resource stock at the start of the first decision period be x_1. Decisions on input or output levels u_i are made at the beginning of each of n decision periods (subscripted by i) in the planning horizon. Any stock remaining at the end of the n-th decision period has a final value $F\{x_{n+1}\}$. The period gain resulting from decision u_i is denoted by $a_i\{x_i, u_i\}$. The discount factor per decision period which applies over all decision periods is α, and equals $1/(1+r)$ where r is the discount rate.

TABLE 1.2
Examples of Functions in Some Resource Control Problems

Natural resource	x_i	u_i	$g_i\{x_i\}$	$h_i\{x_i, u_i\}$	$a_i\{x_i, u_i\}$[a]
Mine	Resource mass	Level of extraction	—	Level of extraction	Net returns from extraction = $p_i h_i - C_i\{x_i, u_i\}$
Irrigated crop	Crop biomass	Level of irrigation	—	Irrigation induced growth ($h_i < 0$)	Net returns from irrigation = $-p_n h_i(1+r)^{n-i} - c_i u_i$
Beef cattle	Liveweight	Level of feed	—	Weight gain ($h_i < 0$)	Net returns from feeding = $-p_n h_i(1+r)^{n-i} - c_i u_i$
Timber thinning	Timber biomass	Level of thinning	Autonomous growth	Level of thinning	Net returns from thinning = $(p_i - c_i)u_i$
Fishery	Fish biomass	Level of fishing effort	Autonomous growth	Level of harvesting	Net returns from harvesting = $p_i h_i - C_i\{x_i, u_i\}$

[a] p_i = price of product; c_i = cost per unit of control; $C_i\{\cdot\}$ = total control cost; r = rate of discount.

The present value of the initial resource stock, denoted by lower-case v_1, is

$$v_1\{x_1, u_1, \ldots, u_n\} = \sum_{i=1}^{n} \alpha^{i-1} a_i\{x_i, u_i\} + \alpha^n F\{x_{n+1}\} \tag{1.1}$$

The problem is to maximize (1.1) with respect to $u_1, \ldots, u_n$ subject to an initial stock level $\bar{x}_1$ and the stock dynamics equations

$$x_{i+1} = x_i + g_i\{x_i\} - h_i\{x_i, u_i\} \qquad (i = 1, \ldots, n) \tag{1.2}$$

where $g_i\{x_i\}$ is the autonomous growth per period of the resource and $h_i\{x_i, u_i\}$ is the per period reduction in stock level consequent on x_i and u_i.

The nature of the functions $a_i\{x_i, u_i\}$, $g_i\{x_i\}$ and $h_i\{x_i, u_i\}$ depends upon the natural resource to be managed. Some examples are shown in Table 1.2.

The formulation applies most immediately to timber thinning and the fishery, for which u_i is the level of harvesting effort in period i. The functions $g_i\{x_i\}$ and $h_i\{x_i, u_i\}$ are usually distinguished in the case of the fishery to deal separately with the natural growth of the fish biomass and the harvest of biomass, respectively.

In the cases of an irrigated crop and beef cattle, there is no need for both $g_i\{\cdot\}$ and $h_i\{\cdot\}$, so $g_i\{\cdot\}$ is dropped. Here the u_i's are levels of inputs, and h_i's are negative representing *increases* in the resource stock. In the simple mining problem, u_i is the level of extraction equal to h_i. However, u_i could instead represent exploration effort, in which case h_i would be negative, representing discoveries of new resource stocks.

In many cases there are restrictions on u_i. There may be a limit to the amount of water available across periods for irrigating a crop, or the appetites of livestock (a function of liveweight) may limit the level of feed input. Any constraints on u_i are easily taken care of when solving problems numerically using dynamic programming. They are ignored for ease of exposition in the rest of this section.

Equation (1.1) states that the value of the initial resource stock equals the present value of all period gains arising from the management of the resource plus the final value of the resource. Alternatively, the value of the initial resource stock is the sum of the gain in period 1 and the value of the resource at the beginning of period 2 discounted one period. In symbols:

$$v_1\{x_1, u_1, \ldots, u_n\} = a_1\{x_1, u_1\} + \sum_{i=2}^{n} \alpha^{i-1} a_i\{x_i, u_i\} + \alpha^n F\{x_{n+1}\}$$

$$= a_1\{x_1, u_1\} + \alpha v_2\{x_2, u_2, \ldots, u_n\}$$

Let u_i^* denote the optimal level of u_i, and upper-case $V_i\{x_i\}$ denote the value of the resource stock at the beginning of period i if controls $u_i^*, \ldots, u_n^*$ are implemented. Suppose $V_2\{x_2\}$ has already been determined. Then the management problem at the beginning of the first period can be reformulated as finding

$$V_1\{x_1\} = v_1\{x_1, u_1^*, \ldots, u_n^*\}$$

$$= a_1\{x_1, u_1^*\} + \alpha v_2\{x_2, u_2^*, \ldots, u_n^*\}$$

$$= a_1\{x_1, u_1^*\} + \alpha V_2\{x_2\}$$

$$= \max_{u_1} [a_1\{x_1, u_1\} + \alpha V_2\{x_2\}] \qquad (1.3)$$

Note that by the transformation equation (1.2), x_2 is the following function of x_1 and u_1:

$$x_2 = x_1 + g_1\{x_1\} - h_1\{x_1, u_1\}$$

This is the approach of dynamic programming. A one-decision variable problem is abstracted from the original n-decision variable problem, although clearly the solution depends on $V_2\{x_2\}$ which is initially unknown. It may appear that the cart is being placed before the horse, but $V_2\{x_2\}$ can be determined through the process of backward induction described in Chapter 2.

Assuming that $a_1\{x_1, u_1\}$, $V_1\{x_1\}$, and $V_2\{x_2\}$ can be differentiated, and for simplicity that the solution is an interior solution, from (1.3) and (1.2) a necessary condition for optimality is

$$\partial a_1/\partial u_1 + \alpha(\mathrm{d}V_2/\mathrm{d}x_2)(\partial x_2/\partial u_1) = 0 \qquad (1.4)$$

The term $\mathrm{d}V_2/\mathrm{d}x_2$ denotes the rate of change in the optimal value of the resource stock *in situ* at the beginning of period 2 with respect to x_2. It is the value of a marginal change in x_2 after allowing for optimal reaction in the control variables $u_2, \ldots, u_n$. It is conveniently written as

λ_2 because it has the same interpretation as the Lagrange multiplier λ_2 used in the alternative derivation of the intertemporal optimality conditions in Appendix 1.A. Thus (1.4) can be written as

$$\partial a_1/\partial u_1 = -\alpha\lambda_2(\partial x_2/\partial u_1) = \alpha\lambda_2(\partial h_1/\partial u_1) \tag{1.5}$$

Equation (1.5) shows that immediate gains (losses) must be balanced against the present value of future losses (gains) in determining u_1^*. It is required that u_1 be increased whilst the immediate marginal gains (present value of future losses) more than offset the present value of future losses (immediate gains), until (1.5) holds.

If u_1 is the usage of a resource then the right-hand side of (1.5) represents the marginal user cost referred to in Section 1.1. For a one-period problem (or if $n = 1$), the right-hand side of (1.5) is zero.

Although x_1 is not a control variable, if the stock is optimally managed across all periods then the resource owner should be indifferent between leaving the marginal unit of the resource in place or extracting it. To see that this condition does indeed hold, differentiate both sides of (1.3) with respect to x_1 to obtain:

$$\mathrm{d}V_1/\mathrm{d}x_1 = \partial a_1/\partial x_1 + \alpha(\mathrm{d}V_2/\mathrm{d}x_2)(\partial x_2/\partial x_1)$$

or

$$\lambda_1 = \partial a_1/\partial x_1 + \alpha\lambda_2(1 + \mathrm{d}g_1/\mathrm{d}x_1 - \partial h_1/\partial x_1) \tag{1.6}$$

Augmenting the current resource base by one unit has two impacts. The period gain is altered, and the resource stock at the beginning of period 2 is changed. Thus, the increase in the optimal value of the resource stock at the beginning of period 1 equals the additional period gain plus the increase in the optimal value of the resource stock at the beginning of period 2, discounted one period.

The generalized natural resource problem subsumes not only the first-period problem (1.3) but also the remaining sequence of single-period problems for $i = 2, \ldots, n$. The two conditions for optimality, (1.5) and (1.6), generalize to

$$\partial a_i/\partial u_i = -\alpha\lambda_{i+1}(\partial x_{i+1}/\partial u_i) \qquad (i = 1, \ldots, n) \tag{1.7}$$

$$\lambda_i = \partial a_i/\partial x_i + \alpha\lambda_{i+1}(\partial x_{i+1}/\partial x_i) \qquad (i = 1, \ldots, n) \tag{1.8}$$

with boundary conditions

$$x_1 = \bar{x}_1 \tag{1.9}$$

$$\lambda_{n+1} = \mathrm{d}F/\mathrm{d}x_{n+1} \tag{1.10}$$

The last boundary condition holds because the value of the resource after all decisions have been made, $V_{n+1}\{x_{n+1}\}$, is set equal to $F\{x_{n+1}\}$.

Equations (1.7) to (1.10) represent a special case of the discrete maximum principle, although (1.8) is usually rewritten to give the between period change in λ. Equation (A1.2) in Appendix 1.A shows the alternative version of (1.8). The results are related to the Hamiltonian used in control theory in Appendix 1.B.

It may be possible to use equations (1.7) to (1.10) to solve a resource problem analytically. However, the solution of a set of difference equations with two-point boundaries is not always straightforward. In Chapter 2 it is explained how numerical solutions to the problem formulated in (1.1) can be obtained directly using dynamic programming.

1.4.2 The general resource problem with replacement

The generalized resource problem dealt with so far has as decision variables the levels of maintenance inputs and harvesting outputs, but not whether to replace the resource itself. Some resource problems, such as those of the mine and the fishery, do not ordinarily include questions about replacement. However, as noted in Section 1.3, many resource problems in agriculture and forestry do.

Perhaps the simplest resource replacement problem is the determination of the optimal rotation period for a forest, assuming no inputs or outputs other than planting inputs and final-felling outputs. The condition for optimality is known as the Faustmann formula after the German forester who published a correct method of calculating the value of a forest stand in 1849. In Chapter 9 the conditions for optimality are derived analytically using the logic of dynamic programming. The condition for optimality is also given for the case for which early replacement may be forced because of some random factor, such as a forest fire.

A more complex replacement problem is one which requires decisions at the beginning of each period not only on whether to cull and replace, but also on the levels of maintenance inputs or continual harvesting of produce. Examples are the forest thinning and rotation problem, and the livestock feeding and replacement problem. There are two methods of solving this type of replacement problem. One is a two-stage process. The optimality conditions presented in Section 1.4 for the maintenance or harvesting sub-problem for a range of planning horizons, T, are used to determine optimal returns. The optimality conditions for the simple replacement problem can then be used to determine the optimal T.

A second more direct method is to solve the problem numerically using dynamic programming. The decision to be made at the beginning of each decision period is expanded from an input or output level to include the possibility of replacement.

1.5 Numerical Dynamic Programming

1.5.1 Types of resource problem

An important feature of numerical dynamic programming is the ability to allow for the decision set to consist of qualitatively quite different decisions. For example, in the case of managing livestock subject to drought, the decision set might be: (a) graze; (b) agist; (c) sell; and (d) slaughter. Another example is the tractor replacement problem, for which the decision set might be replacement with a range of alternative tractor capacities.

Numerical solution by dynamic programming is particularly appropriate when for one reason or another simple optimality conditions such as those presented in the previous section cannot be derived, or when obtaining a solution based on them is difficult. Derivatives of the functions $a_i\{x_i, u_i\}$, $g_i\{x_i\}$ and $h_i\{x_i, u_i\}$ may not exist. The functions may be stochastic. Although optimality conditions for stochastic control problems may be obtained following the same logic as before, finding a solution based on them may be difficult.

Numerical dynamic programming is well suited to solving resource problems involving decisions with irreversible outcomes. For example, resource problems may have irreversible aspects if decisions over time could threaten the existence of a species, or if the development of an amenity area could not be undone. Krutilla and Fisher (1975) show how optimal control theory may be used to draw conclusions about the optimal timing of irreversible development under certainty. Arrow and Fisher (1974) show how the principles of sequential decision making under uncertainty can be applied to the irreversible development decision under uncertainty. Dynamic programming formulations can capture directly changing decision sets over time. The discontinuities of decision or state variables arising from irreversibilities can be readily handled by using numerical dynamic programming. This topic is considered in more detail in Section 8.3.

1.5.2 Links with simulation

Compared with many other solution techniques, numerical dynamic programming shares with simulation great flexibility in accommodating functions $a_i\{x_i, u_i\}$, $g_i\{x_i\}$ and $h_i\{x_i, u_i\}$ that are discontinuous and stochastic. Numerical dynamic programming is not only a flexible modelling device but has the added advantage of being an optimizing algorithm. Not surprisingly, dynamic programming models have been developed which incorporate simulation submodels. This is not to say that dynamic programming is always to be preferred to simulation. Simulation can be used to find solutions to a high degree of accuracy by adding search routines, and there are certain restrictions on the dynamic programming formulation of a problem which are absent in simulation modelling. These restrictions, that the system being modelled possess the Markov property, and that the objective function be separable, are explained in the next chapter. Another restriction is the limit on the number of state variables which can be included if a numerical solution is required. This problem is discussed further in Chapter 4.

1.5.3 Solution procedures

Many dynamic programming problems are solved numerically, with time and the state and decision variables formulated as discrete variables. The calculations required to obtain a solution are generally lengthy and repetitive, and are therefore best left to the computer. However, partly because of the flexibility that dynamic programming allows in specifying the functions $a_i\{x_i, u_i\}$, $g_i\{x_i\}$ and $h_i\{x_i, u_i\}$, until recently there have not been many general-purpose computer algorithms available for solving dynamic programming problems. This distinguishes dynamic programming from many other operations research techniques such as linear programming. Nevertheless, whilst most dynamic programming problems will require some computing input by the user, the routine work common to most dynamic programming problems can be performed using a general-purpose computer program. General purpose algorithms written in BASIC are presented in this book.

1.5.4 Types of dynamic programming problem

However, it should not be thought that all dynamic programming problems are solved numerically. For some problems it may be possible to specify the variables as continuous variables and to obtain solutions analytically. The solution procedure need not make direct use of the optimality conditions derived analytically in the previous section. Examples of such problems are discussed in Chapters 4 and 6. An important advantage of the analytical approach is that the restriction on the number of state variables specified may be overcome.

Besides numerical versus analytical dynamic programming, there are several other categories of dynamic programming problem, requiring different solution procedures. These are: deterministic versus stochastic and adaptive; finite-stage versus infinite-stage; and without-discounting versus with-discounting of stage returns. Examples of the formulation and solution of these different types of problem are presented in Part II.

1.6 References

Anderson, J.R., Dillon, J.L. and Hardaker, J.B. (1977), *Agricultural Decision Analysis*, Iowa State University Press, Ames.

Arrow, K.J. and Fisher, A.C. (1974), Environmental preservation, uncertainty, and irreversibility, *Quarterly Journal of Economics*, **88**(2), 312–9.

Benavie, A. (1970), The economics of the maximum principle, *Western Economic Journal*, **8**(4), 426–30.

Boulding, K.E. (1966), The economics of the coming spaceship earth, in *Environmental Quality in a Growing Economy*, H. Jarrett (ed.), Johns Hopkins University Press, Baltimore.

Chappelle, D.E. and Nelson, T.C. (1964), Estimation of optimal stocking levels and rotation ages of loblolly pine, *Forest Science*, **10**, 471–83.

Clark, C.W. (1976), *Mathematical Bioeconomics: The Optimal Management of Renewable Resources*, J. Wiley, New York.

Dillon, J.L. (1977), *The Analysis of Response in Crop and Livestock Production*, 2nd edition, Pergamon Press, Oxford.

Dorfman, R. (1969), An economic interpretation of optimal control theory, *American Economic Review*, **59**(5), 817–31.

Faustmann, G. (1849), On the determination of the value which forest land and immature stands possess for forestry, reprinted in English in *Oxford Institute Papers*, **42**, 1968.

Georgescu-Roegen, N. (1971), *The Entropy Law and the Economic Process*, Harvard University Press, Cambridge, Massachusetts.

Krutilla, J.V. and Fisher, A.C. (1975), *The Economics of Natural Environments*, Johns Hopkins University Press, Baltimore.

Magnusson, G. (1969), *Production under Risk: a Theoretical Study*, Almqvist and Wiksell's, Uppsala.

Martin, L.R., ed. (1977), *A Survey of Agricultural Economics Literature*, Vol.2, University of Minnesota Press, Minneapolis.

McInerney, J.P. (1976), The simple analytics of natural resource economics, *Journal of Agricultural Economics*, **27**(1), 31–52.

McInerney, J.P. (1978), On the optimal policy for exploiting renewable resource stocks, *Journal of Agricultural Economics*, **29**(2), 183–8.

McInerney, J.P. (1981), Natural resource economics: the basic analytical principles, in *Economics and Resources Policy*, J. A. Butlin (ed.), Longman, London.

Sethi, S.P. and Thompson, G.L. (1981), *Optimal Control Theory: Applications to Management Science*, Martinus Nijhoff, Boston.
Throsby, C.D. (1974), New methodologies in agricultural production economics: A review, in *The Future of Agriculture*, Papers and reports, 15th International Conference of Agricultural Economists, São Paulo, Brazil, 150–69.

1.A Appendix: A Lagrangian Derivation of the Discrete Maximum Principle

A general control problem for the management of a natural resource was presented in Chapter 1. The conditions which must hold for intertemporal optimality were derived using a dynamic programming approach. The derivation introduced the idea of recursive induction which is the heart of the dynamic programming approach. In this appendix the same conditions are derived by finding the conditions for which the relevant Lagrangian expression is maximized. The alternative derivation is included for comparison of the two methods and for completeness.

A Lagrangian expression for the problem studied in Section 1.4 is

$$L = \sum_{i=1}^{n} \alpha^{i-1} a_i\{x_i, u_i\} + \lambda_1(\bar{x}_1 - x_1) + \alpha^n F\{x_{n+1}\}$$

$$+ \sum_{i=1}^{n} \alpha^i \lambda_{i+1}(x_i + g_i\{x_i\} - h_i\{x_i, u_i\} - x_{i+1})$$

$$= \sum_{i=1}^{n} \alpha^{i-1}[a_i\{x_i, u_i\} + \alpha\lambda_{i+1}(x_i + g_i\{x_i\} - h_i\{x_i, u_i\} - x_{i+1})]$$

$$+ \lambda_1(\bar{x}_1 - x_1) + \alpha^n F\{x_{n+1}\}$$

where the Lagrange multiplier or costate variable, λ_i, is the contribution which an additional unit of the resource stock would make to the value of the resource stock at the beginning of period i. If a small change in u_i or x_i changes x_{i+1} by Δx_{i+1}, the value at the beginning of period i of the change is $\alpha\lambda_{i+1}\Delta x_{i+1}$. In some derivations of the discrete maximum principle, the Lagrangian expression is formulated so that what is represented by $\alpha\lambda_{i+1}$ in this analysis is represented instead by λ_i.

Necessary conditions for an interior solution are

$$\partial L/\partial u_i = \alpha^{i-1}[\partial a_i/\partial u_i - \alpha\lambda_{i+1}(\partial h_i/\partial u_i)] = 0$$

which implies

$$\partial a_i/\partial u_i = \alpha\lambda_{i+1}(\partial h_i/\partial u_i) \qquad (i = 1, \ldots, n) \qquad \text{(A1.1)}$$

$$\partial L/\partial x_i = \alpha^{i-1}[\partial a_i/\partial x_i + \alpha\lambda_{i+1}(1 + \mathrm{d}g_i/\mathrm{d}x_i - \partial h_i/\partial x_i) - \lambda_i] = 0$$

which implies

$$\lambda_i = \partial a_i/\partial x_i + \alpha\lambda_{i+1}(1 + \mathrm{d}g_i/\mathrm{d}x_i - \partial h_i/\partial x_i)$$

or, after rearrangement and noting that $\alpha = 1/(1 + r)$ where r is the rate of discount

$$\lambda_{i+1} - \lambda_i = r\lambda_i - \lambda_{i+1}(\mathrm{d}g_i/\mathrm{d}x_i - \partial h_i/\partial x_1) - (1 + r)\partial a_i/\partial x_i \qquad (i = 1, \ldots, n) \qquad \text{(A1.2)}$$

$$x_1 = \bar{x}_1 \qquad \text{(A1.3)}$$

$$\partial L/\partial x_{n+1} = \alpha^n[-\lambda_{n+1} + \mathrm{d}F/\mathrm{d}x_{n+1}] = 0$$

which implies

$$\lambda_{n+1} = \mathrm{d}F/\mathrm{d}x_{n+1} \qquad \text{(A1.4)}$$

No allowance has been made for any constraints on u_i. If there are constraints on the u_i, they can be incorporated in the Lagrangian expression. Another extension is to allow for the possibility of boundary solutions. Some of the conditions become inequalities, and the Kuhn–Tucker conditions apply.

Dorfman (1969) and Benavie (1970) present the derivation of the discrete maximum principle with useful economic interpretations. Clark (1976) presents derivations of both the continuous and discrete maximum principles, and applies them to a wide variety of problems encountered in the management of natural resources.

1.B Appendix: A Note on the Hamiltonian used in Control Theory

The conditions for the optimal management of a resource through time are often derived from optimal control theory. Although not immediately obvious, they can also be derived intuitively from the basic equation of dynamic programming. The derivation here is restricted to cases where all functions are continuously differentiable. The discrete maximum principle for the general resource problem formulated in Chapter 1 can be stated in terms of the expression

$$H_i\{x_i, u_i, \lambda_{i+1}\} \equiv a_i\{x_i, u_i\} + \alpha\lambda_{i+1}(g_i\{x_i\} - h_i\{x_i, u_i\}) \qquad (i = 1, \ldots, n) \qquad \text{(B1.1)}$$

The discrete maximum principle states that a necessary condition for optimality is that u_i maximize $H_i\{\cdot\}$ at each stage i. If the solution is an interior solution, the requirement is that

$$\partial H_i/\partial u_i = 0 \qquad (i = 1, \ldots, n) \qquad \text{(B1.2)}$$

The associated adjoint equation

$$\lambda_{i+1} - \lambda_i = r\lambda_i - (1 + r)\partial H_i/\partial x_i \qquad (i = 1, \ldots, n) \qquad \text{(B1.3)}$$

relates the rate of change of the adjoint variable or dynamic Lagrange multiplier to $H_i\{\cdot\}$ and the rate of interest r, where $\alpha = 1/(1 + r)$.

The fundamental dynamic programming equation for optimal control is the general form of (1.3):

$$\begin{aligned} V_i\{x_i\} &= \max_{u_i} [a_i\{x_i, u_i\} + \alpha V_{i+1}\{x_{i+1}\}] \\ &= \max_{u_i} [a_i\{x_i, u_i\} + \alpha V_{i+1}\{x_i + g_i\{x_i\} - h_i\{x_i, u_i\}\}] \end{aligned} \qquad (i = 1, \ldots, n) \qquad \text{(B1.4)}$$

The purpose of this appendix is to show that (B1.4) implies (B1.2) and (B1.3). A necessary condition for an interior solution is

$$\partial a_i/\partial u_i - \alpha(\mathrm{d}V_{i+1}/\mathrm{d}x_{i+1})\partial h_i/\partial u_i = 0 \qquad (i = 1, \ldots, n) \qquad \text{(B1.5)}$$

which is the same as (B1.2) after substituting λ_{i+1} for $\mathrm{d}V_{i+1}/\mathrm{d}x_{i+1}$ evaluated at x_{i+1}.

Denoting optimal u_i by u_i^*, (B1.4) can be rewritten as

$$V_i\{x_i\} = a_i\{x_i, u_i^*\} + \alpha V_{i+1}\{x_i + g_i\{x_i\} - h_i\{x_i, u_i^*\}\} \qquad (i = 1, \ldots, n) \qquad \text{(B1.6)}$$

Because u_i^* is a function of x_i, (B1.6) can be totally differentiated with respect to x_i to give

$$\mathrm{d}V_i/\mathrm{d}x_i = \partial a_i/\partial x_i + \alpha \mathrm{d}V_{i+1}/\mathrm{d}x_{i+1}(1 + \mathrm{d}g_i/\mathrm{d}x_i - \partial h_i/\partial x_i) \qquad (i = 1, \ldots, n) \qquad \text{(B1.7)}$$

After rearrangement and substitution for $\mathrm{d}V_i/\mathrm{d}x_i$ (B1.7) becomes

$$\lambda_i(1+r) - \lambda_{i+1} = (1+r)\partial a_i/\partial x_i + \lambda_{i+1}(\mathrm{d}g_i/\mathrm{d}x_i - \partial h_i/\partial x_i)$$

or

$$\lambda_{i+1} - \lambda_i = r\lambda_i - (1+r)\partial H_i/\partial x_i \qquad (i = 1, \ldots, n) \qquad \text{(B1.8)}$$

which is the same as (B1.3). The term $H_i\{\cdot\}$ introduced in (B1.1) is the current value Hamiltonian of control theory, discounted one period.

PART 2

The Methods of Dynamic Programming

Chapter 2

INTRODUCTION TO DYNAMIC PROGRAMMING

In this chapter the solution of deterministic, finite-stage dynamic programming problems using backward recursion is explained. The solution procedure is valid for any problem for which Bellman's Principle of Optimality holds. The necessary structure and properties of a dynamic programming problem are discussed. The process of backward recursion is illustrated in the solution of two problems by compiling solution tables. The way in which the general-purpose dynamic programming programs listed in Appendix A2 may be used to solve the problems is also explained.

2.1 Backward Recursion Applied to the General Resource Problem

In Chapter 1 an n-stage general resource problem was introduced. It was shown how the problem could be reformulated as one of maximizing, with respect to the initial decision variable u_1, an objective function subsuming an $(n-1)$-stage problem. The same type of decision problem exists at all subsequent decision stages. The overall problem of finding $u_1^*, \ldots, u_n^*$ can be solved by finding u_i^* which satisfies the following recursive functional equation:

$$V_i\{x_i\} = \max_{u_i} [a_i\{x_i, u_i\} + \alpha V_{i+1}\{x_i + g_i\{x_i\} - h_i\{x_i, u_i\}\}] \qquad (i = n, \ldots, 1) \qquad (2.1)$$

with

$$V_{n+1}\{x_{n+1}\} = F\{x_{n+1}\}$$

$$x_1 = \bar{x}_1$$

where

$V_i\{x_i\}$ = value derived from implementing $u_i^*, \ldots, u_n^*$ given the level of the resource stock is x_i;
u_i = decision on level of input or output;
$a_i\{x_i, u_i\}$ = stage return;
$g_i\{x_i\}$ = autonomous addition to the resource stock between stage i and stage $i+1$;
$h_i\{x_i, u_i\}$ = controlled reduction in the resource stock between stage i and stage $i+1$;
$\bar{x}_1$ = stock of the resource available at stage 1;
$F\{x_{n+1}\}$ = final value of the resource stock remaining at stage $n+1$; and
α = discount factor.

Equation (2.1) is a functional equation because functions appear on both sides. It is recursive because determining $V_{i+1}\{x_{i+1}\}$ enables $V_i\{x_i\}$ to be determined. The equation also reflects Bellman's Principle of Optimality which is expounded more fully in Section 2.2 in connection with the numerical solution of problems.

Because (2.1) is recursive the general resource problem can be solved by a process of backward induction. The term $V_{i+1}\{\cdot\}$ on the right-hand side is initially unknown for all i except for $i = n$. In the case of $i = n$, $V_{n+1}\{x_{n+1}\} = F\{x_{n+1}\}$. Equation (2.1) can therefore be solved for $i = n$ to give u_n^* and $V_n\{x_n\}$. The solution to the general resource problem is obtained by repeating the process for all i from $i = n$ to $i = 1$.

In practice the solution of functional equations may not be straightforward. In Chapter 1 it was assumed that $V_i\{x_i\}$ and $a_i\{x_i, u_i\}$ were differentiable functions and that solutions were always interior solutions. It was also pointed out that many resource problems do not have these features, in which case it still may be possible to obtain solutions numerically.

A numerical formulation of the general resource problem restricts the values of x_i and u_i to finite sets for all i. A further restriction is that any feasible combination of x_i and u_i must imply access to a value of x_{i+1}. The dynamic programming problem can then be interpreted as one of finding the optimal path through a network of nodes. A perhaps obvious but exploitable characteristic of the optimal path is described in Bellman's Principle of Optimality.

2.2 The Principle of Optimality

The Principle of Optimality is illustrated with reference to a simple mining problem. Suppose x_i represents the number of units of a mineral in the ground at stage i, and always takes one of the nine values between 0 and 80 shown in the grid in Fig. 2.1. The decision variable u_i represents the number of units mined and can take one of the values 0, 10 or 20, subject to $u_i \leq x_i$. The amount to mine at each of four stages has to be decided. The net returns from mining, $a\{x_i, u_i\}$, are shown in Table 2.1.

Any stock left unmined at the end of the fourth period has zero value. The problem is

$$\max_{u_1 \text{ to } u_4} \sum_{i=1}^{4} a\{x_i, u_i\}$$

subject to

$$x_1 = 80$$

$$x_{i+1} = x_i - u_i$$

and to the constraints on u_i.

The feasible stock levels at each stage are bordered by the broken lines in Fig. 2.1. At each decision stage, the three levels of mining 0, 10 and

TABLE 2.1
Mining Stage Returns $a\{x, u\}$

Stock (x)	*Mined* (u)		
	0	*10*	*20*
0	−10	—	—
10	−10	−60	—
20	−10	−40	−60
30	−10	−25	−40
40	−10	−15	−20
50	−20	0	−10
60	−20	10	10
70	−20	20	35
80	−20	30	70

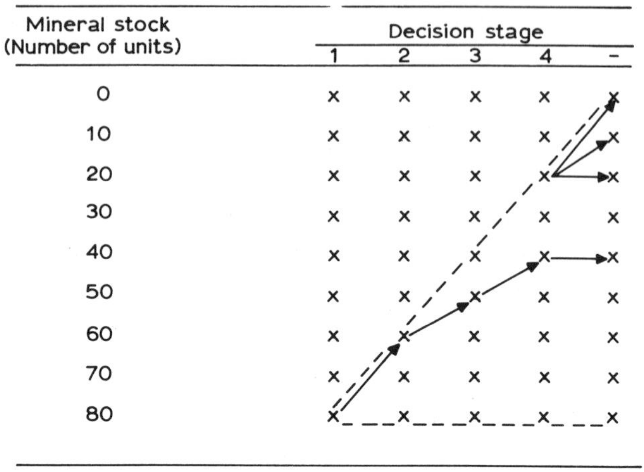

FIG. 2.1 Grid of feasible stock levels.

20 are possible at each feasible state node. As an example, the decisions are illustrated in Fig. 2.1 at stage 4 for stock level 20.

The optimal path through state nodes is 80, 60, 50, 40, 40 and is shown in Fig. 2.1. The corresponding optimal mining policy is 20, 10, 10, 0 which gives a total return $(70 + 10 + 0 - 10) = 70$. The method of identifying the optimal path is not dwelt on at this stage. The point to be made here is that for the policy to be optimal, it must conform with Bellman's (1957, p.83) Principle of Optimality:

> An optimal policy has the property that, whatever the initial state and optimal first decision may be, the remaining decisions constitute an optimal policy with regard to the state resulting from the first decision.

In other words, if 80, 60, 50, 40, 40 really is an optimal path starting with $x_1 = 80$, then 60, 50, 40, 40 must be an optimal path starting with $x_2 = 60$; 50, 40, 40 must be an optimal path starting with $x_3 = 50$; and 40, 40 must be an optimal path starting with $x_4 = 40$. It is simple to check the validity of these statements.

Note that the Principle of Optimality is not a definition of an optimal policy, for the principle states that an optimal policy consists of a nesting of shorter optimal policies. Rather it is a statement about the recursive nature of an optimal policy which suggests that sequential decision problems can be solved by a process of backward recursion. Referring to the example, it was argued by a process of contraction along the optimal path starting with $x_1 = 80$ that 40, 40 must be the optimal final path starting with $x_4 = 40$. Conversely, an optimal path over any number of decision stages can be found by a reverse process of expansion. Optimal final paths are determined from all possible values of x_4 (20 to 80) in the knowledge that one of them must be part of each of the optimal paths starting from x_3 (40 to 80). This enables the optimal paths from each possible value of x_3 to be determined in the knowledge that one of them must be part of the optimal path starting from each possible value of x_2. In this way the optimal path starting with $x_1 = 80$ is found.

In the process of finding the overall optimal path, information is generated about many possible optimal sub-paths, only a few of which are actually embedded in the overall optimal path. Such information may be redundant. On the other hand, it may be useful for sensitivity analysis, or it may be useful for pursuing a feedback control strategy when transitions from one stock level to another are subject to uncertainty. In any event, the number of decision consequences explored using backward recursion is much less than the number using total enumeration for problems with many decision stages. In the example problem, $3^4 = 81$ decision consequences must be evaluated using total enumeration compared with $(1 \times 3) + (3 \times 3) + (5 \times 3) + (7 \times 3) = 48$ using backward recursion. In general, if the number of decision stages is n and the initial stock is large enough, the number of decision consequences which must be evaluated is 3^n by total enumeration and $3n^2$ by backward recursion. For $n > 3$, backward recursion requires the evaluation of fewer decision consequences.

2.3 The Structure of Dynamic Programming Problems

In Chapter 1 it was stressed that one of the advantages of dynamic programming was that few constraints are placed on the functions $a_i\{x_i, u_i\}$, $g_i\{x_i\}$ and $h_i\{x_i, u_i\}$. The functions could be non-linear, discontinuous and/or stochastic. However, for it to be possible to solve a problem by dynamic programming, it must have a particular structure. The structure must be the same or similar to that for the general resource problem already encountered.

To describe the required structure, the following terms are employed: decision, decision stage, state, transformation function, stage return function and objective function. The problem must consist of a sequence of *decisions*, $u_1, \ldots, u_n$. A point in time at which a decision is made is a *decision stage*, often referred to merely as a stage. Any decision u_i made at the i-th decision stage has two consequences. First, it results in a change in the *state* of the decision system from x_i at stage i to x_{i+1} at stage $i+1$. The change is expressed by the *transformation function* which for the general resource problem was written as

$$x_{i+1} = x_i + g_i\{x_i\} - h_i\{x_i, u_i\}$$

but which can be more generally written as

$$x_{i+1} = t_i\{x_i, u_i\}$$

Secondly, the decision results in a return at each decision stage given by the *stage return function*, $a_i\{x_i, u_i\}$. The overall objective of the problem must be to select the decision sequence $u_1, \ldots, u_n$ so that a separable *objective function* of the n stage returns is optimized. The objective functions most frequently encountered are the sum of stage returns, or the present value of stage returns

$$\sum_{i=1}^{n} \alpha^{i-1} a_i\{x_i, u_i\}$$

The final decision to be taken, u_n, determines the terminal state of the system, x_{n+1}. There may be some final value $F\{x_{n+1}\}$ associated with the terminal state, in which case it is included in the objective function.

Although the additive objective function is the most commonly used one, there are other separable objective functions for which the Principle of Optimality holds. Nemhauser (1966) following Mitten (1964) has presented a sufficient condition on the objective function for the application of the Principle of Optimality, besides separability. If the objective function is separable, then the objective function for the last two decision stages can be written

$$f_{n-1} = f_{n-1}\{a_{n-1}, f_n\{a_n, F\}\}$$

where a_n is the stage return at stage n and F is the terminal value. In general

$$f_i = f_i\{a_i, f_{i+1}\} \qquad (i = 1, \ldots, n-1)$$

The sufficient condition is that

$$f_i\{a_i, f'_{i+1}\} \geq f_i\{a_i, f''_{i+1}\} \quad \text{if} \quad f'_{i+1} > f''_{i+1} \text{ for all } a_i \qquad (i = 1, \ldots, n-1)$$

In words, f_i must be a monotonically non-decreasing function of f_{i+1} for all a_i. This condition is satisfied if the objective function is additive in stage returns, or the product of stage returns provided all stage returns are non-negative. The latter objective function has been used in crop-irrigation problems because it implies zero total return if any stage return is zero as a result of crop failure. Cooper and Cooper (1981) have pointed out that the monotonicity condition is not a necessary condition, but has also commented that so far there has been no satisfactory statement of the necessary conditions on the objective function for the Principle of Optimality to hold.[1]

It is important to note that the decision system must be fully described at any stage i by the state of the system, x_i, in the sense that x_{i+1} and a_i depend only on x_i, u_i and any exogenous variables. Only to this extent may the behaviour of the system be dependent on the history of the system prior to stage i. In other words, the decision system must possess the Markov property.[2]

To sum up, a dynamic programming problem with an additive objective function has the following form:

$$\max_{u_1,\ldots,u_n} \sum_{i=1}^{n} \alpha^{i-1} a_i\{x_i, u_i\} + \alpha^n F\{x_{n+1}\}$$

subject to

$$x_1 = \bar{x}_1$$

$$x_{i+1} = t_i\{x_i, u_i\} \qquad (i = 1, \ldots, n)$$

and any constraints on u_i.

The corresponding recursive equation for solving the problem is

$$V_i\{x_i\} = \max_{u_i} [a_i\{x_i, u_i\} + \alpha V_{i+1}\{t_i\{x_i, u_i\}\}] \qquad (i = n, \ldots, 1) \qquad (2.2)$$

with

$$V_{n+1}\{x_{n+1}\} = F\{x_{n+1}\}$$

$$x_1 = \bar{x}_1$$

and any constraints on u_i.

[1] Hastings (1973, pp.25–33) proves that two conditions, which he terms separability and optimality conditions, are necessary and sufficient for solving a multistage decision problem by dynamic programming.

[2] Problems with causality lags greater than one period can be formulated to conform with the first-order Markov requirement by suitably defining the state variable. For example, if x_{i+1} and a_i depend on x_{i-1} as well as x_i and u_i, the requirement is met by defining both x_i and x_{i-1} as state variables.

2.4 A Numerical Example

Consider the problem of finding the least-cost path through the network shown in Fig. 2.2. Nodes have been designated (i, J_i) where i is the decision stage and J_i is the state number. The optimal path must start at (1, 1) and end at (5, 1). Inter-node costs, or negative stage returns, are shown beside the linking decision arrows. The problem is

$$\min_{J_2, J_3, J_4} \sum_{i=1}^{4} a_i\{J_i, J_{i+1}\} + F\{J_4\}$$

where $a_i\{J_i, J_{i+1}\}$ is the cost of linking (i, J_i) and $(i+1, J_{i+1})$; and

$$F\{J_4\} = \begin{vmatrix} 8 & \text{if } J_4 = 1 \\ 5 & \text{if } J_4 = 2 \end{vmatrix}$$

The recursive equation is

$$V_i\{J_i\} = \min_{J_{i+1}} [a_i\{J_i, J_{i+1}\} + V_{i+1}\{J_{i+1}\}] \qquad (i = 3 \text{ to } 1)$$

with

$$V_4\{J_4\} = F\{J_4\}$$

A useful system for solving dynamic programming problems by hand is the preparation of a series of tables, one for each stage, starting with the final decision stage. Each table has a row for each feasible state. Against each feasible state, the total return to the end of the planning

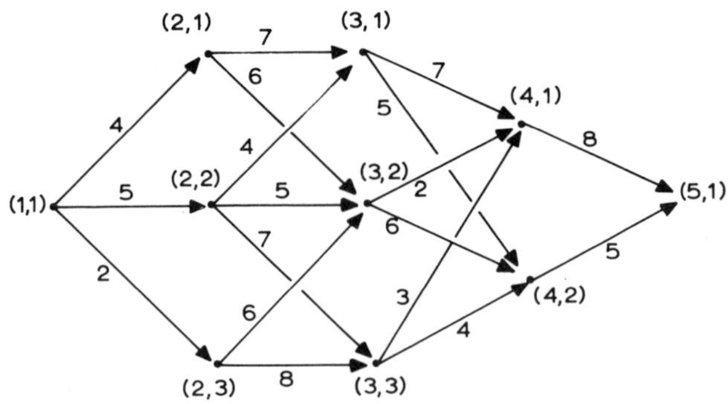

FIG. 2.2 A least-cost network problem.

TABLE 2.2
Costs to (5, 1) from Stage-3 Nodes

Node	*Next node*		*Least cost*	*Optimal next node*
	(4, 1)	*(4, 2)*	$V_3\{J_3\}$	
(3, 1)	7 + 8 = 15	5 + 5 = 10*	10	(4, 2)
(3, 2)	2 + 8 = 10*	6 + 5 = 11	10	(4, 1)
(3, 3)	3 + 8 = 11	4 + 5 = 9*	9	(4, 2)

horizon for each feasible decision is shown. Total return is the sum of the stage return and the optimal return to the planning horizon from the state accessed at the next stage. The last two columns of the table show the optimal total return and associated optimal decision. The procedure is demonstrated for the network problem of Fig. 2.2 in Tables 2.2 to 2.4.

The first entry in Table 2.2 consists of the cost of linking nodes (3, 1) and (4, 1), ($a_3\{1, 1\} = 7$) plus ($V_4\{1\} = 8$) which equals 15. The second entry consists of the cost of linking nodes (3, 1) and (4, 2), ($a_3\{1, 2\} = 5$) plus ($V_4\{2\} = 5$) which equals 10. The least cost of moving from (3, 1) to (5, 1) is therefore min [15, 10] = 10 via (4, 2), as recorded in the last two columns of row one. Other rows are determined in the same way. Optimal decisions are indicated by asterisks in the 'next node' column.

TABLE 2.3
Costs to (5, 1) from Stage-2 Nodes

Node	*Next node*			*Least cost*	*Optimal next node*
	(3, 1)	*(3, 2)*	*(3, 3)*	$V_2\{J_2\}$	
(2, 1)	7 + 10 = 17	6 + 10 = 16*		16	(3, 2)
(2, 2)	4 + 10 = 14*	5 + 10 = 15	7 + 9 = 16	14	(3, 1)
(2, 3)		6 + 10 = 16*	8 + 9 = 17	16	(3, 2)

TABLE 2.4
Costs to (5, 1) from (1, 1)

Node	*Next node*			*Least cost*	*Optimal next node*
	(2, 1)	*(2, 2)*	*(2, 3)*	$V_1\{J_1\}$	
(1, 1)	4 + 16 = 20	5 + 14 = 19	2 + 16 = 18*	18	(2, 3)

Table 2.4 shows that the least cost path from (1,1) to (5,1) incurs a cost of 18. The least cost path itself is found by tracking forward through the tables. Table 2.4 shows that (2,3) should succeed (1,1), Table 2.3 that (3,2) should succeed (2,3), and Table 2.2 that (4,1) should succeed (3,2). The optimal sequence of nodes is therefore (1,1) (2,3) (3,2) (4,1) (5,1) with associated cost $2+6+2+8=18$.

2.5 Forward Recursion and Stage Numbering

The network problem in Section 2.4 was that of finding the least cost path from (1,1) to (5,1). It is clear that a second problem of finding the least cost path from (5,1) to (1,1) would have yielded the same path in reverse. Thus the solution to the first problem could have been found by solving the second problem, even if the direction to be taken in practice is (1,1) to (5,1). In order to solve the second problem by backward recursion, the first stage considered would have been stage 3, then stage 4, and finally stage 5. The arrows in Fig. 2.2 would have been reversed.

Instead of describing the second procedure as a method of solving the first problem with the decision sequence reversed, using backward recursion, it could be described as a method of solving the first problem with the decision sequence as originally given, using forward recursion. Stage 3 would be taken to be the first decision stage, the decision at any node being the optimal node *to have come from*. Stage 4 nodes and then (5,1) would be examined in sequence in the usual way. The recursive equation would be

$$V_i\{J_i\} = \min_{J_{i-1}} [a_i\{J_{i-1}, J_i\} + V_{i-1}\{J_{i-1}\}] \qquad (i = 3 \text{ to } 5)$$

with

$$V_2\{J_2\} = \left| \begin{array}{ll} 4 & \text{if } J_2 = 1 \\ 5 & \text{if } J_2 = 2 \\ 2 & \text{if } J_2 = 3 \end{array} \right.$$

In general it is possible to solve numerical dynamic programming problems which are deterministic by forward as well as by backward recursion. Forward recursion has an advantage over backward recursion for problems with uncertain planning horizons. The relevant planning horizon may be judged to be the one for which, if it were increased, there would be no change in the optimal first decision, or first few decisions.

The relevant decision horizon must be found by experimenting with successively longer horizons until the requisite criterion is met. If backward recursion were used, a fresh problem would have to be solved for each planning horizon. If forward recursion were used, once calculations had been performed for a planning horizon of n stages, only calculations for one additional stage would need to be performed for a decision horizon of $n+1$ stages.

At this point, it is probably as well to point out that the ordering of stage subscripts in dynamic programming recursive equations is conventionally the reverse of that used here. That is, if backward recursion is used, stage subscript i denotes the number of decision stages remaining, instead of the number of the stage reached in the sequence in which decisions are actually taken. The convention of reverse ordering is perhaps useful when the process of backward recursion is actually being worked through. It is not followed here because temporal ordering of stage subscripts makes the interpretation of the recursion equations more straightforward.

2.6 A Simple Crop-Irrigation Problem

In this section, for exposition purposes, a highly simplified problem in the management of a growing resource is introduced. The problem is formulated as a dynamic programming problem and solved numerically by hand. In Section 2.7 the same solution is obtained using a computer program. Other versions of the problem are formulated in subsequent chapters. In Chapter 3 the problem is modified to illustrate the solution of stochastic and infinite-stage problems, and in Chapter 4 to illustrate the analytical solution of a problem with continuous state and decision variables. More realistic crop-irrigation problems are discussed in Chapter 6.

2.6.1 The formulation of the problem

A farmer grows three horticultural crops in successive seasons over one year on 100 ha. Each crop takes four months, or one season, to reach maturity from the time of planting. The yield of each crop (in hundreds of tonnes per 100 ha) is given by

$$y_i = w_i - 0{\cdot}1 w_i^2 \qquad (i = 1 \text{ to } 3)$$

where w_i is the depth of water in centimetres received by the crop grown in the i-th season. The depth of water received depends on the height of water released from storage at the beginning of each season (u_i in metres) and rainfall received during each season (q_i in centimetres). The area of the dam is 1 ha, so

$$w_i = u_i + q_i$$

The dam is full at the beginning of the first season with a water height of 3 m. The amount of water which can be released at the beginning of any season is limited to integer values of metres of water, and by the amount in storage. Rainfall augments the water in storage. The catchment area is 100 ha, so 1 cm of rainfall raises the level of the dam (x_i) by 1 m, provided the dam is not full.

The farmer's objective is to select integer values of u_i so that the present value of receipts from sale of the crops is maximized. The price (in dollars per tonne) received for the *i*-th season crop is b_i. The problem can be formulated as:

$$\max_{u_1, u_2, u_3} \sum_{i=1}^{3} \alpha^{i-1} b_i (w_i - 0{\cdot}1 w_i^2)$$

subject to

$$0 \leq u_i \leq x_i \leq 3, \quad u_i, x_i \text{ integer}$$

$$x_1 = 3$$

$$x_{i+1} = \min(x_i - u_i + q_i, 3)$$

with data

$$b = [50 \quad 100 \quad 150]$$

$$q = [2 \quad 1 \quad 1]$$

$$\alpha = 0{\cdot}95$$

2.6.2 The solution procedure

The backward recursive equation used to solve this problem is

$$V_i\{x_i\} = \max_{0 \leqslant u_i \leqslant x_i} [b_i(w_i - 0{\cdot}1 w_i^2) + \alpha\, V_{i+1}\{x_i - u_i + q_i\}] \qquad (i = 3 \text{ to } 1) \qquad (2.3)$$

with

$$w_i = u_i + q_i$$
$$V_4\{x_4\} = 0$$

The recursive equation is used in the compilation of Table 2.5. The layout is the same as in Table 2.2 for the network problem. Consider the

TABLE 2.5
Optimal Irrigation Returns ($000)

Stage 3

Dam water level x_3 (m)	Water released (ha m)				$V_3\{x_3\}$	$u_3^*\{x_3\}$
	$u_3 = 0$	$u_3 = 1$	$u_3 = 2$	$u_3 = 3$		
	$q_3 = 1;\ w_3 = 1$	$q_3 = 1;\ w_3 = 2$	$q_3 = 1;\ w_3 = 3$	$q_3 = 1;\ w_3 = 4$		
0	13·5 + 0·0 = 13·5				13·5	0
1	13·5 + 0·0 = 13·5	24·0 + 0·0 = 24·0			24·0	1
2	13·5 + 0·0 = 13·5	24·0 + 0·0 = 24·0	31·5 + 0·0 = 31·5		31·5	2
3	13·5 + 0·0 = 13·5	24·0 + 0·0 = 24·0	31·5 + 0·0 = 31·5	36·0 + 0·0 = 36·0	36·0	3

Stage 2

Dam water level x_2 (m)	Water released (ha m)				$V_2\{x_2\}$	$u_2^*\{x_2\}$
	$u_2 = 0$	$u_2 = 1$	$u_2 = 2$	$u_2 = 3$		
	$q_2 = 1;\ w_2 = 1$	$q_2 = 1;\ w_2 = 2$	$q_2 = 1;\ w_2 = 3$	$q_2 = 1;\ w_2 = 4$		
0	9·0 + 22·8 = 31·8				31·8	0
1	9·0 + 29·9 = 38·9	16·0 + 22·8 = 38·8			38·9	0
2	9·0 + 34·2 = 43·2	16·0 + 29·9 = 45·9	21·0 + 22·8 = 43·8		45·9	1
3	9·0 + 34·2 = 43·2	16·0 + 34·2 = 50·2	21·0 + 29·9 = 50·9	24·0 + 22·8 = 46·8	50·9	2

Stage 1

Dam water level x_1 (m)	Water released (ha m)				$V_1\{x_1\}$	$u_1^*\{x_1\}$
	$u_1 = 0$	$u_1 = 1$	$u_1 = 2$	$u_1 = 3$		
	$q_1 = 2;\ w_1 = 2$	$q_1 = 2;\ w_1 = 3$	$q_1 = 2;\ w_1 = 4$	$q_1 = 2;\ w_1 = 5$		
0	8·0 + 43·6 = 51·6				51·6	0
1	8·0 + 48·4 = 56·4	10·5 + 43·6 = 54·1			56·4	0
2	8·0 + 48·4 = 56·4	10·5 + 48·4 = 58·9	12·0 + 43·6 = 55·6		58·9	1
3	8·0 + 48·4 = 56·4	10·5 + 48·4 = 58·9	12·0 + 48·4 = 60·4	12·5 + 43·6 = 56·1	60·4	2

options which can be taken at stage 2 if the dam water level is 3 m. If no water is released, the only water received by the crop is 1 cm of rainfall. The value of the additional crop yield is \$9000. The dam water level will still be 3 m at stage 3 despite the rainfall because of overflow. The optimal return at stage 3 from a water level of 3 m has been calculated to be \$36 000 in the stage-3 section of Table 2.2. The return discounted one stage is $\$(0{\cdot}95 \times 36\,000 = 34\,200)$. The value of releasing no water at stage 2 with 3 m of water is therefore $\$(9{,}000 + 34\,200 = 43\,200)$.

The value of releasing all of the available dam water is the sum of the value of the additional crop yield (\$24 000), and the discounted value of returns from the dam with 1 m of water depth remaining at stage 3 ($\$(0{\cdot}95 \times 24\,000 = 22\,800)$), which equals \$46 800. The optimal amount of water to release can be seen to be 2 m, giving a total return of \$50 900.

The table shows that if the dam is full at the start of the growing season, the maximum return is \$60 400. The optimal decision path, shown in Table 2.6, was derived by tracking forward through time from stage 3 to stage 1. The three crops receive a total of ten units of water. Six are supplied from the dam, and four are received as rainfall. The dam is emptied at the beginning of the third season, though subsequent rainfall leaves the dam with a water height of 1 m at the end of the third season.

TABLE 2.6
Optimal Irrigation Sequence Starting With the Dam Full

Season	*Dam water level (m)*	*Water released (ha m)*	*Season return (\$000)*	*Discounted total return (\$000)*
1	3	2	12·0	60·4
2	3	2	21·0	50·9
3	2	2	31·5	31·5

Table 2.5 can be used to find the optimal sequence of irrigation starting with other water levels and seasons.

2.7 A General-purpose Computer Program for Solving Dynamic Programming Problems

The repetitive nature of the procedure for solving numerical problems is apparent from solving the network and crop-irrigation problems. For problems of any realistic size computer programs are written to find solutions. Usually the computer program is written specifically for one dynamic programming problem. The necessity of writing a computer program to solve a dynamic programming problem may be one of the barriers to the wider adoption of dynamic programming. As noted in Chapter 1, there has been little use of general-purpose computer programs because of the lack of restrictions on the specification of the return function $a_i\{x_i, u_i\}$ and transformation function $t_i\{x_i, u_i\}$. One possibility, followed by Gillett (1976), is to write computer programs for solving specific classes of dynamic programming problem having the same types of return and transformation function—such as network, replacement and inventory problems.

However, it is possible to go further and develop routines which will perform the recursive numerical optimization, storage and forward tracking procedures which are common to the solution of any numerical dynamic programming problem. Such routines have to be given as input data the values of $a_i\{x_i, u_i\}$ and x_{i+1} for all feasible values of x_i and u_i for all decision stages i. For simple problems, it may be straightforward to determine this information and enter it directly. For larger problems it will be necessary to write computer programs to find $a_i\{x_i, u_i\}$ and x_{i+1}.

Hastings (1975) has developed a general-purpose dynamic programming package which is available commercially, known as DYNACODE. Morin (1977) refers to some other codes which are available. A simple set of routines for solving general-purpose dynamic programming problems (referred to as GPDP) is presented in this book. The programs are written in BASIC to make them easily accessible for owners of microcomputers. In this section, the way in which GPDP can be used to solve the simple network and crop-irrigation problems introduced earlier is explained. Instructions for using GPDP to solve more complex problems are given in Chapters 3 and 4.

2.7.1 An introduction to the GPDP programs

There are four GPDP programs, with the following names and functions:

Program	*Purpose*
DPD	To write the data for a DP problem to file
FDP	To read data from file and to determine the *finite-stage* DP solution
IDP	To read data from file and to determine the *infinite-stage* DP solution
DIM	To redimension arrays in FDP and IDP to solve problems of different size

The programs are listed with further background information in Appendix A2. They have been written to run on any computer system with as few modifications as possible. When the programs are run, data are entered directly via a terminal in response to prompting by the programs. To solve the two example problems only DPD and FDP are required. The operation of these programs is explained below. The use of IDP is discussed in Section 3.5.2, and of DIM is Section 4.7.2.

2.7.2 Data entry using DPD

Before data for a dynamic programming problem can be written to file using DPD, the problem data have to be structured. The basic components of the problem—the decision stages, states, decisions, return functions and transformation functions—must be identified. For each stage, each possible state of the system receives an identification number I, starting with $I = 1$ and proceeding upwards. For each decision which can be made at each state number and stage, the state number J which is reached at the next stage is identified, together with the associated stage return, $R(I, J)$.

Data are entered at the prompting of DPD. Up to five blocks of data are required.To make clear the sequence of data entry and to facilitate data preparation, coding sheets are presented in Appendix A1. The relevant coding sheets for a deterministic problem are those for data blocks A and B, CD, D and ED.

Block A data are the data which generally describe the problem: the problem name, the number of decision stages, the rate of discount, whether the problem is deterministic or stochastic, whether the problem is stationary,[3] and whether labels for states and decisions are available.

Block B data are the values of terminal states. DPD asks for the entry, line by line, of terminal state number I followed by value $V(I)$, starting with $I = 1$. Block B is terminated by entering '0, 0'. If $V(I) = 0$ for all I, only the terminating line '0, 0' need be entered.

The next block of data is CD if the problem is deterministic. The data concern all of the decisions which can be made at the last decision stage. The first line consists of the first state number ($I = 1$), the succeeding state number J if the first decision for $I = 1$ were implemented, and the associated stage return $R(I, J)$. Similar lines are entered for all other decisions which can be made for $I = 1$. Lines are then entered for $I = 2$, and so on. Block CD is terminated by entering '0, 0, 0'.

[3] The problem is stationary if the return and transformation functions are the same for all decision stages.

```
PROBLEM NAME ? ROUTE

NO. OF DECISION STAGES ? 4

RATE OF DISCOUNT - PER CENT? 0

PROBLEM TYPE - D(DETERMINISTIC) OR S(STOCHASTIC) ? D

ARE STATES,DECISIONS AND RETURNS THE SAME FOR ALL
STAGES - Y(YES) OR N(NO)? N

DO YOU HAVE LABELS FOR STATES AND DECISIONS -
(THE SAME FOR ALL STAGES) - Y(YES) OR N(NO) ? N
--------------------------------------------------------
VALUES OF TERMINAL STATES AT STAGE  5 :
YOU WILL BE ASKED TO ENTER I,V(I) FOR ALL I
WHERE  I    = TERMINAL STATE NO.
      V(I) = VALUE OF 'I'
ENTER '0,0' AFTER FINAL ENTRY
--------------------------------------------------------
ENTER I,V(I)? 0,0

--------------------------------------------------------
STAGE RETURNS FOR EACH POSSIBLE DECISION:
YOU WILL BE ASKED TO ENTER I,J,R(I,J) FOR ALL I
WHERE  I    = CURRENT STATE NO.
       J    = NEXT STATE NO. (DEPENDENT ON DECISION)
    R(I,J) = STAGE RETURN
ENTER '0,0,0' AFTER FINAL ENTRY FOR EACH STAGE
--------------------------------------------------------
FOR STAGE 4  ENTER I,J,R(I,J)? 1,1,-8
FOR STAGE 4  ENTER I,J,R(I,J)? 2,1,-5
FOR STAGE 4  ENTER I,J,R(I,J)? 0,0,0


FOR STAGE 3  ENTER I,J,R(I,J)? 1,1,-7
FOR STAGE 3  ENTER I,J,R(I,J)? 1,2,-5
FOR STAGE 3  ENTER I,J,R(I,J)? 2,1,-2
FOR STAGE 3  ENTER I,J,R(I,J)? 2,2,-6
FOR STAGE 3  ENTER I,J,R(I,J)? 3,1,-3
FOR STAGE 3  ENTER I,J,R(I,J)? 3,2,-4
FOR STAGE 3  ENTER I,J,R(I,J)? 0,0,0

FOR STAGE 2  ENTER I,J,R(I,J)? 1,1,-7
FOR STAGE 2  ENTER I,J,R(I,J)? 1,2,-6
FOR STAGE 2  ENTER I,J,R(I,J)? 2,1,-4
FOR STAGE 2  ENTER I,J,R(I,J)? 2,2,-5
FOR STAGE 2  ENTER I,J,R(I,J)? 2,3,-7
FOR STAGE 2  ENTER I,J,R(I,J)? 3,2,-6
FOR STAGE 2  ENTER I,J,R(I,J)? 3,3,-8
FOR STAGE 2  ENTER I,J,R(I,J)? 0,0,0

FOR STAGE 1  ENTER I,J,R(I,J)? 1,1,-4
FOR STAGE 1  ENTER I,J,R(I,J)? 1,2,-5
FOR STAGE 1  ENTER I,J,R(I,J)? 1,3,-2
FOR STAGE 1  ENTER I,J,R(I,J)? 0,0,0


*** DATA WRITTEN TO ROUTE.DAT ***
```

FIG. 2.3 Data entry using DPD for the problem 'ROUTE'.

FDP and IDP are designed to maximize the present value of the stream of $R(I, J)$ across all stages. If the present value of the stream of $R(I, J)$ is to be minimized instead, negative values of $R(I, J)$ should be entered.

Block D data are the labels of each state number and decision number if they are available. The user may find them useful for interpreting the solution tables printed by FDP. However, they can only be used if the labels for each state number and decision number are the same at all stages. If in the block A data the answer was 'N' for no to the question 'Are labels available?', no information for labels is requested by DPD.

If the problem is stationary, no further data are requested. However, for non-stationary deterministic problems, block ED data are required. Block ED data consist of data for the remaining decision stages, entered

```
-----FINITE STAGE DP PROBLEM-----
PROBLEM NAME? ROUTE


--------------------------------------------------------------
         DETERMINISTIC DYNAMIC PROGRAMMING SOLUTION
                     PROBLEM ---ROUTE---

PROBLEM PARAMETERS:
NO. OF DECISION STAGES         =  4
RATE OF DISCOUNT (PER CENT)    =  0
--------------------------------------------------------------

OPTIMAL STATE SEQUENCE FOR ALL INITIAL STATES
--------------------------------------------------------------
            STAGE                  VALUE
 1     2    3    4    5
--------------------------------------------------------------
 1     3    2    1    1          -18
--------------------------------------------------------------


DETAILS OF OPTIMAL PATH
ENTER INITIAL STATE NO. (OR '0' TO FINISH) ? 1

--------------------------------------------------------------
STAGE      STATE          DECISION        DISCOUNTED        VALUE
 NO.    ----------       ----------       STAGE RETURN
           NO.              NO.
--------------------------------------------------------------
  1         1                3              -2               -18
  2         3                1              -6               -16
  3         2                1              -2               -10
  4         1                1              -8                -8
  5         1                0               0                 0
--------------------------------------------------------------


DETAILS OF OPTIMAL PATH
ENTER INITIAL STATE NO. (OR '0' TO FINISH) ? 0
```

FIG. 2.4 Solution to the problem 'ROUTE' using FDP.

in the reverse of the temporal order. The input format is the same as for the CD block data.

2.7.3 Using GPDP to solve the least-cost network problem

The way data for the network problem are entered for writing to file using DPD is shown in Fig. 2.3. The entries made by the user are the data following each '?'. The problem has been reformulated as consisting of four instead of three decision stages to facilitate interpretation of the solution tables produced by FDP. The fourth decision stage is a dummy decision stage in that for both states 1 and 2 there is only one decision option—namely to go to terminal state 1 at stage 5. By giving the problem the name 'ROUTE' (Fig. 2.3, line 1), DPD writes the problem data to a file 'ROUTE.DAT'.

Running FDP causes the data to be read from ROUTE.DAT and a solution printed. The results are shown in Fig. 2.4. The last Table in Fig. 2.4 confirms that the optimal sequence of nodes is (1,1), (2,3), (3,2), (4,1), (5,1) with associated total cost of 18.

One reason for writing the problem data to file with DPD before obtaining a solution using FDP is that it is simple to alter any of the data by editing the data file, and then run FDP again, without having to repeat data entry. Another reason is that if the problem is stationary it may be possible to find the infinite-stage solution by running IDP which uses the same data file.

2.7.4 Using GPDP to solve the crop-irrigation problem

The first task is to structure the data for writing to file using DPD. Although the problem is not stationary, the possible states of the system and decisions which can be made at each stage are the same, and can be given the same labels. Table 2.7 shows the system of numbering and labelling states and decisions used.

TABLE 2.7
Identification of States and Decisions

Dam level (m) x	*State number* I	*Label*	*Water released (ha m)* u	*Decision number* D	*Label*
0	1	L0	0	1	R0
1	2	L1	1	2	R1
2	3	L2	2	3	R2
3	4	L3	3	4	R3

```
PROBLEM NAME ? CIDF

NO. OF DECISION STAGES ? 3

RATE OF DISCOUNT - PER CENT? 5.2632

PROBLEM TYPE - D(DETERMINISTIC) OR S(STOCHASTIC) ? D

ARE STATES,DECISIONS AND RETURNS THE SAME FOR ALL
STAGES - Y(YES) OR N(NO)? N

DO YOU HAVE LABELS FOR STATES AND DECISIONS -
(THE SAME FOR ALL STAGES) - Y(YES) OR N(NO) ? Y
----------------------------------------------------
VALUES OF TERMINAL STATES AT STAGE  4 :
YOU WILL BE ASKED TO ENTER I,V(I) FOR ALL I
WHERE  I    = TERMINAL STATE NO.
     V(I) = VALUE OF 'I'
ENTER '0,0' AFTER FINAL ENTRY
----------------------------------------------------
ENTER I,V(I)? 0,0

----------------------------------------------------
STAGE RETURNS FOR EACH POSSIBLE DECISION:
YOU WILL BE ASKED TO ENTER I,J,R(I,J) FOR ALL I
WHERE  I    = CURRENT STATE NO.
       J    = NEXT STATE NO. (DEPENDENT ON DECISION)
    R(I,J) = STAGE RETURN
ENTER '0,0,0' AFTER FINAL ENTRY FOR EACH STAGE
----------------------------------------------------
FOR STAGE 3  ENTER I,J,R(I,J)? 1,2,13.5
FOR STAGE 3  ENTER I,J,R(I,J)? 2,3,13.5
FOR STAGE 3  ENTER I,J,R(I,J)? 2,2,24
FOR STAGE 3  ENTER I,J,R(I,J)? 3,4,13.5
FOR STAGE 3  ENTER I,J,R(I,J)? 3,3,24
FOR STAGE 3  ENTER I,J,R(I,J)? 3,2,31.5
FOR STAGE 3  ENTER I,J,R(I,J)? 4,4,13.5
FOR STAGE 3  ENTER I,J,R(I,J)? 4,4,24
FOR STAGE 3  ENTER I,J,R(I,J)? 4,3,31.5
FOR STAGE 3  ENTER I,J,R(I,J)? 4,2,36
FOR STAGE 3  ENTER I,J,R(I,J)? 0,0,0
STATE NAME ? LEVEL
LABEL FOR STATE NO.  1 ? L0
LABEL FOR STATE NO.  2 ? L1
LABEL FOR STATE NO.  3 ? L2
LABEL FOR STATE NO.  4 ? L3
DECISION NAME ? RELEASE
LABEL FOR DECISION NO.  1 ? R0
LABEL FOR DECISION NO.  2 ? R1
LABEL FOR DECISION NO.  3 ? R2
LABEL FOR DECISION NO.  4 ? R3

FOR STAGE 2  ENTER I,J,R(I,J)? 1,2,9
FOR STAGE 2  ENTER I,J,R(I,J)? 2,3,9
FOR STAGE 2  ENTER I,J,R(I,J)? 2,2,16
FOR STAGE 2  ENTER I,J,R(I,J)? 3,4,9
FOR STAGE 2  ENTER I,J,R(I,J)? 3,3,16
FOR STAGE 2  ENTER I,J,R(I,J)? 3,2,21
FOR STAGE 2  ENTER I,J,R(I,J)? 4,4,9
FOR STAGE 2  ENTER I,J,R(I,J)? 4,4,16
FOR STAGE 2  ENTER I,J,R(I,J)? 4,3,21
FOR STAGE 2  ENTER I,J,R(I,J)? 4,2,24
FOR STAGE 2  ENTER I,J,R(I,J)? 0,0,0
```

FIG. 2.5. Data entry using DPD for the crop-irrigation problem.

```
FOR STAGE 1  ENTER I,J,R(I,J)? 1,3,8
FOR STAGE 1  ENTER I,J,R(I,J)? 2,4,8
FOR STAGE 1  ENTER I,J,R(I,J)? 2,3,10.5
FOR STAGE 1  ENTER I,J,R(I,J)? 3,4,8
FOR STAGE 1  ENTER I,J,R(I,J)? 3,4,10.5
FOR STAGE 1  ENTER I,J,R(I,J)? 3,3,12
FOR STAGE 1  ENTER I,J,R(I,J)? 4,4,8
FOR STAGE 1  ENTER I,J,R(I,J)? 4,4,10.5
FOR STAGE 1  ENTER I,J,R(I,J)? 4,4,12
FOR STAGE 1  ENTER I,J,R(I,J)? 4,3,12.5
FOR STAGE 1  ENTER I,J,R(I,J)? 0,0,0

*** DATA WRITTEN TO CIDF.DAT ***
```

FIG. 2.5.—*contd.*

```
-----FINITE STAGE DP PROBLEM-----
PROBLEM NAME? CIDF


-------------------------------------------------------------
      DETERMINISTIC DYNAMIC PROGRAMMING SOLUTION
                  PROBLEM ---CIDF---

PROBLEM PARAMETERS:
NO. OF DECISION STAGES        =  3
RATE OF DISCOUNT (PER CENT)   =  5.2632
-------------------------------------------------------------

OPTIMAL STATE SEQUENCE FOR ALL INITIAL STATES
-------------------------------------------------------------
      STAGE                VALUE
 1    2   3   4
-------------------------------------------------------------
 1    3   3   2            51.6287
 2    4   3   2            56.3787
 3    4   3   2            58.8787
 4    4   3   2            60.3787
-------------------------------------------------------------


DETAILS OF OPTIMAL PATH
ENTER INITIAL STATE NO. (OR '0' TO FINISH) ? 4

-------------------------------------------------------------
STAGE     STATE       DECISION       DISCOUNTED      VALUE
 NO.   -----------  -----------    STAGE RETURN
        NO. LEVEL    NO. RELEASE
-------------------------------------------------------------
  1      4  L3        3  R2          12              60.3787
  2      4  L3        3  R2          19.95           50.925
  3      3  L2        3  R2          28.4287         31.5
  4      2  L1        0              0               0
-------------------------------------------------------------


DETAILS OF OPTIMAL PATH
ENTER INITIAL STATE NO. (OR '0' TO FINISH) ? 0
```

FIG. 2.6. Solution to the crop-irrigation problem using FDP.

The stage-3 data for data block CD and the stage-2 and -1 data for data block ED were drawn up as shown in Table 2.8. The layout is similar to that used in Table 2.5. The data were transferred to copies of the coding sheets in Appendix A1. Because the crop-irrigation problem is deterministic and has finite stages, CIDF was chosen as the problem name. Fig. 2.5 shows the print-out obtained by running DPD and entering the data. Finally, Fig. 2.6 shows the solution obtained by running FDP for the problem CIDF. The first part of the print-out from FDP gives the optimal sequence of state numbers starting from each possible state at stage 1. The second part of the print-out gives further details of the

TABLE 2.8
Stage Data for the Crop-Irrigation Problem

Stage 3

I	*D = 1*		*D = 2*		*D = 3*		*D = 4*	
	J	*R(I, J)*	*J*	*R(I, J)*	*J*	*R(I, J)*	*J*	*R(I, J)*
1	2	13·5						
2	3	13·5	2	24·0				
3	4	13·5	3	24·0	2	31·5		
4	4	13·5	4	24·0	3	31·5	2	36·0

Stage 2

I	*D = 1*		*D = 2*		*D = 3*		*D = 4*	
	J	*R(I, J)*	*J*	*R(I, J)*	*J*	*R(I, J)*	*J*	*R(I, J)*
1	2	9·0						
2	3	9·0	2	16·0				
3	4	9·0	3	16·0	2	21·0		
4	4	9·0	4	16·0	3	21·0	2	24·0

Stage 1

I	*D = 1*		*D = 2*		*D = 3*		*D = 4*	
	J	*R(I, J)*	*J*	*R(I, J)*	*J*	*R(I, J)*	*J*	*R(I, J)*
1	3	8·0						
2	4	8·0	3	10·5				
3	4	8·0	4	10·5	3	12·0		
4	4	8·0	4	10·5	4	12·0	3	12·5

optimal decision sequence for any nominated initial-state number, in this case the state number 4 representing a full dam. The results agree with those presented in Table 2.6 based on the calculations in Table 2.5.

References

Bellman, R.E. (1957), *Dynamic Programming*, Princeton University Press, Princeton, New Jersey.

Cooper, L. and Cooper, M.W. (1981), *Introduction to Dynamic Programming*, Pergamon Press, Oxford.

Gillett, B.E. (1976), *Introduction to Operations Research: A Computer Oriented Approach*, McGraw-Hill, New York.

Hastings, N.A.J. (1973), *Dynamic Programming with Management Applications*, Butterworths, London.

Hastings, N.A.J. (1975), *Dynacode: Dynamic Programming System Handbook*, Albany Interactive, 76 Kooringal Road, Upwey, Victoria, Australia, 3158.

Mitten, L.G. (1964), Composition principles for synthesis of optimal multistage processes, *Operations Research*, **12**, 610–19.

Morin, T.L. (1977), Computational advances in dynamic programming, in *Dynamic Programming and Its Applications*, M.L. Puterman (ed.), Academic Press, New York.

Nemhauser, G.L. (1966), *Introduction to Dynamic Programming*, J.Wiley, New York.

Chapter 3

STOCHASTIC AND INFINITE-STAGE DYNAMIC PROGRAMMING

All decision making is based on the outcomes of alternative actions which can only be anticipated, and is therefore subject to some degree of uncertainty. Deterministic formulations of dynamic programming problems are satisfactory if the uncertainty is sufficiently low or if a certainty-equivalent problem exists. Otherwise, a stochastic formulation is required. The formulation and solution of stochastic problems are explained in Section 3.1.

For some problems the return and transformation functions are the same at all decision stages. Such problems are described as *stationary* problems. As previously, let n denote the number of decision stages of the problem, or the *decision horizon*, and let $u_i^*\{x_i\}$ refer to the vector of optimal decisions corresponding to the vector of states at stage i. For many stationary problems there exists a decision horizon of n' decision stages which results in a first-stage state-decision profile identical to that for any decision horizon larger than n'. The state-decision profile remains unchanged as n' is increased towards infinity, and for this reason is known as the *infinite-stage* decision vector. The infinite-stage decision vector may be of interest because n is known to be large, or in order to discover the maximum number of decision stages which are relevant in the stationary problem.

Different solution procedures are required for the infinite-stage problem, depending on whether stage returns are discounted. The solution of problems with discounting is described in Section 3.2, and of problems without discounting in Section 3.3. The scope for applying infinite-stage methods to problems in agriculture and natural resources is considered in Section 3.4. Solution procedures are illustrated for the crop-irrigation problem. The chapter concludes with instructions for the use of GPDP routines to solve stochastic and infinite-stage problems.

3.1 Stochastic Dynamic Programming

In many decision problems, the state transformation depends not only on the state of the system and the decision taken, but also on unpredictable events outside the control of the decision maker. Likewise, the stage return may partly depend on unpredictable factors. Provided the only unpredictable events affecting the stage return and state transformation at stage i are those occurring at stage i and not any earlier, the problem may be formulated as a stochastic dynamic programming problem without additional state variables. For example, the crop irrigation problem introduced in Chapter 2 may be reformulated as a stochastic dynamic programming problem if rainfall q_i is regarded as unpredictable or as a random variable. With q_i random, the change in the water level of the dam (state transformation) and the financial contribution to total returns (stage return) are uncertain. A worked example for this problem is presented in Section 3.1.2.

3.1.1 Formulation of the stochastic problem

In many deterministic dynamic programming problems the objective function is the present value of stage returns. An alternative objective function is required for stochastic problems. The most straightforward objective function is the present value of *expected* stage returns. This may be a reasonable objective function if the decision maker is risk-neutral, and is the one considered in this chapter. Consideration of alternative objectives under uncertainty is left to Chapter 4.

Let the random variable at stage i which affects the state transformation and possibly also the stage return be independent and denoted by r_i. For a discrete formulation, r_i is limited to a domain of m values. Allow the domain of r_i to be mapped into k_i, the range of integers 1 to m. The probability that r_i takes the k-th discrete value in its domain is denoted by $p_i\{k_i\}$. Instead of referring to r_i as the value of the random variable, it is convenient to refer to k_i for the purposes of describing the stochastic process.

Suppose the state of the system at stage $i+1$ and the stage return at stage i depend on x_i, u_i and k_i only. The decision process is described as a *Markov decision process*. The stage return and state transformation functions can be respecified as $a_i\{x_i, u_i, k_i\}$ and $t_i\{x_i, u_i, k_i\}$, respectively. If the objective function for a problem is the expected present value of stage returns the recursive equation for the stochastic dynamic programming problem is a simple extension of that for the deterministic problem. It is

$$V_i\{x_i\} = \max_{u_i} \left[\sum_{k=1}^{m} p_i\{k_i\}(a_i\{x_i, u_i, k_i\} + \alpha V_{i+1}\{t_i\{x_i, u_i, k_i\}\}) \right] \quad (i = n, \dots, 1) \qquad (3.1)$$

subject to

$$\sum_{k=1}^{m} p_i\{k_i\} = 1$$

with

$$V_{n+1}\{x_{n+1}\} = F\{x_{n+1}\}$$

Recursive solution of (3.1) results in the determination of $V_i\{x_i\}$ and the optimal decision vector $u_i^*\{x_i\}$ for $i = n$ to 1. However, in the absence of knowledge of the sequence k_1 to k_n, it is not possible to find the optimal decision path by tracking forward from the initial state.

Note that there is no difficulty in allowing the probability distribution of k_i to depend on x_i and u_i. Notationally, $p_i\{k_i \mid x_i, u_i\}$ replaces $p_i\{k_i\}$ in (3.1).

3.1.2 A stochastic crop-irrigation problem

As an example of a stochastic dynamic programming problem, consider a stochastic version of the deterministic crop-irrigation problem introduced in Chapter 2. Suppose rainfall is now to be treated as a random

TABLE 3.1
Rainfall Probabilities

k	*Decision stage*					
	1		*2*		*3*	
	$p_1\{k_1\}$	q_1^k	$p_2\{k_2\}$	q_2^k	$p_3\{k_3\}$	q_3^k
1	0·25	1	0·25	0	0·25	0
2	0·50	2	0·50	1	0·50	1
3	0·25	3	0·25	2	0·25	2

variable, q_i^k, where $k = 1$, 2 or 3. In the i-th season, q_i^k occurs with probability $p_i\{k_i\}$ shown in Table 3.1. Expected rainfall in each season is the same as for the deterministic problem.

Suppose the objective is to find the optimal decision vector $u_i^*\{x_i\}$ which maximizes expected net return. The recursive equation is

$$V_i\{x_i\} = \max_{u_i}\left[\sum_{k=1}^{3} p_i\{k_i\}(b_i(u_i + q_i^k - 0{\cdot}1(u_i + q_i^k)^2) + \alpha V_{i+1}\{x_i - u_i + q_i^k\})\right] \qquad (i = 3 \text{ to } 1)$$

subject to

$$0 \leq u_i \leq x_i \leq 3, \quad u_i, x_i \text{ integer}$$

$$\sum_{k=1}^{3} p_i\{k_i\} = 1$$

with

$$V_4\{x_4\} = 0$$

The results from applying the equation recursively are shown in Table 3.2. Comparing Table 3.2 for the stochastic problem with Table 2.5 for the deterministic problem, it is evident that the optimal decision vectors, $u_i^*\{x_i\}$, are the same for both problems. This means that in this example some of the solution information for the deterministic problem, which was redundant for finding the optimal irrigation sequence starting with a full dam and rainfall known, could have been used (optimally, as it turns out) for the stochastic problem. For example, for the deterministic problem, the optimal irrigation sequence, starting with a full dam, leads to a storage level of 3 m at the beginning of season 2. However, if rainfall in season 1 had been only 1 instead of 2 cm, a storage level of 2 m would have resulted. Stage-2 calculations in Table 2.5 show the optimal decision would have been 'release 1 ha m of water', but predicated on rainfall from then on being known. Stage-2 calculations in Table 3.2 show the same decision is optimal for the stochastic case. This result might be expected given the certainty equivalence theorem discussed in Section 4.3.2, although the theorem is not strictly applicable to problems with decision variables subject to integer constraints.

TABLE 3.2
Optimal Irrigation Returns ($000)

Stage 3

Dam water level x_3 (m)	Water released (ha m)				$V_3\{x_3\}$	$u_3^*\{x_3\}$
	$u_3=0$	$u_3=1$	$u_3=2$	$u_3=3$		
0	12·75				12·75	0
1	12·75	23·25			23·25	1
2	12·75	23·25	30·75		30·75	2
3	12·75	23·25	30·75	35·25	35·25	3

Stage 2

Dam water level x_2(m)	Water released (ha m)				$V_2\{x_2\}$	$u_2^*\{x_2\}$
	$u_2=0$	$u_2=1$	$u_2=2$	$u_2=3$		
0	29·88				29·88	0
1	37·00	36·88			37·00	0
2	40·92	44·00	41·88		44·00	1
3	41·99	47·92	49·00	44·88	49·00	2

Stage 1

Dam water level x_1 (m)	Water released (ha m)				$V_1\{x_1\}$	$u_1^*\{x_1\}$
	$u_1=0$	$u_1=1$	$u_1=2$	$u_1=3$		
0	49·08				49·08	0
1	53·11	51·58			53·11	0
2	54·30	55·61	53·08		55·61	1
3	54·30	56·80	57·11	53·58	57·11	2

Table 3.2 cannot be used to find the optimal irrigation sequence until the actual rainfall in each season is known. Thus at the beginning of the first season, if the dam is full, it is known that the present value of expected returns is maximized by releasing 2 ha m of water, but the optimal decision for the second growing interval is unknown until rainfall in the first growing interval has been observed. The present value of expected returns is $57 110.

TABLE 3.3
Ex Post Irrigation Sequence for Two Rainfall Profiles

Season	*Most-likely rainfall*			*Drought*		
	Rainfall (cm)	*Dam level (m)*	*Water released (ha m)*	*Rainfall (cm)*	*Dam level (m)*	*Water released (ha m)*
1	2	3	2	1	3	2
2	1	3	2	0	2	1
3	1	2	2	0	1	1
Present value of returns		\$60 400			\$33 000	

Table 3.3 shows the optimal irrigation sequence for the stochastic case for two possible rainfall profiles which could eventuate—a most likely rainfall profile, and a drought profile. The irrigation sequences are derived from Table 3.2. If the most-likely rainfall pattern occurs, the decision sequence which follows from the decision vectors results in an *ex post* present value of returns equal to \$60 400. This compares with \$57 110 for the present value of expected returns at the beginning of season 1 under uncertainty.

Note that if the drought sequence could have been forecast at the beginning of season 1, a more profitable irrigation sequence could have been followed. The release sequence 0 1 3 instead of 2 1 1 would have resulted in a present value of \$45 000 instead of \$33 000.

3.2 Infinite-stage Dynamic Programming for Problems with Discounting

The optimal decision vector for stationary problems with infinite decision stages is generally invariant with respect to stage. An invariant decision vector does not mean that a single equilibrium state or decision is optimal for all stages. It does mean that at any stage the optimal decision depends only on the state of the system and not on the decision stage. An intuitive rationale is that the optimal decision for any state should not alter if the number of decision stages is very large and is increased by one.

3.2.1 Formulation of the problem

Any stationary decision vector started from one of a finite number of states and applied over a large enough number of decision stages leads to the repetitive visiting of either one state or a sequence of states. In other words, the decision process results eventually in a constant stream of finite returns. The present value of the infinite stream of stage returns therefore converges to a finite sum for any positive discount rate. It follows that the value of each state at the i-th decision stage with an unlimited number of decision stages remaining, $V_i\{x_i\}$, is finite and equal to $V_{i+1}\{x_{i+1}\}$. Consequently the basic recursive equation for the stationary, infinite-stage problem with discounting is the basic recursive equation for the general dynamic programming problem (eqn (2.2)) but without stage subscripts. That is

$$V\{x\} = \max_u [a\{x, u\} + \alpha V\{t\{x, u\}\}] \qquad (3.2)$$

In the nature of the problem, there are no terminal values. Note that the function $V\{x\}$ appears on both sides of the equation. This feature is exploited in solution methods. Two basic solution methods which are explained in this chapter are value iteration and policy iteration. Linear programming as a solution method is discussed in Chapter 4.

3.2.2 Solution by value iteration

The infinite-stage problem is solved using value iteration by applying a modified version of the basic recursive equation for the general dynamic programming problem (eqn (2.2)). Stage subscripts are used to denote iteration number instead of decision-stage number. The recursive equation is

$$V_i\{x_i\} = \max_{u_i} [a\{x_i, u_i\} + \alpha V_{i-1}\{t\{x_i, u_i\}\}] \qquad (i = 1, \ldots, n) \quad (3.3)$$

The idea of value iteration is that after a large number of iterations, n^*, the optimal decision vector obtained for iteration n^* and all subsequent iterations is the same as the optimal infinite-stage decision vector. To initiate the iteration process, the vector $V_0\{x_0\}$ is required. The vector is a guess at $V\{x\}$, which could be the null vector.

The difficulty with value iteration is the determination of n^*. A good indication that the n^*-th iteration has been reached is the constancy of the optimal decision vector over several successive iterations. However, to be certain that the optimal infinite-stage decision vector has indeed been found by the i-th iteration, $V_i\{x_i\}$ must be a close approximation to $V\{x\}$. The closeness of the approximation can be judged by the size of $V_{i+1}\{x_{i+1}\} - V_i\{x_i\}$. The larger the discount factor, α, the less rapid is the reduction in size of the difference between one vector and the next. Hastings (1973, pp.138–43) discusses how the difference vector can be used to assess at what stage satisfactory convergence has been reached. Hastings and Mello (1973) show how suboptimal decisions can be identified and discarded during the solution process.

3.2.3 Solution by policy iteration

The advantage of policy iteration over value iteration is that both the optimal decision vector and $V\{x\}$ can be found exactly within a finite number of iterations. The disadvantage is that large numbers of simultaneous equations have to be solved many times for problems with a large number of state variables. The algorithm for policy iteration requires the iteration of two steps, a value-determination step followed by a policy-improvement step.

The value-determination step is best explained with matrix notation. At each stage, let there be m possible states. For the j-th state, let there be d_j decisions which could be made. Then the total number of possible decision vectors is:

$$z = \prod_{j=1}^{m} d_j$$

Arrange for the numbering of the decision vectors from 1 to z, and for any particular decision vector to be referred to by superscript l. The l-th decision vector implies an m-vector of returns, a^l, and an $m \times m$ transformed-state matrix, B^l. The vector a^l and matrix B^l are defined as follows: the j-th element of a^l is the stage return from going from state j to state k set by the l-th decision vector. The j-th row of B^l consists of zero elements except for the k-th element which is '+1'.

The limiting state-value vector, $V^l\{x\}$, which would result from the application of the l-th decision vector $u^l\{x\}$ at each stage indefinitely, follows from (3.2) and is

$$V^l = a^l + \alpha B^l V^l \tag{3.4}$$

V^l can be found by solving

$$V^l = (I - \alpha B^l)^{-1} a^l \tag{3.5}$$

where I is an identity matrix.

The second step, policy improvement, is performed by applying (3.3), the same recursive equation as is used for value iteration.

In summary, the policy-iteration algorithm is as follows:

(1) *Initialization*
Select $u^l\{x\}$.
(Under policy iteration, $u^l\{x\}$ is a guess at the optimal policy; if a check for optimality is being made under value iteration at stage n, l is the l appropriate for $u_n\{x\}$.)

(2) *Value determination*
Set $V^l = (I - \alpha B^l)^{-1} a^l$.

(3) *Policy improvement*
Set $\hat{V}^l\{x\} = \max\limits_u \; [a\{x, u\} + \alpha V^l\{t\{x, u\}\}]$.

(4) *Optimality check*
If the maximizing $u^*\{x\}$ for 3 corresponds to the l-th decision vector, it is the optimal decision vector and $V\{x\}$ in (3.2) equals $\hat{V}^l\{x\} = V^l\{x\}$. Otherwise, set l equal to the appropriate l for $u^*\{x\}$ and go to 2.

A worked example of the policy-iteration routine with discounting is given in Section 3.4.3.

3.3 Infinite-stage Dynamic Programming for Problems Without Discounting

3.3.1 Formulation of the problem

A different approach is required to find the infinite-stage decision vector if $\alpha = 1$. In general, as $n \to \infty$ in (3.3), $V_i\{x_i\}$ does not converge to a finite $V\{x\}$, but approaches plus or minus infinity. For example, if $a\{x, u\}$ is always positive, $V_i\{x_i\}$ approaches plus infinity as $n \to \infty$.

The solution of infinite-stage problems without discounting is not so straightforward as the solution of those with discounting. For simplicity, only problems which are *unichain* will be considered. Any decision vector implemented indefinitely starting at any one of a finite number of states will eventually lead to the continuous repetition of either states or chains of states. A problem is unichain if any decision vector implemented indefinitely leads to the continuous repetition of one chain of states only. For example, a typical infinite-stage machine replacement problem is a unichain problem. The decision for any age of the machine is either to keep the machine at least one more year or to replace it with a new one, except that a machine which is the oldest possible must always be replaced.

The machine must therefore eventually be replaced, whereupon the system goes to the state representing the lowest-aged machine, say state 1. The state number is incremented by one for every 'keep' decision, but reverts to state 1 once a replacement decision is made. This one chain of states will repeat indefinitely.

Suppose repetition of a chain of t states results in a flow of $\sum_{i=1}^{t} a_i$ after every t periods. If the chain is repeated indefinitely, an average flow per period of $(1/t)\sum a_i = g$ is obtained. Assuming that the objective of the finite-stage problem is to maximize the sum of stage returns, the objective of the corresponding infinite-stage problem is to find that chain of states which maximizes the average flow per period, g. The return b from passing through states not in the chain on the way to reaching a state in the chain is of secondary importance, because it is finite.

It can be shown that as the number of stages in the planning horizon, n, goes to infinity, the average flow per period is g, regardless of the initial state. After n stages the average flow per period is

$$\mathrm{AF} = \left(b + \mathrm{int}\,(n/t)\sum_{i=1}^{t} a_i + \sum_{i=1}^{r} a_i\right)\Big/ n$$

$$= \left(b + \sum_{i=1}^{r} a_i\right)\Big/ n + \sum_{i=1}^{t} a_i/(t + r/\mathrm{int}\,(n/t)) \tag{3.6}$$

where $\mathrm{int}\,(z)$ is the integer transform of z which rounds downwards, and $r = n - (\mathrm{int}\,(n/t))t$.

As $n \to \infty$, the first term of (3.6) approaches zero, and the second term approaches $\sum_{i=1}^{t} a_i/t$ which equals g. Therefore AF approaches g as n approaches infinity.

This result can be used to relate the value of starting at one state to the value of starting at other states. Suppose the initial state is the i-th state, and implementing the stationary decision vector leads to the j-th state at the next stage with return $a\{i,j\}$. The value of starting at the i-th state with n (approaching infinity) stages in the planning horizon is $V\{i\}$. The same average flow per period is obtained from j as from i, though $V\{j\}$ need not equal $V\{i\}$. If the planning horizon is extended by one period there are two equivalent ways of expressing the value of starting at state i: first, as $V\{i\}$ plus the average flow per period, g; secondly, as the value of starting at state j with n periods in the planning horizon, plus the stage return associated with the first transition from state i to state j. That is

$$V\{i\} + g = a\{i,j\} + V\{j\} \tag{3.7}$$

Values of $V\{i\}$ cannot be found for $n = \infty$ because they are unbounded. However, it is possible to find values of $V\{i\}$ relative to the value of an arbitrarily selected state, say the first. The relative-value function is defined

$$V_r\{i\} \equiv V\{i\} - V\{1\} \qquad (i = 1, \ldots, m)$$

which implies

$$V_r\{1\} = 0$$

Subtracting $V\{1\}$ from both sides of (3.7) leads to

$$V_r\{i\} + g = a\{i,j\} + V_r\{j\} \qquad (i = 1, \ldots, m) \tag{3.8}$$

Note that g depends on the decision vector $j\{i\}$. Values of the $(m-1)$ relative values $V_r\{i\}$, $i = 2, \ldots, m$ and of g can be determined given the m equations (3.8). The problem is to find the optimal decision vector $j^*\{i\}$ which maximizes

$$V_r\{i\} = a\{i, j^*\{i\}\} + V_r\{j^*\{i\}\} - g\{j^*\} \qquad (i = 1, \ldots, m)$$

Reverting to the notation of previous sections, the problem is to find

$$V_r\{x\} + g = \max_u \left[a\{x,u\} + V_r\{t\{x,u\}\}\right] \tag{3.9}$$

where g is the maximal average flow per period and depends on $u^*\{x\}$.

3.3.2 Solution by value iteration

The procedure is similar to value iteration with discounting. Equation (3.3) is applied iteratively, starting with $i = 1$ and a guess at $V_0\{x_0\}$. Iterations continue until the optimal decision vector appears to have been determined, judged by the constancy of the decision vector over successive iterations. Without discounting, $V_{i+1}\{x_{i+1}\} - V_i\{x_i\}$ converges to g instead of to zero as in the case of the discounted problem. Hastings (1973, pp. 136–8) shows how the accuracy of $V_{i+1}\{x_{i+1}\}$ as an estimate of $V\{x\}$ can be judged by the closeness of $V_{i+1}\{x_{i+1}\} - V_i\{x_i\}$ to optimal g. He also shows how the method can be used in conjunction with tests which allow nonoptimal decisions to be progressively eliminated (Hastings, 1976).

3.3.3 Solution by policy iteration

The procedure is similar to policy iteration with discounting. The same notation is used. The equation for value determination for the l-th decision vector is

$$V_r^l = a^l + B^l V_r^l - g^l$$

where g^l is a vector with each element equal to the gain for the l-th decision vector. V_r^l can be found by solving

$$V_r^l = (I - B^l)^{-1}(a^l - g^l)$$

where I is an identity matrix.

At the policy improvement step a check is made to see if the decision vector can be improved. Following (3.9), this is accomplished by finding the decision vector which satisfies

$$R^l\{x\} = \max_u [a\{x, u\} + V_r^l\{t\{x, u\}\}]$$

If the decision vector is the same as the l-th, the l-th decision vector is optimal. If it is not, the value-determination step is repeated for the new decision vector.

The algorithm for policy iteration is:

(1) *Initialization*
Select $u^l\{x\}$.

(2) *Gain and value determination*
Calculate g^l given $u^l\{x\}$.
Set $V_r^l = (I - B)^{-1}(a^l - g^l)$.

(3) *Policy improvement*
Set $R^l\{x\} = \max_u [a\{x, u\} + V_r^l\{t\{x, u\}\}]$.

(4) *Optimality check*
If the maximizing $u^*\{x\}$ for 3 corresponds to the l-th decision vector, it is the optimal decision vector and $V_r\{x\}$ in (3.9) equals $V_r^l\{x\}$.
Otherwise, set l equal to the l appropriate for $u^*\{x\}$ and go to 2.

A worked example of the policy-iteration routine without discounting is given in Section 3.4.4.

3.4 Solving Infinite-stage Problems in Practice

3.4.1 Applications to agriculture and natural resources

One reason for finding the infinite-stage solution may be that decisions are expected to continue indefinitely. Such expectations are more likely in the management of biological resources which grow and reproduce than in the management of exhaustible resources such as minerals. However, even in the case of the management of a biological resource such as a stock of fish, the decision sequence may be finite because it may be optimal to eventually exhaust or extinguish the resource.

A more likely reason for interest in infinite-stage solutions is the knowledge that, for stationary problems with planning horizons beyond a finite, often small, number of stages, the optimal decision vector is stationary. Burt and Allison (1963, pp. 135–6) have argued that, even for problems which are non-stationary in the short run, because forecasts of changes from stage to stage can be made, an infinite-stage solution for the uncertain longer run may be relevant for making immediate decisions. That is, the problem of the longer run may be characterized as stationary for lack of information about future changes. The infinite-stage solution for the longer-run problem gives state-value vectors which may be used as the terminal values for the short-run non-stationary problem.

For a stationary decision vector to be optimal, the problem must be stationary. That is, the return and transformation functions must be the same for each decision stage. How restrictive is an assumption that unit costs, prices and production relationships do not change from stage to stage?

An assumption of unchanging production relationships between decision stages may be reasonable for those resources from which an annual harvest such as cereals, wool or fruit is taken. However, biological input–output relationships may change from stage to stage because biomass changes, or because of aging. Such changes can be allowed for within a stationary framework if the state variable is defined to include biomass or age.

An assumption of constant prices, even if they are in real rather than money terms, may be less satisfactory for planning horizons greater than a year or so. If the analyst has forecasts of price changes for several periods into the future, infinite-stage solutions may still be useful in two ways. First, sensitivity analysis may be conducted by obtaining infinite-stage solutions for a range of prices. Secondly, the suggestion of Burt and Allison of obtaining infinite-stage solutions for the uncertain longer run may be relevant.

Another approach to price variability is to treat price at stage i as a random variable dependent on price at stage $i-1$. A stochastic dynamic programming problem may be formulated with previous-stage price included as a state variable. The problem is stationary if the autoregressive price process is stationary. An example involving stochastic livestock prices is considered in Section 7.4.3.

Agricultural problems with several decision stages within the year such as multicropping, cattle raising and dairying are often complicated by seasonal change. Seasonal weather patterns may change the levels of important inputs such as water and solar energy between decision stages. The availability and hence price of certain feeds may vary throughout the year. However, seasonal variation can be incorporated within a stationary framework by adding season as a state variable. If there are n_1 decision stages within the year subject to seasonal change, and n_2 possible biomass levels at each stage, the decision system can be fully described by one of $n_1 \times n_2$ possible states. As an example, the crop-irrigation problem is formulated as a stationary problem in the next section.

The solution procedure for policy iteration is different depending on whether stage returns are discounted. If stage returns are not discounted, the problem must be unichain for all decision vectors if the policy-iteration procedure described above is to be applicable. The requirement that a problem be unichain can be quite restrictive. For example, problems which allow decisions for the state to remain unchanged from one stage to the next are not unichain. Water level in a dam remains constant if the water taken out for irrigation is balanced by the input of water from rainfall. The biomass of a fish stock remains constant if the weight of fish harvested is the same as the weight of natural net increase in the stock.

The rate of interest appropriate for a stationary problem is the real, and not the money, rate of interest. If there are several decision stages within the year, the discount factor, α, may be so close to one that there may be no significant difference between results for the discounted and those for the undiscounted problem. The matrix $(I - \alpha B^l)$ in (3.5) may even be singular, making it impossible to use the policy-iteration method described in Section 3.2.3. If the problem is unichain, the policy-iteration method described in Section 3.3.3 is relevant. If the problem is not unichain, solution by value iteration is always possible.

3.4.2 The infinite-stage crop-irrigation problem

The crop-irrigation problem introduced in Chapter 2 had three decision stages for each season within one year. The return and transformation functions were different for each stage, so the problem was non-stationary. In this section the problem is extended to finding the optimal irrigation decisions at each decision stage across many years. This means that in order to determine how much water to use in a year, the amount of water that will be left for the start of the following year must be taken into account.

The extended problem could be solved for a finite planning horizon of many years in straightforward fashion by applying (2.3) recursively. Suppose, however, that the infinite-stage solution is required. Although the problem as formulated in Chapter 2 is non-stationary, the return and transformation functions for each season through all years are the same. The infinite-stage solution could be obtained using value iteration in the same sort of way as for a finite horizon of many years. Iterations would cease once it was evident that the decision vectors for each season were stationary. Alternatively, the problem can be reformulated to make it stationary, and solved by policy iteration. This approach is demonstrated below.

The state of the system at any decision stage can be completely specified by both the water level of the dam x and the season s, which takes the value 1, 2 or 3. The additional state variable s is used to describe the system in the reformulated recursive equation. For any combination of x and s, the return and transformation functions are independent of the decision stage. The state transformation function in the reformulated problem makes x_{i+1} a function of x_i and u_i as before, but s_{i+1} is updated automatically from s_i without reference to x_i and u_i.

The recursive equation for the infinite-stage problem is

$$V\{x, s\} = \max_u [b_s(u + q_s - 0{\cdot}1(u + q_s)^2) + \alpha\, V\{x - u + q_s, t\{s\}\}]$$

where

$$t\{s\} = \begin{vmatrix} s+1 & \text{if } s = 1 \text{ or } 2 \\ 1 & \text{if } s = 3 \end{vmatrix}$$

3.4.3 Solution to the crop-irrigation problem with discounting

The infinite-stage solution is obtained by value iteration after the second iteration. The solution tables for value iteration are not shown because the solution procedure is similar to that explained in Chapter 2. Instead, tables are shown for solution by policy iteration.

There are 12 combinations of the four dam levels and three seasons which describe all possible states of the stationary irrigation problem. The system for numbering states is shown in Table 3.4. The combination of x (dam level) and s (season) is denoted by the state number I. Following the policy iteration algorithm detailed in Section 3.2.3, an initial decision vector is selected $u^l\{I\}$ which is a guess at the optimal decision vector. The trial initial decision vector $u^1\{I\}$ shown in Table 3.4 results in all of the water in the dam being released.

In the value-determination step the present value of the returns from following the policy $u^l\{I\}$ for an infinite number of stages is calculated for all 12 possible states. The method suggested in Section 3.2.3 for calculating $V^l\{I\}$ requires the solution of 12 simultaneous equations. Because the problem is deterministic, a simpler method of calculation can be followed. First the recurring chain of states is identified by starting at any state and following through the subsequent states dictated by the decision vector. For example, starting with state 1 and implementing $u^l\{I\}$, the sequence of states is

$$1 \quad 7\ 10\ 2 \quad 7\ 10\ 2 \quad 7\ldots$$

The chain 7 10 2 is reached after starting at any other state. Given the values of $a\{x, s, u\}$ shown in the first tableau of Table 3.4, the present value of repeating the chain an infinite number of times is

$$\begin{aligned} V^1\{7\} &= \$1000(21{\cdot}0 + 24{\cdot}0(0{\cdot}95) + 10{\cdot}5(0{\cdot}95)^2) \\ &\quad \times (1{\cdot}166\,35/0{\cdot}166\,35) \\ &= \$373\,500 \end{aligned}$$

where 0·166 35 is the annual rate of interest obtained by compounding the rate per season ($1/0{\cdot}95 - 1 = 0{\cdot}052\,63$) over three seasons.

$V^1\{10\}$ and $V^1\{2\}$ can be calculated in similar fashion. It is straightforward to find the remaining $V^l\{I\}$. For example,

$$\begin{aligned} V^1\{1\} &= \$(8000 + 0{\cdot}95(V^1\{7\})) \\ &= \$362\,900 \end{aligned}$$

The policy-improvement calculations for the first iteration are shown in the second tableau of Table 3.4. The calculations are similar to those performed in value iteration. For the first and second iterations l is arbitrarily set equal to 1 and 2, respectively, for convenience. The final column shows the decision vector $u_1^*\{I\}$ which maximizes $\hat{V}^1\{I\}$. Because $u_1^*\{I\}$ differs from the initial guess $u^1\{I\}$ it cannot be optimal. Calculations proceed with another round of value determination and policy improvement. Because $u_2^*\{I\} = u^2\{I\}$, $u_2^*\{I\}$ is optimal. This is confirmed by noting that $\hat{V}^2\{I\} = V^2\{I\}$.

TABLE 3.4
Solution Tableaux for the Crop-Irrigation Problem with Discount Factor $\alpha = 0{\cdot}95$

Iteration 1: Value determination given $u^1\{I\}$

Current state number I	Dam water level x (m)	Season s	Water released $u^1\{I\}$ (ha m)	Next state number J	Season return $a\{x, s, u\}$ ($000)	Value $V^1\{I\}$ ($000)
1	0	1	0	7	8·0	362·9
2	1	1	1	7	10·5	365·4
3	2	1	2	7	12·0	366·9
4	3	1	3	7	12·5	367·4
5	0	2	0	10	9·0	361·5
6	1	2	1	10	16·0	368·5
7	2	2	2	10	21·0	373·5
8	3	2	3	10	24·0	376·5
9	0	3	0	2	13·5	360·6
10	1	3	1	2	24·0	371·1
11	2	3	2	2	31·5	378·6
12	3	3	3	2	36·0	383·1

Iteration 1: Policy improvement given $V^1\{I\}$

Current state number I	Dam water level x (m)	Season s	Water released (ha m)				$\hat{V}^1\{I\}$ ($000)	$u_1^*\{I\}$
			$u=0$	$u=1$	$u=2$	$u=3$		
1	0	1	362·8				362·8	0
2	1	1	365·7	365·3			365·7	0
3	2	1	365·7	368·2	366·8		368·2	1
4	3	1	365·7	368·2	369·7	367·3	369·7	2
5	0	2	361·5				361·5	0
6	1	2	368·7	368·5			368·7	0
7	2	2	372·9	375·7	373·5		375·7	1
8	3	2	372·9	379·9	380·7	376·5	380·7	2
9	0	3	360·6				360·6	0
10	1	3	362·1	371·1			371·1	1
11	2	3	360·9	372·6	378·6		378·6	2
12	3	3	360·9	373·0	380·1	383·1	383·1	3

TABLE 3.4—*contd.*

Iteration 2: Value determination given $u^2\{I\} = u_1^*\{I\}$

Current state number I	Dam water level x (m)	Season s	Water released $u^2\{I\}$ (ha m)	Next state number J	Season return $a\{x, s, u\}$ (\$000)	Value $V^2\{I\}$ (\$000)
1	0	1	0	7	8·0	390·5
2	1	1	0	8	8·0	395·3
3	2	1	1	8	10·5	397·8
4	3	1	2	8	12·0	399·3
5	0	2	0	10	9·0	388·6
6	1	2	0	11	9·0	395·7
7	2	2	1	11	16·0	402·7
8	3	2	2	11	21·0	407·7
9	0	3	0	2	13·5	389·0
10	1	3	1	2	24·0	399·5
11	2	3	2	2	31·5	407·0
12	3	3	3	2	36·0	411·5

Iteration 2: Policy improvement given $V^2\{I\}$

Current state number I	Dam water level x (m)	Season s	Water released (ha m)				$\hat{V}^2\{I\}$ (\$000)	$u_2^*\{I\}$
			$u = 0$	$u = 1$	$u = 2$	$u = 3$		
1	0	1	390·6				390·6	0
2	1	1	395·3	393·1			395·3	0
3	2	1	395·3	397·8	394·6		397·8	1
4	3	1	395·3	397·8	399·3	395·1	399·3	2
5	0	2	388·5				388·5	0
6	1	2	395·7	395·5			395·7	0
7	2	2	399·9	402·7	400·5		402·7	1
8	3	2	399·9	406·9	407·7	403·5	407·7	2
9	0	3	389·0				389·0	0
10	1	3	391·4	399·5			399·5	1
11	2	3	392·8	401·9	407·0		407·0	2
12	3	3	392·8	403·3	409·4	411·5	411·5	3

TABLE 3.5
Solution Tableaux for the Crop-Irrigation Problem without Discounting

Iteration 1: Value determination given $u^1\{I\}$

Current state number I	*Dam water level* x *(m)*	*Season* s	*Water released* $u^1\{I\}$ *(ha m)*	*Next state number* J	*Season return* $a\{x, s, u\}$ *($000)*	*Value* $V^1\{I\}$ *($000)*
1	0	1	0	7	8·0	0·0
2	1	1	1	7	10·5	2·5
3	2	1	2	7	12·0	4·0
4	3	1	3	7	12·5	4·5
5	0	2	0	10	9·0	−1·5
6	1	2	1	10	16·0	5·5
7	2	2	2	10	21·0	10·5
8	3	2	3	10	24·0	13·5
9	0	3	0	2	13·5	−2·5
10	1	3	1	2	24·0	8·0
11	2	3	2	2	31·5	15·5
12	3	3	3	2	36·0	20·0

Average flow per period $g^1 = \$18\,500$

Iteration 1: Policy improvement given $V_r^1\{I\}$

Current state number I	*Dam water level* x *(m)*	*Season* s	*Water released (ha m)*				$R^1\{I\}$ *($000)*	$u_1^*\{I\}$
			$u = 0$	$u = 1$	$u = 2$	$u = 3$		
1	0	1	18·5				18·5	0
2	1	1	21·5	21·0			21·5	0
3	1	1	21·5	24·0	22·5		24·0	1
4	3	1	21·5	24·0	25·5	23·0	25·5	2
5	0	2	17·0				17·0	0
6	1	2	24·5	24·0			24·5	0
7	2	2	29·0	31·5	29·0		31·5	1
8	3	2	29·0	36·0	36·5	32·0	36·5	2
9	0	3	16·0				16·0	0
10	1	3	17·5	26·5			26·5	1
11	2	3	18·0	28·0	34·0		34·0	2
12	3	3	18·0	28·5	35·5	38·5	38·5	3

TABLE 3.5—*contd.*

Iteration 2: Value determination given $u^2\{I\} = u_1^*\{I\}$

Current state number I	*Dam water level* x *(m)*	*Season* s	*Water released* $u^2\{I\}$ *(ha m)*	*Next state number* J	*Season return* $a\{x, s, u\}$ *($000)*	*Value* $V^2\{I\}$ *($000)*
1	0	1	0	7	8·0	0·0
2	1	1	0	8	8·0	5·0
3	2	1	1	8	10·5	7·5
4	3	1	2	8	12·0	9·0
5	0	2	0	10	9·0	−23·3
6	1	2	0	11	9·0	51·7
7	2	2	1	11	16·0	121·7
8	3	2	2	11	21·0	171·7
9	0	3	0	2	13·5	−16·7
10	1	3	1	2	24·0	88·3
11	2	3	2	2	31·5	163·3
12	3	3	3	2	36·0	208·3

Average flow per period $g^2 = \$20\,170$

Iteration 2: Policy improvement given $V_r^2\{I\}$

Current state number I	*Dam water level* x *(m)*	*Season* s	*Water released (ha m)* $u = 0$	$u = 1$	$u = 2$	$u = 3$	$R^2\{I\}$ *($000)*	$u_2^*\{I\}$
1	0	1	20·17				20·17	0
2	1	1	25·17	22·67			25·17	0
3	2	1	25·17	27·67	24·17		27·67	1
4	3	1	25·17	27·67	29·17	24·67	29·17	2
5	0	2	17·83				17·83	0
6	1	2	25·33	24·83			25·33	0
7	2	2	29·83	32·33	29·83		32·33	1
8	3	2	29·83	36·83	37·33	32·83	37·33	2
9	0	3	18·50				18·50	0
10	1	3	21·00	29·00			29·00	1
11	2	3	22·50	31·50	36·50		36·50	2
12	3	3	22·50	33·00	39·00	41·00	41·00	3

The optimal infinite-stage decision vector gives the same decisions for each season as those for the three-stage problem shown in Table 2.5. This in itself suggests that the infinite-stage solution would have been found in only a few iterations using value iteration, starting with $V_0\{x_0\}$ set equal to zero in (3.3) for $i = 1$. As already mentioned, the solution is in fact obtained after the second iteration. However, for larger problems, the infinite-stage solution will generally be found in fewer iterations using policy iteration.

3.4.4 Solution to the crop-irrigation problem without discounting

The infinite-stage solution is found for the problem without discounting using the policy-iteration algorithm detailed in Section 3.3.3. The gain and state relative-value vector are calculated in Table 3.5 for the same initial guess $u^1\{I\}$ as before. For demonstration purposes, these values were not found by the matrix algebra detailed in the algorithm. It has already been shown that the chain of states which is eventually repeated indefinitely given $u^1\{I\}$ is 7 10 2. The gain is therefore

$$g^1 = \$1000(21{\cdot}0 + 24{\cdot}0 + 10{\cdot}5)/3 = \$18\,500$$

$V^1\{1\}$ is set equal to zero. The remaining $V^1\{I\}$ values for $I > 0$ are found using (3·7). For example,

$$V^1\{1\} + g^1 = \$(8000 + V^1\{7\})$$

from which it follows that

$$V^1\{7\} = \$1000(0{\cdot}0 + 18{\cdot}5 - 8{\cdot}0) = \$10\,500$$

The policy-improvement step alternates with the value-determination step as before. Calculations are shown in Table 3.5. The optimal decision vector is again identified at the end of the second policy-improvement stage. It turns out that for the crop-irrigation problem the solution for $\alpha = 1{\cdot}00$ is the same as that for $\alpha = 0{\cdot}95$.

3.5 Using GPDP to Solve Stochastic and Infinite-stage Problems

3.5.1 Stochastic problems

In a stochastic dynamic programming problem there is more than one state number J which may be reached at the next stage given the current state and decision number. The probability of reaching state number J is represented by $P(J)$. In order to write the data to file using DPD, stages, states and decisions are numbered in the same way as for a deterministic problem. Data can be brought together for entry using DPD by filling up the coding sheets shown in Appendix A1 for data blocks A and B, CS, D and ES.

The decision number *D* must be entered for the stochastic problem. DPD prompts for data on a new decision once the sum of the *P*(*J*)'s for the current decision equals 1.000. This means that the problem must be formulated so that one of the specified state numbers is always reached at the next stage, even if the state number represents a dummy state such as extinction.

By way of illustration, the DPD and FDP routines were used to solve the stochastic, finite-stage crop-irrigation problem introduced in Section 3.1.2. The way in which data were entered for writing to the file CISF.DAT using DPD is shown in Fig. 3.1. The solution obtained by then running FDP, entering CISF as the problem name, is shown in Fig. 3.2. Optimal decisions are presented for all states for all decision stages, and agree with the results obtained earlier by calculation in Table 3.2.

3.5.2 Infinite-stage problems

Data for the infinite-stage problem can be written to file using DPD in exactly the same way as for the problem with a finite number of stages. When entering block A data, a value must be entered in answer to the query about the number of decision stages. Whatever the finite value entered, it is not subsequently used in calculations performed by IDP. For block B data, only the terminating line '0, 0' is required because there are no terminal states.

Solutions are obtained by running IDP. Different routines within IDP are performed, dependent on whether the rate of discount is positive or zero.

For deterministic problems, after the infinite-stage solution is presented, the user is asked if the optimal sequence of states and decisions for any nominated initial state is required. If it is, the optimal cycle of states which would be eventually reached and repeated indefinitely is identified.

For stochastic problems, the user can opt to obtain for each state the long-run probabilities of the system assuming that state if the optimal policy is followed indefinitely.

To illustrate the print-out from running IDP, data for the deterministic infinite-stage crop-irrigation problem with discounting were first written to a file CIDI.DAT using DPD. The results from running IDP for the problem 'CIDI' are shown in Fig. 3.3, and agree with those obtained earlier by calculation in Table 3.4.

```
PROBLEM NAME ? CISF

NO. OF DECISION STAGES ? 3

RATE OF DISCOUNT - PER CENT? 5.2632

PROBLEM TYPE - D(DETERMINISTIC) OR S(STOCHASTIC) ? S

ARE STATES,DECISIONS AND RETURNS THE SAME FOR ALL
STAGES - Y(YES) OR N(NO)? N

DO YOU HAVE LABELS FOR STATES AND DECISIONS -
(THE SAME FOR ALL STAGES) - Y(YES) OR N(NO) ? Y
-----------------------------------------------------------
VALUES OF TERMINAL STATES AT STAGE  4 :
YOU WILL BE ASKED TO ENTER I,V(I) FOR ALL I
WHERE  I    = TERMINAL STATE NO.
      V(I) = VALUE OF 'I'
ENTER '0,0' AFTER FINAL ENTRY
-----------------------------------------------------------
ENTER I,V(I)? 0,0

-----------------------------------------------------------

RETURNS AND TRANSITION PROBABILITIES FOR EACH POSSIBLE DECISION:
YOU WILL BE ASKED TO ENTER I,D,R(I,D) FOR ALL I
WHERE  I    = CURRENT STATE NO.
       D    = DECISION NO.
    R(I,D) = STAGE RETURN
AFTER ENTERING I,D,R(I,D) YOU WILL BE ASKED TO ENTER
J,PROB(I,D,J) FOR ALL J
WHERE      J    = POSSIBLE STATE NO. REACHED AT NEXT STAGE
  PROB(I,D,J) = PROBABILITY OF REACHING J
ENTER '0,0,0' AFTER FINAL ENTRY FOR EACH STAGE
-----------------------------------------------------------
FOR STAGE 3 , ENTER I,D,R(I,D)? 1,1,12.75
FOR I=  1 , D=  1 , ENTER J,PROB(I,D,J)? 1,.25
FOR I=  1 , D=  1 , ENTER J,PROB(I,D,J)? 2,.5
FOR I=  1 , D=  1 , ENTER J,PROB(I,D,J)? 3,.25
FOR STAGE 3 , ENTER I,D,R(I,D)? 2,1,12.75
FOR I=  2 , D=  1 , ENTER J,PROB(I,D,J)? 2,.25
FOR I=  2 , D=  1 , ENTER J,PROB(I,D,J)? 3,.5
FOR I=  2 , D=  1 , ENTER J,PROB(I,D,J)? 4,.25
FOR STAGE 3 , ENTER I,D,R(I,D)? 2,2,23.25
FOR I=  2 , D=  2 , ENTER J,PROB(I,D,J)? 1,.25
FOR I=  2 , D=  2 , ENTER J,PROB(I,D,J)? 2,.5
FOR I=  2 , D=  2 , ENTER J,PROB(I,D,J)? 3,.25
FOR STAGE 3 , ENTER I,D,R(I,D)? 3,1,12.75
FOR I=  3 , D=  1 , ENTER J,PROB(I,D,J)? 3,.25
FOR I=  3 , D=  1 , ENTER J,PROB(I,D,J)? 4,.5
FOR I=  3 , D=  1 , ENTER J,PROB(I,D,J)? 4,.25
FOR STAGE 3 , ENTER I,D,R(I,D)? 3,2,23.25
FOR I=  3 , D=  2 , ENTER J,PROB(I,D,J)? 2,.25
FOR I=  3 , D=  2 , ENTER J,PROB(I,D,J)? 3,.5
FOR I=  3 , D=  2 , ENTER J,PROB(I,D,J)? 4,.25
FOR STAGE 3 , ENTER I,D,R(I,D)? 3,3,30.75
FOR I=  3 , D=  3 , ENTER J,PROB(I,D,J)? 1,.25
FOR I=  3 , D=  3 , ENTER J,PROB(I,D,J)? 2,.5
FOR I=  3 , D=  3 , ENTER J,PROB(I,D,J)? 3,.25
```

FIG. 3.1. Data entry using DPD for the problem 'CISF'.

```
FOR STAGE 3 , ENTER I,D,R(I,D)? 4,1,12.75
FOR I=  4 , D=  1 , ENTER J,PROB(I,D,J)? 4,.25
FOR I=  4 , D=  1 , ENTER J,PROB(I,D,J)? 4,.5
FOR I=  4 , D=  1 , ENTER J,PROB(I,D,J)? 4,.25
FOR STAGE 3 , ENTER I,D,R(I,D)? 4,2,23.25
FOR I=  4 , D=  2 , ENTER J,PROB(I,D,J)? 3,.25
FOR I=  4 , D=  2 , ENTER J,PROB(I,D,J)? 4,.5
FOR I=  4 , D=  2 , ENTER J,PROB(I,D,J)? 4,.25
FOR STAGE 3 , ENTER I,D,R(I,D)? 4,3,30.75
FOR I=  4 , D=  3 , ENTER J,PROB(I,D,J)? 2,.25
FOR I=  4 , D=  3 , ENTER J,PROB(I,D,J)? 3,.5
FOR I=  4 , D=  3 , ENTER J,PROB(I,D,J)? 4,.25
FOR STAGE 3 , ENTER I,D,R(I,D)? 4,4,35.25
FOR I=  4 , D=  4 , ENTER J,PROB(I,D,J)? 1,.25
FOR I=  4 , D=  4 , ENTER J,PROB(I,D,J)? 2,.5
FOR I=  4 , D=  4 , ENTER J,PROB(I,D,J)? 3,.25
FOR STAGE 3 , ENTER I,D,R(I,D)? 0,0,0


STATE NAME ? LEVEL
LABEL FOR STATE NO.  1 ? L0
LABEL FOR STATE NO.  2 ? L1
LABEL FOR STATE NO.  3 ? L2
LABEL FOR STATE NO.  4 ? L3
DECISION NAME ? RELEASE
LABEL FOR DECISION NO.  1 ? R0
LABEL FOR DECISION NO.  2 ? R1
LABEL FOR DECISION NO.  3 ? R2
LABEL FOR DECISION NO.  4 ? R3


FOR STAGE 2 , ENTER I,D,R(I,D)? 1,1,8.5
FOR I=  1 , D=  1 , ENTER J,PROB(I,D,J)? 1,.25
FOR I=  1 , D=  1 , ENTER J,PROB(I,D,J)? 2,.5
FOR I=  1 , D=  1 , ENTER J,PROB(I,D,J)? 3,.25
FOR STAGE 2 , ENTER I,D,R(I,D)? 2,1,8.5
FOR I=  2 , D=  1 , ENTER J,PROB(I,D,J)? 2,.25
FOR I=  2 , D=  1 , ENTER J,PROB(I,D,J)? 3,.5
FOR I=  2 , D=  1 , ENTER J,PROB(I,D,J)? 4,.25
FOR STAGE 2 , ENTER I,D,R(I,D)? 2,2,15.5
FOR I=  2 , D=  2 , ENTER J,PROB(I,D,J)? 1,.25
FOR I=  2 , D=  2 , ENTER J,PROB(I,D,J)? 2,.5
FOR I=  2 , D=  2 , ENTER J,PROB(I,D,J)? 3,.25
FOR STAGE 2 , ENTER I,D,R(I,D)? 3,1,8.5
FOR I=  3 , D=  1 , ENTER J,PROB(I,D,J)? 3,.25
FOR I=  3 , D=  1 , ENTER J,PROB(I,D,J)? 4,.5
FOR I=  3 , D=  1 , ENTER J,PROB(I,D,J)? 4,.25
FOR STAGE 2 , ENTER I,D,R(I,D)? 3,2,15.5
FOR I=  3 , D=  2 , ENTER J,PROB(I,D,J)? 2,.25
FOR I=  3 , D=  2 , ENTER J,PROB(I,D,J)? 3,.5
FOR I=  3 , D=  2 , ENTER J,PROB(I,D,J)? 4,.25
FOR STAGE 2 , ENTER I,D,R(I,D)? 3,3,20.5
FOR I=  3 , D=  3 , ENTER J,PROB(I,D,J)? 1,.25
FOR I=  3 , D=  3 , ENTER J,PROB(I,D,J)? 2,.5
FOR I=  3 , D=  3 , ENTER J,PROB(I,D,J)? 3,.25
FOR STAGE 2 , ENTER I,D,R(I,D)? 4,1,8.5
FOR I=  4 , D=  1 , ENTER J,PROB(I,D,J)? 4,.25
FOR I=  4 , D=  1 , ENTER J,PROB(I,D,J)? 4,.5
FOR I=  4 , D=  1 , ENTER J,PROB(I,D,J)? 4,.25
```

(continued)

FIG. 3.1.—*contd.*

```
FOR STAGE 2 , ENTER I,D,R(I,D)? 4,2,15.5
FOR I=  4 , D=  2 , ENTER J,PROB(I,D,J)? 3,.25
FOR I=  4 , D=  2 , ENTER J,PROB(I,D,J)? 4,.5
FOR I=  4 , D=  2 , ENTER J,PROB(I,D,J)? 4,.25
FOR STAGE 2 , ENTER I,D,R(I,D)? 4,3,20.5
FOR I=  4 , D=  3 , ENTER J,PROB(I,D,J)? 2,.25
FOR I=  4 , D=  3 , ENTER J,PROB(I,D,J)? 3,.5
FOR I=  4 , D=  3 , ENTER J,PROB(I,D,J)? 4,.25
FOR STAGE 2 , ENTER I,D,R(I,D)? 4,4,23.5
FOR I=  4 , D=  4 , ENTER J,PROB(I,D,J)? 1,.25
FOR I=  4 , D=  4 , ENTER J,PROB(I,D,J)? 2,.5
FOR I=  4 , D=  4 , ENTER J,PROB(I,D,J)? 3,.25
FOR STAGE 2 , ENTER I,D,R(I,D)? 0,0,0

FOR STAGE 1 , ENTER I,D,R(I,D)? 1,1,7.75
FOR I=  1 , D=  1 , ENTER J,PROB(I,D,J)? 2,.25
FOR I=  1 , D=  1 , ENTER J,PROB(I,D,J)? 3,.5
FOR I=  1 , D=  1 , ENTER J,PROB(I,D,J)? 4,.25
FOR STAGE 1 , ENTER I,D,R(I,D)? 2,1,7.75
FOR I=  2 , D=  1 , ENTER J,PROB(I,D,J)? 3,.25
FOR I=  2 , D=  1 , ENTER J,PROB(I,D,J)? 4,.5
FOR I=  2 , D=  1 , ENTER J,PROB(I,D,J)? 4,.25
FOR STAGE 1 , ENTER I,D,R(I,D)? 2,2,10.25
FOR I=  2 , D=  2 , ENTER J,PROB(I,D,J)? 2,.25
FOR I=  2 , D=  2 , ENTER J,PROB(I,D,J)? 3,.5
FOR I=  2 , D=  2 , ENTER J,PROB(I,D,J)? 4,.25
FOR STAGE 1 , ENTER I,D,R(I,D)? 3,1,7.75
FOR I=  3 , D=  1 , ENTER J,PROB(I,D,J)? 4,.25
FOR I=  3 , D=  1 , ENTER J,PROB(I,D,J)? 4,.5
FOR I=  3 , D=  1 , ENTER J,PROB(I,D,J)? 4,.25
FOR STAGE 1 , ENTER I,D,R(I,D)? 3,2,10.25
FOR I=  3 , D=  2 , ENTER J,PROB(I,D,J)? 3,.25
FOR I=  3 , D=  2 , ENTER J,PROB(I,D,J)? 4,.5
FOR I=  3 , D=  2 , ENTER J,PROB(I,D,J)? 4,.25
FOR STAGE 1 , ENTER I,D,R(I,D)? 3,3,11.75
FOR I=  3 , D=  3 , ENTER J,PROB(I,D,J)? 2,.25
FOR I=  3 , D=  3 , ENTER J,PROB(I,D,J)? 3,.5
FOR I=  3 , D=  3 , ENTER J,PROB(I,D,J)? 4,.25
FOR STAGE 1 , ENTER I,D,R(I,D)? 4,1,7.75
FOR I=  4 , D=  1 , ENTER J,PROB(I,D,J)? 4,.25
FOR I=  4 , D=  1 , ENTER J,PROB(I,D,J)? 4,.5
FOR I=  4 , D=  1 , ENTER J,PROB(I,D,J)? 4,.25
FOR STAGE 1 , ENTER I,D,R(I,D)? 4,2,10.25
FOR I=  4 , D=  2 , ENTER J,PROB(I,D,J)? 4,.25
FOR I=  4 , D=  2 , ENTER J,PROB(I,D,J)? 4,.5
FOR I=  4 , D=  2 , ENTER J,PROB(I,D,J)? 4,.25
FOR STAGE 1 , ENTER I,D,R(I,D)? 4,3,11.75
FOR I=  4 , D=  3 , ENTER J,PROB(I,D,J)? 3,.25
FOR I=  4 , D=  3 , ENTER J,PROB(I,D,J)? 4,.5
FOR I=  4 , D=  3 , ENTER J,PROB(I,D,J)? 4,.25
FOR STAGE 1 , ENTER I,D,R(I,D)? 4,4,12.25
FOR I=  4 , D=  4 , ENTER J,PROB(I,D,J)? 2,.25
FOR I=  4 , D=  4 , ENTER J,PROB(I,D,J)? 3,.5
FOR I=  4 , D=  4 , ENTER J,PROB(I,D,J)? 4,.25
FOR STAGE 1 , ENTER I,D,R(I,D)? 0,0,0


*** DATA WRITTEN TO CISF.DAT ***
```

FIG. 3.1.—*contd.*

```
-----FINITE STAGE DP PROBLEM-----
PROBLEM NAME? CISF

------------------------------------------------------------
         STOCHASTIC DYNAMIC PROGRAMMING SOLUTION
                  PROBLEM ---CISF---

PROBLEM PARAMETERS:
NO. OF DECISION STAGES          =  3
RATE OF DISCOUNT (PER CENT)     =  5.2632
------------------------------------------------------------

STOCHASTIC SOLUTION FOR STAGE  3
------------------------------------------------------------
   STATE           DECISION         STAGE           EXPECTED
-----------      -----------        RETURN           VALUE
 NO. LEVEL        NO. RELEASE
------------------------------------------------------------
  1  L0            1   R0            12.75            12.75
  2  L1            2   R1            23.25            23.25
  3  L2            3   R2            30.75            30.75
  4  L3            4   R3            35.25            35.25
------------------------------------------------------------

STOCHASTIC SOLUTION FOR STAGE  2
------------------------------------------------------------
   STATE           DECISION         STAGE           EXPECTED
-----------      -----------        RETURN           VALUE
 NO. LEVEL        NO. RELEASE
------------------------------------------------------------
  1  L0            1   R0            8.5              29.875
  2  L1            1   R0            8.5              37
  3  L2            2   R1            15.5             44
  4  L3            3   R2            20.5             49
------------------------------------------------------------

STOCHASTIC SOLUTION FOR STAGE  1
------------------------------------------------------------
   STATE           DECISION         STAGE           EXPECTED
-----------      -----------        RETURN           VALUE
 NO. LEVEL        NO. RELEASE
------------------------------------------------------------
  1  L0            1   R0            7.75             49.075
  2  L1            1   R0            7.75             53.1125
  3  L2            2   R1            10.25            55.6125
  4  L3            3   R2            11.75            57.1125
------------------------------------------------------------
```

FIG. 3.2. Solution to the problem 'CISF' using FDP.

```
-----INFINITE STAGE DP PROBLEM-----
PROBLEM NAME ? CIDI


----------------------------------------------------------
        DETERMINISTIC DYNAMIC PROGRAMMING SOLUTION
                    PROBLEM ---CIDI---

PROBLEM PARAMETERS:
INFINITE NO. OF STAGES
RATE OF DISCOUNT (PER CENT)  =  5.2632
----------------------------------------------------------

NO. OF ITERATIONS TO FIND SOLUTION =  3


----------------------------------------------------------
  CURRENT       DECISION        NEXT            PRESENT
   STATE                        STATE            VALUE
-----------   -----------   -----------
 NO. S/LEVEL   NO. RELEASE   NO. S/LEVEL
----------------------------------------------------------
  1  S1/L0      1   R0        7  S2/L2          390.541
  2  S1/L1      1   R0        8  S2/L3          395.291
  3  S1/L2      2   R1        8  S2/L3          397.791
  4  S1/L3      3   R2        8  S2/L3          399.291
  5  S2/L0      1   R0       10  S3/L1          388.55
  6  S2/L1      1   R0       11  S3/L2          395.675
  7  S2/L2      2   R1       11  S3/L2          402.675
  8  S2/L3      3   R2       11  S3/L2          407.675
  9  S3/L0      1   R0        2  S1/L1          389.026
 10  S3/L1      2   R1        2  S1/L1          399.526
 11  S3/L2      3   R2        2  S1/L1          407.026
 12  S3/L3      4   R3        2  S1/L1          411.526
----------------------------------------------------------


OPTIMAL STAGE CYCLE
ENTER INITIAL STATE NO. (OR '0' TO FINISH) ? 4
------------------------------------
  STAGE        STATE         DECISION
   NO.      -----------    -----------
            NO. S/LEVEL     NO. RELEASE
------------------------------------
   1         4  S1/L3        3  R2

   2         8  S2/L3        3  R2
   3        11  S3/L2        3  R2
   4         2  S1/L1        1  R0
   5         8  S2/L3        3  R2
------------------------------------

OPTIMAL STAGE CYCLE
ENTER INITIAL STATE NO. (OR '0' TO FINISH) ? 0
```

FIG. 3.3. Solution to the problem 'CIDI' using IDP.

3.6 References

Burt, O.R. and Allison, J.R. (1963), Farm management decisions with dynamic programming, *Journal of Farm Economics*, **45**(1), 121–36.

Hastings, N.A.J. (1973), *Dynamic Programming with Management Applications*, Butterworth, London.

Hastings, N.A.J. (1976), A test for nonoptimal actions in undiscounted finite Markov decision chains, *Management Science*, **23**(1), 87–92.

Hastings, N.A.J. and Mello, J.M.C. (1973), Tests for suboptimal actions in discounted Markov programming, *Management Science*, **19**(9), 1019–22.

Chapter 4

EXTENSIONS TO THE BASIC FORMULATION

The principles of dynamic programming have been explained and illustrated in Chapters 2 and 3. Specialist texts on dynamic programming deal with extensions to the basic framework of various kinds. Some extensions which may be useful in the context of managing agricultural and natural resources are introduced in this chapter. The aim is to provide some insight into the formulation and application of the extensions, rather than to treat each topic with full rigour.

A natural topic to follow on from Chapter 2 is the dynamic programming problem which is both stochastic and infinite-stage. Such a problem does not raise any new conceptual issues, and may be solved in the ways already discussed. An interesting alternative solution method which is explained in Section 4.1 is through the use of linear programming. An obvious advantage is the ready availability of standard computer software for solving linear programming problems. However, for problems of any size, users will probably want to write their own programs to form the linear programming matrix.

In Section 4.2 another extension of stochastic dynamic programming, adaptive dynamic programming, is dealt with. This applies to problem situations in which the probabilities of the stochastic events are themselves uncertain. Estimates of the probabilities are revised from one decision stage to the next in the light of additional information on the stochastic variables.

So far dynamic programming has been seen as a technique for deducing optimality conditions (Chapter 1) or as an algorithm for obtaining numerical solutions (Chapters 2 and 3). In Section 4.3 it is shown how dynamic programming can also be used to obtain analytical solutions to certain problems. A major limitation of numerical dynamic programming is that because the number of possible ways of describing the state of a decision system increases exponentially with the number of state variables, problems with many state variables may impose excessive or infeasible computing and memory requirements. If an analytical approach is feasible, solutions may not be bound by the number of state variables to be considered. One important class of problem which may be solved analytically consists of problems with quadratic objective functions and linear equality constraints. This class of problem is also of interest because certain stochastic versions of the problem can be readily solved using the principle of certainty equivalence.

Burt and Cummings (1977) have suggested a useful analytical approach for obtaining approximate infinite-stage solutions. The approach is described in Section 4.4.

Decision makers are usually assumed to be interested in finding decisions which maximize some objective function measured in terms say of profit or utility. However, it is quite possible that an agricultural or natural resource problem may need to be evaluated by more than one criterion. The criteria may be efficiency and equity, or private profit and pollution. Methods of formulating dynamic programming models to take account of multiple objectives are considered in Section 4.5.

Methods of alleviating the limitation of numerical dynamic programming mentioned above (referred to by Bellman and Dreyfus (1962, p. 323) as 'the curse of dimensionality') are discussed in Section 4.6.

Finally, in Section 4.7, the format of the data files read by the computer programs FDP and IDP is explained. This information is necessary if the data for a dynamic programming problem are to be generated by a user-written computer program rather than entered directly using the DPD routine. The way in which the arrays in FDP and IDP can be redimensioned for solving large problems is also explained.

4.1 Linear Programming for Solving Stochastic, Infinite-stage Problems

Stochastic, infinite-stage problems may be solved by value iteration or policy iteration in the ways discussed in Chapter 3. For policy iteration, the transformed-state matrix B^l required for the stochastic case is a generalized version of that for the deterministic case. It consists of elements b^l_{ij} equal to the probability of going from state number I to state number J under the l-th decision vector.

The policy-iteration procedure is similar to the simplex procedure of sequential improvement of bases followed in solving linear programming problems. An initial decision vector is selected which must be feasible, and which implies an initial $m \times m$ transformed-state matrix B^l. The steps of value determination and policy improvement results in an improved decision vector, and a revised $m \times m$ B^l matrix. The process continues until no further improvement is possible. As Hadley (1970, Ch. 11) and others have pointed out, the infinite-stage dynamic programming problem can be solved using linear programming. The method is illustrated by applying it to the stochastic crop-irrigation problem formulated as an infinite-stage problem. It could be just as readily applied to the deterministic problem, which can be viewed as a special case of the stochastic problem. The discounted and non-discounted cases are treated in turn.

4.1.1 Linear programming formulations of problems with discounting

Consider the irrigation decisions which can be made at the beginning of season 1 when the dam level is 2 m. The state of the system for $x = 2$ and $s = 1$ in the infinite-stage problem is identified by $I = 3$ (see Table 3.4). The possible decisions are $u_1 = 0$, 1 or 2 ha m with corresponding expected stage returns of 7·75, 10·25 and 11·75 thousand dollars. Let $V\{I\}$ denote the expected return from state number I over infinite stages. If $V\{3\}$ is to be as large as possible, the following inequalities must hold:

$$V\{3\} \geq 7{\cdot}74 + \alpha V\{8\}$$
$$V\{3\} \geq 10{\cdot}25 + 0{\cdot}25\alpha V\{7\} + 0{\cdot}75\alpha V\{8\}$$
$$V\{3\} \geq 11{\cdot}75 + 0{\cdot}25\alpha V\{6\} + 0{\cdot}50\alpha V\{7\} + 0{\cdot}25\alpha V\{8\}$$

The inequalities correspond to the three decisions which can be made from state number $I = 3$. The inequality corresponding to the optimal decision holds as an equality. The other inequalities remain as strict inequalities, indicating that the corresponding decisions are inferior to the optimal decision.

An inequality constraint applies for each decision which can be made at each of the m possible states of the system. If d_j is the number of decisions which can be made at the j-th state, the total number of inequalities is $\sum_{j=1}^{m} d_j$. For the crop-irrigation problem, 30 inequalities apply. The objective is to find the decision set which maximizes all values of $V\{I\}$, $I=1,\ldots,12$. This can be achieved with linear programming by maximizing the objective function $\sum_{I=1}^{12} V\{I\}$.

In short, the solution to the crop-irrigation problem may be found by solving the linear programming problem

$$\max c'v$$

subject to

$$Av \le b$$
$$v \ge 0$$

where v is the vector of $V\{I\}$'s; c is a vector of ones; and the A matrix and b vector are as shown in Table 4.1 in partial form. The linear programming solution vector gives the optimal $V\{I\}$ values. The optimal decision for each of the 12 possible states is given by the 12 inequalities which are binding constraints. The solution obtained for $\alpha = 0{\cdot}95$ is shown in Table 4.2.

The order of matrix A is 30×12. Because computation required to solve a linear programming problem increases more rapidly with the number of rows than with the number of columns, it is computationally more efficient to solve the dual problem

$$\min b'y$$

subject to

$$A'y \ge c$$
$$y \ge 0$$

The 30×1 solution vector y gives the shadow prices of the 30 constraints in the primal problem. The optimal decision vector for the dynamic programming problem is therefore deduced by observing which 'y' values are positive. The v vector is given by the shadow prices of the 12 constraints in the dual problem.

The optimal v vector cannot be obtained using linear programming if any element of v is negative. This restriction on linear programming as a solution device can be readily overcome by adding a sufficiently large constant θ to each term of the b vector. The solution vector v is obtained by subtracting $\theta/(1-\alpha)$ from each element of the linear programming solution vector.

4.1.2 Linear programming formulations of problems without discounting

The approach for solving infinite-stage, non-discounted problems has been described in Section 3.3. The objective is to maximize the average

TABLE 4.1

'*A*' and '*b*' in the Linear Programming Model of the Crop-irrigation Problem with Discount Factor α

Row number		Column number 1	2	3	4	5	6	7	8	9	10	11	12		
1		−1·0					0·25α	0·50α	0·25α						−7·75
2			−1·0					0·25α	0·75α						−7·75
3			−1·0				0·25α	0·50α	0·25α						−10·25
4				−1·0					1·00α						−7·75
5	$A =$			−1·0				0·25α	0·75α					$b =$	−10·25
6				−1·0			0·25α	0·50α	0·25α						−11·25
⋮		⋮											⋮		⋮
30		0·25α	0·50α	0·25α									−1·0		−35·25

TABLE 4.2
The Linear Programming Solution to the Crop-irrigation Problem with Discount Factor = 0·95

Current state number I	*Dam water level* x *(m)*	*Season* s	*Binding constraint number* C	*Optimal decision (from C) (ha m)*	*Value* $V\{I\}$ *($000)*
1	0	1	1	0	366
2	1	1	2	0	370
3	2	1	5	1	373
4	3	1	9	2	374
5	0	2	11	0	364
6	0	2	12	0	371
7	2	2	15	1	378
8	3	2	19	2	383
9	0	3	21	0	364
10	1	3	23	1	374
11	2	3	26	2	382
12	3	3	30	3	386

flow per period, g, subject to constraints based on (3.8), such as the following which must hold for $I = 3$:

$$V_r\{3\} \geq 7{\cdot}75 + V_r\{8\} - g$$
$$V_r\{3\} \geq 10{\cdot}25 + 0{\cdot}25\, V_r\{7\} + 0{\cdot}75\, V_r\{8\} - g$$
$$V_r\{3\} \geq 11{\cdot}75 + 0{\cdot}25\, V_r\{6\} + 0{\cdot}50\, V_r\{7\} + 0{\cdot}25\, V_r\{8\} - g$$

where $V_r\{I\}$ is the value of state I relative to the value of state 1. Only the inequality representing the optimal decision holds as an equality in the solution.

Let the activity vector in the linear programming problem be q, with $q\{1\} = g$ and $q\{I\} = V_r\{I\}$ for $I = 2, \ldots, 12$. The reference value $V\{1\}$ is set large enough to ensure all $q\{I\} \geq 0$. The linear programming formulation is

$$\max\ c'q$$

subject to

$$Aq \leq b$$
$$q \geq 0$$

where c is a vector of zeroes except for the first element which is one. The A matrix and b vector are shown in Table 4.3 in partial form, with

TABLE 4.3

'*A*' and '*b*' in the Linear Programming Model of the Crop-irrigation Problem Without Discounting

Row number		*Column number* 1	2	3	4	5	6	7	8	9	10	11	12		*b*
1		−1·0					0·25	0·50	0·25						−7·75 + 1000
2		−1·0	−1·0					0·25	0·75						−7·75
3		−1·0	−1·0				0·25	0·50	0·25						−10·25
4		−1·0		−1·0					1·00						−7·75
5	*A* =	−1·0		−1·0				0·25	0·75					*b* =	−10·25
6		−1·0		−1·0			0·25	0·50	0·25						−11·25
⋮		⋮											⋮		⋮
30		−1·0	0·50	0·25									−1·0		−35·25 − 250

TABLE 4.4
The Linear Programming Solution to the Crop-irrigation Problem Without Discounting

Current state number I	*Dam water level* x *(m)*	*Season* s	*Binding constraint number* C	*Optimal decision (from C) (ha m)*	*Value* $q\{I\}$ *($000)*
1	0	1	1	0	18·85
2	1	1	2	0	1 004·25
3	2	1	5	1	1 006·75
4	3	1	9	2	1 008·25
5	0	2	11	0	997·10
6	1	2	12	0	1 004·60
7	2	2	15	1	1 011·60
8	3	2	19	2	1 016·60
9	0	3	21	0	997·71
10	1	3	23	1	1 008·21
11	2	3	26	2	1 015·71
12	3	3	30	3	1 020·21

$V\{1\} = \$1$ million. Note that entries for state 1 appear as additional constants in the b vector. The optimal decisions for each of the 12 possible states are again given by the 12 inequalities which are binding constraints. The solution obtained is shown in Table 4.4. Whilst the optimal decision vector is the same as that for the corresponding deterministic problem (shown in Table 3.5), the maximum expected gain ($18 850) is lower than the maximum gain for the deterministic problem ($20 170).

4.2 Adaptive Dynamic Programming

In the stochastic formulation of the dynamic programming problem the stage return and state transformation depend on the outcome of a random event. In Section 3.1.1 the random event was numbered k, and could take any integer value from 1 to m. It was assumed that, whilst k_i was unknown before a decision at stage i was made, the probability distribution of k_i was known. In practice, the probability distribution may not be known with certainty. However, it may be that after observing the occurrence of the random events over a number of stages, an improved estimate of the probability distribution can be made. If so, it may be desirable to allow for subsequent learning in determining the optimal decision to be made at any stage. Dynamic programming which makes such allowance is referred to as adaptive.

The types of random variables encountered in agricultural and natural resource problems are: economic variables such as input and output prices, and the rate of interest; weather variables such as temperature and rainfall; and technical parameters particular to the resource system, such as growth rates for biological systems and exploration success in the case of mining. Of the three types, there is probably greatest scope for revising technical parameters particular to the resource system, over the same period as decisions are being implemented. For example, in the case of feeding beef cattle, there may be uncertainty about the quality of feed. The probability distribution of feed quality may be revised after analysing the weight gain of a group of cattle and the amount of feed consumed by the cattle. The gain potential of individual cattle may be revised after analysing the weight gain of individual cattle and the amount of feed they have consumed. In the case of the fishery, harvest levels relative to fishing effort may be used to revise estimates of stock levels.

It is likely that for the adaptive approach to be worthwhile there must be little change in the probability distribution throughout the decision periods. Some price distributions may be constantly changing in response to economic shocks which alter the way supply and demand interact. Parameters of an econometric forecasting model may be updated as prices and explanatory variables are revealed, but dynamic programming is severely limited in the number of variables which can be used for revising distributions. It is also likely that for an adaptive approach to be successful the probability distribution should not have existed or remained stable for long periods before the decision periods. For example, there may be little point in taking an adaptive approach towards revising the probability distribution of rainfall for a particular area if rainfall records over many past periods exist and there is no reason to suppose that there has been a change in the rainfall pattern. In both of these cases, either where there is no constancy in a probability distribution, or where the constancy has been so great that there is little scope for subsequent learning, a straightforward stochastic dynamic programming approach is probably the most suitable.

Amongst the possible methods of revising probabilities in the light of new observations are the use of relative frequencies and of Bayes' theorem. The two methods are discussed taking a simplified version of the crop-irrigation problem as an example. The problem is simplified by disallowing seasonality, so that the rainfall probability distribution and the yield and return functions are the same for all stages. Suppose the probabilities that are to be revised are the probabilities of rainfall. Rainfall probabilities are chosen to provide a simple illustration which extends the stochastic crop-irrigation problem already considered, rather than an illustration which is likely to be important in practice.

Suppose rainfall over each season always falls in the range 0·50 to 3·49 cm, and that the events $k = 1$, 2 and 3 correspond to rainfalls 0·50–1·49, 1·50–2·49, 2·50–3·49 cm. If the relative frequency approach is taken to estimating probabilities, no particular probability distribution for rainfall is assumed. At the first decision stage, assume that rainfall observations over the previous s periods show $k = 1$ recorded f_0^1 times, $k = 2$ recorded f_0^2 times and $k = 3$ recorded $f_0^3 = s - f_0^1 - f_0^2$ times. Estimated probabilities for $k = 1$, 2 and 3 at stage 1 would be f_0^1/s, f_0^2/s and f_0^3/s. If during the first season before stage 2 rainfall is such that $k = 1$, the probabilities for $k = 1$, 2 and 3 at stage 2 would be revised to $(f_0^1 + 1)/(s + 1)$, $f_0^2/(s + 1)$ and $f_0^3/(s + 1)$. Probabilities would be revised in this way at each decision stage. The adaptive decision problem can be viewed as a stochastic dynamic programming problem with the set of probabilities for k included as an additional state variable. The expected return at each stage depends on the current probability set. The probability set at the next stage depends only on the current probability set and the k which is observed before the next decision stage.

To follow through in the spirit of dynamic programming, all of the

TABLE 4.5
All Possible Observed Frequencies of Rainfall Category at Stage 4

f_3^1	f_3^2	f_3^3
$f_0^1 + 0$	$f_0^2 + 0$	$f_0^3 + 3$
$f_0^1 + 1$	$f_0^2 + 0$	$f_0^3 + 2$
$f_0^1 + 2$	$f_0^2 + 0$	$f_0^3 + 1$
$f_0^1 + 3$	$f_0^2 + 0$	$f_0^3 + 0$
$f_0^1 + 0$	$f_0^2 + 1$	$f_0^3 + 2$
$f_0^1 + 1$	$f_0^2 + 1$	$f_0^3 + 1$
$f_0^1 + 2$	$f_0^2 + 1$	$f_0^3 + 0$
$f_0^1 + 0$	$f_0^2 + 2$	$f_0^3 + 1$
$f_0^1 + 1$	$f_0^2 + 2$	$f_0^3 + 0$
$f_0^1 + 0$	$f_0^2 + 3$	$f_0^3 + 0$

probability sets that could have eventuated by each stage must be considered at each decision stage during backward recursion. For example, consider stage 4. All possible frequencies which could have been observed over the previous three periods are listed in Table 4.5 using f_i^k to refer to the number of rainfall observations by stage i in the k-th category. The probability $p_i\{k_i\}$ is given by $f_i^k/(s-1+i)$.

For $i=4$, there are 10 possible frequency observations. In general, the number of frequency observations is $i(i+1)/2$. For an adaptive dynamic programming formulation, an additional state variable is required to record the frequency observations f_i^1, f_i^2. The probabilities for $k=1$, 2 and 3 at stage i follow directly from f_i^1, f_i^2, i and s. The problem can be formulated as

$$V_i\{x_i, f_i^1, f_i^2\} = \max_{u_i}\left[\sum_{k=1}^{3} p_i\{k_i\}(a_i\{x_i, u_i, k_i\} + \alpha V_{i+1}\{x_i + k_i - u_i, t_i^1\{f_i^1, k_i\}, t_i^2\{f_i^2, k_i\}\})\right] \qquad (i = n, \ldots, 1)$$

where

$$t_i^1\{f_i^1, k_i\} = \begin{cases} f_{i+1}^1 & \text{if } k_i = 1 \\ f_i^1 & \text{otherwise} \end{cases}$$

$$t_i^2\{f_i^2, k_i\} = \begin{cases} f_{i+1}^{2+1} & \text{if } k_i = 2 \\ f_i^2 & \text{otherwise} \end{cases}$$

$$f_i^3 = s - f_i^1 - f_i^2 - 1 + i$$
$$p_i\{k_i\} = f_i^k/(s-1+i)$$

with

$$V_{n+1}\{x_{n+1}, f_{n+1}^1, f_{n+1}^2\} = 0$$

An adaptive formulation of the seasonal crop-irrigation problem would have a prohibitive number of state variables. With three growing seasons in each year, the state variables at stage i would be x_i, s_i, f_i^1, f_i^2, f_{i-1}^1, f_{i-1}^2, f_{i-2}^1, f_{i-2}^2. Thus at stage 10, after three lots of rainfall observations for each season, the state of the system would assume one out of $4 \times 3 \times 10 \times 10 \times 10 = 12\,000$ possible states. This example underscores the computational problems which are likely to be faced in attempting to obtain solutions numerically.

It may be possible to avoid a heavy computational burden. The load need not be so great if: (a) the form of the probability distribution of the random variable is known and can be described by a set of sufficient statistics; and (b) an updating procedure such as the application of Bayes' theorem is used for revising estimates of the sufficient statistics from stage to stage. For example, it may be judged that as a reasonable approximation the rainfall distribution can be treated as a normal distribution. The normal distribution is completely described by its mean and standard deviation. Estimates of the mean and standard deviation may be improved as k_i is observed from stage to stage, using Bayes' theorem. In this case the mean and standard deviation would be the only additional state variables. For numerical solutions, a range of discrete values of the mean and standard deviation would be used. Alternatively, it may be possible to formulate the problem with continuous variables, and to obtain a solution analytically. If so, it is quite possible to extend the formulation to the adaptive case with a normal distribution for the random variable, without the great increase in computation required for the numerical approach. Simple equations have been derived for revising the expected mean and standard deviation of the mean using Bayes' theorem. Further details can be found in the literature on optimal control theory, such as Jacobs (1974) and Rausser and Hochman (1979).

An application of adaptive dynamic programming to the decision on whether to cull dairy cows is described in Chapter 7. The analytical formulation and solution of dynamic programming problems is discussed in the next section.

4.3 Analytical Dynamic Programming

The amount of computation required to solve a dynamic programming problem numerically may be very large if there are more than a few state variables. Certain classes of problem with state and decision variables treated as continuous can be solved analytically by dynamic programming. As a general rule more state variables can be accommodated in problems which are solved analytically than in problems which are solved numerically.

4.3.1 Deterministic, quadratic return, linear transformation problems

An important class of dynamic programming problem which can be solved analytically consists of problems with quadratic stage returns and linear state transformation. The solution procedure for the general problem using vector notation is explained in detail in texts such as Chow (1975, Ch. 8). A simple exposition and an example are given here. Problems of this type may be formulated as

$$\max V_1\{x_1\} = \sum_{i=1}^{n} (a_i + b_i u_i + c_i x_i + d_i u_i^2 + e_i x_i^2 + f_i u_i x_i) \tag{4.1}$$

with respect to $u_1, \ldots, u_n$, subject to

$$x_{i+1} = t_i + w_i x_i + y_i u_i \qquad (i = 1, \ldots, n) \tag{4.2}$$

with

$$x_1 = \bar{x}_1$$

Adopting backward recursion, the subproblem for the final stage is to find

$$\begin{aligned} V_n\{x_n\} &= \max_{u_n} [a_n + b_n u_n + c_n x_n + d_n u_n^2 + e_n x_n^2 + f_n u_n x_n] \\ &= \max_{u_n} [p_n\{x_n\} + q_n\{x_n\} u_n + r_n u_n^2] \end{aligned} \tag{4.3}$$

where

$$p_n\{x_n\} = a_n + c_n x_n + e_n x_n^2 \qquad q_n\{x_n\} = b_n + f_n x_n \qquad r_n = d_n$$

Differentiating the right-hand side with respect to u_n and setting the derivative to zero gives the optimal value of u_n:

$$u_n^* = -q_n\{x_n\}/2r_n$$

provided $r_n < 0$. Substituting u_n^* for u_n in (4.3) gives

$$V_n\{x_n\} = p_n\{x_n\} - (q_n\{x_n\})^2/4r_n \tag{4.4}$$

The subproblem for stage $n-1$ is to find

$$\begin{aligned} V_{n-1}\{x_{n-1}\} = \max_{u_{n-1}} [&a_{n-1} + b_{n-1} u_{n-1} + c_{n-1} x_{n-1} + d_{n-1} u_{n-1}^2 + e_{n-1} x_{n-1}^2 \\ &+ f_{n-1} u_{n-1} x_{n-1} + V_n\{t_{n-1} + w_{n-1} x_{n-1} + y_{n-1} u_{n-1}\}] \end{aligned} \tag{4.5}$$

Because $p_n\{x_n\}$ and hence $V_n\{x_n\}$ are quadratic functions of x_n, and because x_n is a linear function of u_{n-1} and x_{n-1}, $V_n\{x_n\}$ is a quadratic function of u_{n-1} and x_{n-1}. Equation (4.5) can therefore be written in the same form as (4.3) as

$$V_{n-1}\{x_{n-1}\} = \max_{u_{n-1}} [p_{n-1}\{x_{n-1}\} + q_{n-1}\{x_{n-1}\} u_{n-1} + r_{n-1} u_{n-1}^2]$$

In the same way as for stage n, $u^*_{n-1}\{x_{n-1}\}$ and $V_{n-1}\{x_{n-1}\}$ can be found in terms of $p_{n-1}\{x_{n-1}\}$, $q_{n-1}\{x_{n-1}\}$ and r_{n-1}. This procedure can be followed recursively for all stages through to stage 1. In order to solve the original problem for $x_1 = \bar{x}_1$, the functions already calculated, $p_i\{x_i\}$ and $q_i\{x_i\}$, and r_i, $i = 1, \ldots, n$, are then used. Starting with stage 1

$$u^*_1\{x_1\} = -q_1\{x_1\}/2r_1$$

$$V_1\{x_1\} = p_1\{x_1\} - (q_1\{x_1\})^2/4r_1$$

$$x_2 = t_1 + w_1 x_1 + y_1 u^*_1\{x_1\}$$

The calculations are performed, repetitively, to find u^*_i, V_i and x_i for all i.

An important advantage of the analytical approach is that the same type of solution procedure can be followed if x and u are vectors of state and decision variables, thus overcoming the curse of dimensionality. On the other hand, an important limitation is that it may not be easily applied if each stage solution is constrained by restrictions on u_i and x_i. The limitation is apparent in attempting to solve analytically the deterministic three-stage crop-irrigation problem, which consists of quadratic stage returns and linear state transformations. Relaxing the integer constraints, the following constraints still apply:

$$0 \leq u_i \leq x_i \leq 3 \qquad (i = 1, 2, 3) \tag{4.6}$$

The analytical solution ignoring these constraints, for a full dam in the first season, leads to $x_2 = 4{\cdot}64$ m which violates (4.6). However, in this case an analytical solution to the constrained problem can be obtained by recognizing that the marginal product of irrigation water is positive for water releases of up to 5 ha m. It is therefore optimal to release 2 ha m of water in the first season, leaving a water level of 3 m for the start of the second season. The analytical solution for the crop-irrigation problem starting with a full dam in the second season does not violate (4.6) and is therefore applicable. The full solution is shown in Table 4.6 with the integer solution obtained in Chapter 2 shown in brackets.

To permit a further comparison of results obtained from numerical and analytical solution methods, the crop-irrigation model was modified by changing the maximum water level of the dam from 3 to 9 m, increasing the number of decision stages to 6 and assuming an initial water level of 6 m. The results are shown in Table 4.7.

In this solution no inequality constraints on u_i or x_i are violated. It is

TABLE 4.6
Analytical Solution to the Crop-irrigation Problem

Year	*Season*	*Dam level (m)*	*Water released (ha m)*	*Discounted total return ($000)*
1	1	3·00 (3)	2·00 (2)	60·64 (60·4)
1	2	3·00 (3)	1·65 (2)	51·22 (50·9)
1	3	2·35 (2)	2·35 (2)	33·42 (31·5)
2	1	1·00 (1)	— (—)	— (—)

apparent that for this particular problem, the restriction of the state and decision variables to integer values in the numerical solution approach does not lead to any significant fall in discounted total return. The time taken to compute the analytical solution was a fifth that for the numerical solution. However, whereas analytical solutions for alternative initial water levels would have to be computed from scratch, numerical solutions for alternative initial water levels could be obtained with little additional computation. Another important difference concerning the practice of the two approaches is that constraints on x_i and u_i reduce computation if the numerical approach is taken, but greatly increase the complexity of finding the solution analytically.

4.3.2 Stochastic, quadratic return, linear transformation problems

Suppose the problem posed in (4.1) and (4.2) is made stochastic by adding a random disturbance term with zero mean to the right-hand side of the

TABLE 4.7
Analytical Solution to the Modified Crop-irrigation Problem

Year	*Season*	*Dam level (m)*	*Water released (ha m)*	*Discounted total return ($000)*
1	1	6·00 (6)	0·81 (1)	120·33 (120·05)
1	2	7·19 (7)	2·85 (3)	116·03 (115·32)
1	3	5·35 (5)	3·19 (3)	97·22 (96·13)
2	1	3·15 (3)	0·44 (0)	63·90 (63·29)
2	2	4·71 (5)	2·65 (3)	57·55 (58·20)
2	3	3·06 (3)	3·06 (3)	36·16 (36·00)
3	1	1·00 (1)	— (—)	— (—)

linear transformation equation (4.2), and by making the objective the maximizing of *expected* total return. An example would be the crop-irrigation problem without inequality constraints and with rainfall treated as a random variable. Chow (1975, Ch. 7) has shown that the dynamic programming feedback-control solution to the deterministic problem is the same as the feedback-control solution to the stochastic problem. By feedback-control solution is meant the specification of optimal $u_i\{x_i\}$. Chow has termed this a multiperiod certainty-equivalence result, and is an extension of Theil's (1964) *first-period certainty-equivalence theorem.* It can be shown from Theil's theorem that the first-stage solution of the stochastic problem is the same as the first-stage solution of the deterministic problem.

4.3.3 Other problems which can be solved analytically

In many problems the objective function is not the sum of quadratic stage returns. However, it may be possible to approximate a more general objective function by an additive quadratic function using a Taylor series expansion. On the other hand, sometimes the structure of the problem is such that the problem can be solved analytically even though the objective function cannot be specified as additive quadratic. An example is the fertilizer carry-over problem examined in Chapter 6.

4.4 Approximately Optimal Infinite-stage Solutions

In Chapter 1 it was shown how the logic of dynamic programming could be used to derive the necessary conditions for an interior solution to the general resource management problem. Whilst these conditions may be given a useful economic interpretation, they may be difficult to use directly to find optimal management policies. However, the conditions are simpler and therefore more operational in the case of the infinite-stage problem—a problem with stage-independent functions and an unbounded planning horizon. This is particularly so if implementing the infinite-stage decision vector leads to one state which once reached is optimal from then on—that is, if the solution is a steady-state solution. The conditions may be used to find both the steady state and the corresponding decision which is optimal for the steady state. Perhaps more importantly, under certain circumstances, they give approximately optimal infinite-stage decision rules for states not equal to the steady state. They may even give approximately optimal decision rules for problems with infinite-stage solutions which lead to the continuous repetition of chains of states.

Noting that $\alpha = 1/(1+r)$, the optimality conditions (1.7) and (1.8) can be rewritten as

$$\lambda_{i+1}/(1+r) = (\partial a/\partial u)/(\partial h/\partial u) \tag{4.7}$$

$$\partial a/\partial x = (\lambda_{i+1}/(1+r))(r - \mathrm{d}g/\mathrm{d}x + \partial h/\partial x) - (\lambda_{i+1} - \lambda_i) \tag{4.8}$$

where λ_i is $\mathrm{d}V/\mathrm{d}x$ evaluated at x_i. Stage subscripts on the derivatives of $a\{x, u\}$, $g\{x\}$ and $h\{x, u\}$ are dropped because for the infinite-stage problem the functions must be independent of stage. For stage i all derivatives are evaluated at x_i and u_i^* as appropriate, except λ_{i+1} which equals $\mathrm{d}V/\mathrm{d}x$ evaluated at x_{i+1}.

Combining (4.7) and (4.8) gives

$$\frac{\partial a/\partial u}{\partial h/\partial u} = \frac{\partial a/\partial x + (\lambda_{i+1} - \lambda_i)}{r - \mathrm{d}g/\mathrm{d}x + \partial h/\partial x} \tag{4.9}$$

Suppose there is an optimal steady-state stock x^*, with associated optimal decision u^*. It follows that if $x_i = x^*$ and u_i^* is implemented, then $x_{i+1} = x^*$, and $\lambda_i = \lambda_{i+1} = \mathrm{d}V/\mathrm{d}x$ evaluated at x^*. Equation (4.9) for $x_i = x^*$ becomes

$$\frac{\partial a/\partial u}{\partial h/\partial u} = \frac{\partial a/\partial x}{r - \mathrm{d}g/\mathrm{d}x + \partial h/\partial x} \tag{4.10}$$

It also follows that if $x_i = x_{i+1} = x^*$, the net increase in stock from natural growth and harvesting must be zero, or

$$g\{x^*\} - h\{x^*, u^*\} = 0 \tag{4.11}$$

One way to find x^* and u^* is to solve (4.10) and (4.11). However, (4.10) alone can be used as an approximately optimal feedback-control rule for $x_i \neq x^*$. The rule is to set the decision for observed x_i equal to u_i which solves (4.10) evaluated at $x = x_i$. Denote the approximately optimal feedback rule by $\hat{u}\{x_i\}$, and the optimal feedback rule by $u^*\{x_i\}$.

The optimal rule $u^*\{x_i\}$ could be determined from (4.9), but this requires a knowledge of λ_{i+1} and λ_i which are not observable. However, it can be seen from (4.9) that $\hat{u}\{x_i\}$ obtained using (4.10) is a good approximation to $u^*\{x_i\}$ if $(\lambda_{i+1} - \lambda_i)$ is small relative to $\partial a/\partial x$ evaluated at x_i.

Some insight into the size of $(\lambda_{i+1} - \lambda_i)$ can be gained by considering the shape of the $V\{x\}$ schedule. An example schedule with positive but

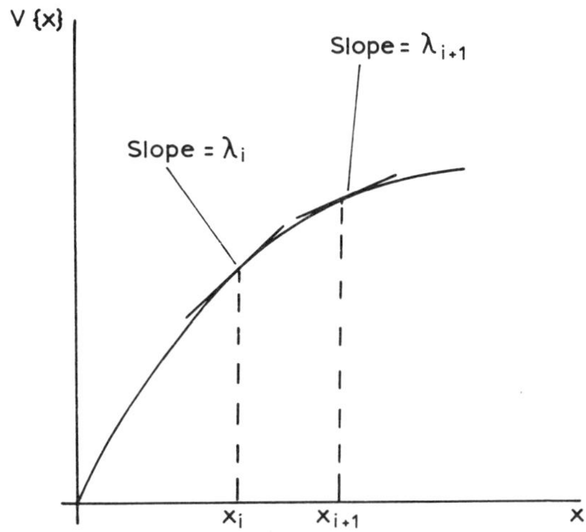

FIG. 4.1 Change in λ related to change in x between stages.

declining slope is shown in Fig. 4.1. This may be assumed typical of $V\{x\}$ for resources in general, but for interest some sufficient conditions for the schedule to have this shape are obtained in Appendix 4.A.

Figure 4.1 indicates the slope of $V\{x\}$ at x_i, and at x_{i+1} which is the resource stock reached after implementing $u^*\{x_i\}$. The slopes are equivalent to λ_i and λ_{i+1}. Clearly λ_i and λ_{i+1} will be closer the closer x_i is to x_{i+1}, and the more nearly linear is $V\{x\}$ over the domain x_i to x_{i+1}.

The closer x_i is to x^*, the closer $g\{x_i\}$ is to $h\{x_i, u_i^*\}$, and hence the closer x_{i+1} is to x_i. Thus (4.10) is likely to provide near-optimal rules for x_i in the vicinity of x^*. Also, if the optimal solution $u^*\{x_i\}$ results in the continuous repetition of states, (4.10) may still provide reasonable rules if $V\{x\}$ is approximately linear in the region of x_i and the repeated states.

Burt and Cummings (1977) reach similar conclusions with more rigour using a different approach. They refer to two evaluations of approximately optimal decision rules derived in the way suggested. One study concerned a groundwater allocation problem (Burt, 1964); the other a soil conservation problem (Burt, 1981). In the first study, the approximation was very close for groundwater stocks large relative to the change in stocks from one period to the next. In the second study, the error in the approximately optimal decision rule was less than one per cent over the relevant range of the resource stock.

If the $V\{x\}$ schedule shows diminishing returns as depicted in Fig. 4.1, statements can be made about the bias of the approximately optimal rule $\hat{u}\{x_i\}$ compared with the optimal rule $u^*\{x_i\}$. To take a simple example, suppose the decision variable u is also the level of harvesting, so that $\partial h/\partial u = 1$ and $\partial h/\partial x = 0$. Equation (4.10) becomes

$$\partial a/\partial u = (\partial a/\partial x)/(r - \mathrm{d}g/\mathrm{d}x) \tag{4.12}$$

Suppose also that $\partial a/\partial u > 0$ and $\partial^2 a/\partial u^2 < 0$. If implementing $\hat{u}\{x_i\}$ calculated from (4.12) results in $x_{i+1} > x_i$, then $(\lambda_{i+1} - \lambda_i) < 0$ by concavity of $V\{x\}$. Ignoring $(\lambda_{i+1} - \lambda_i)$ in the numerator of the right-hand side of (4.12) leads to the overvaluation of $\partial a/\partial u$, and a corresponding $\hat{u}\{x_i\}$ which is too low. Thus if $\hat{u}\{x_i\}$ leads to a rise in the resource stock, $\hat{u}\{x_i\} < u^*\{x_i\}$, and if it leads to a fall in the resource stock, $\hat{u}\{x_i\} > u^*\{x_i\}$. The approximately-optimal policy is biased towards attaining the steady state x^* in too short a time.

Knapp (1983) has extended the analysis of Burt and Cummings by obtaining conditions for steady-state solutions to problems with inequality constraints on the state and decision variables.

4.5 Multiple Objectives

Dynamic programming problems are often formulated with the aim of increasing economic efficiency. For example, objectives may be maximizing the expected present value of net returns in the case of a private interest problem, or maximizing the expected present value of consumers' and producers' surplus in the case of an economy-wide problem. Although an efficiency criterion is often seen to be important, for many problems there are other relevant criteria. In the case of agricultural and natural resource decisions examples of other relevant criteria are resulting levels of environmental quality and risk.

4.5.1 Multi-attribute utility

Whilst net present value is often an important criterion for evaluating resource management, there may be other criteria which are difficult to assess in money terms. In other words, there may be several management goals which are incommensurable, or which cannot be incorporated in one objective function. A farmer may assess a grazing policy not only in terms of the direct returns from grazing but also in terms of possible erosion damage, or in terms of lost leisure. At a more aggregate level, the management of resources may lead to external costs which need to be considered in social policy making. For example, decisions on fertilizer applications may pollute rivers used for drinking or recreation purposes, and decisions to fell trees may result in the water table being raised, and subsequent leaching of farm land if the water is saline. Forestry decisions, such as tree type, density of stands, and thinning and harvesting, may involve aesthetic as well as financial considerations. For some resources there may be extra-market returns from the stock just because of the existence of the stock. People may feel better off in the knowledge that some types of whale do not face extinction, even though they do not experience any contact with whales.

Methods of dealing with multiple objectives in agricultural and natural resource problems have been reviewed by Sinden and Worrell (1979) and Thampapillai (1978). For present purposes three types of strategy for dealing with non-financial criteria are considered. The first two involve transforming measures of the several criteria so that they are commensurable. If this is possible, the multiple objective problem collapses to a single objective problem. One strategy, often pursued in cost–benefit analysis, is to estimate a dollar value for non-market flows. For example, considerable work has been done on valuing such non-market flows as travel time, health, pollution and aesthetic appreciation (see, for example, Pearce, 1978). However, not all those who advocate cost–benefit analysis agree that all costs and benefits can be reduced to money terms.

A second strategy obtains commensurability by measuring all criterion flows in terms of utility instead of money. This type of approach has often been taken in the case of risk criteria, and is discussed in Section 4.5.2. In the case of other criteria, developments have led to multi-attribute utility theory (Keeney and Raiffa, 1976). Suppose the various criteria to be evaluated can be measured by n attributes with levels $z_1, \ldots, z_n$. The decision-making goal becomes maximization of utility given by the function $U(z_1, \ldots, z_n)$. Various resource allocation decisions may be cast as problems requiring the maximization of multi-attribute utility. The resource may be funds, land or water. For example, a public body may face the problem of how best to allocate public land to the following uses: grazing, mining, bush walking, skiing and pleasure driving in four-wheel-drive vehicles. The allocation problem can be formulated as a sequential decision-making problem. The transfer from one stage to another does not, however, represent the elapse of time. The ordering of stages is therefore arbitrary. In the example problem there are five decision stages for allocating available land to each of the five activities. The problem is

$$\max_{z_1,\ldots,z_5} U\{z_1,\ldots,z_5\}$$

subject to

$$\sum_{i=1}^{5} z_i \leq L$$

where z_i is the area allocated to the i-th activity and L is the total land available.

The problem can be solved by dynamic programming if the utility function satisfies the separability and monotonicity requirements outlined in Chapter 2. Keeney and Raiffa (1976) have shown that if there are separate utility functions for each attribute, $U_1\{z_1\},\ldots,U_n\{z_n\}$, and attribute z_i is *utility independent* of all other attributes, then given certain assumptions utility is quasi-separable and takes the form

$$U\{z_1,\ldots,z_n\} = \left(\prod_{i=1}^{n}(Kk_iU_i\{z_i\}+1)-1\right)\Big/K$$

where k_i and K are scaling factors. The U_i functions are scaled to take on values from zero to one. The scaling factors k_i are derived in a way which ensures $0 < k_i < 1$. The objective function is additive or multiplicative depending on $\sum_{i=1}^{n} k_i$ relative to one.

If $\sum k_i = 1$, it can be shown that $K = 0$, and

$$U\{z_1,\ldots,z_n\} = \sum_{i=1}^{n} k_iU_i\{z_i\}$$

which is a straightforward additive objective function.

Alternatively, if $\sum k_i > 1$, then $-1 < K < 0$. Because

$$(U\{z_1,\ldots,z_n\}K)+1 = \prod_{i=1}^{n}(Kk_iU_i\{z_i\}+1)$$

and K is negative, maximizing $U\{z_1,\ldots,z_n\}$ is equivalent to *minimizing*

$$\prod_{i=1}^{n}(Kk_iU_i\{z_i\}+1)$$

The limits of K, k_i and U_i ensure that each term of the product is positive, so the multiplicative objective function can be minimized using dynamic programming. The recursive equation is

$$V_i\{x_i\} = \min_{z_i \leqslant x_i} [(Kk_iU_i\{z_i\} + 1)(V_{i+1}\{x_i - z_i\})] \qquad (i = n, \ldots, 1) \qquad (4.13)$$

with

$$V_{n+1}\{x_{n+1}\} = 1$$

The remaining possibility is that $\sum k_i < 1$, in which case $0 < K < \infty$. Because K is positive, (4.13) would apply after replacing 'min' on the right-hand side by 'max'.

A third strategy for dealing with incommensurable criteria is to present the decision maker with efficiency frontiers showing the trade-off possibilities between criteria. It is left to the decision maker to consult his preference trade-offs and thereby determine an optimal solution. Tauxe *et al.* (1979*a* and *b*) have discussed and exemplified the derivation of such frontiers using what they term *multi-objective dynamic programming*.

Suppose a resource-allocation problem has one decision variable u and one state variable x, but two criteria

$$V = \sum_{i=1}^{n} a_i\{x_i, u_i\} \quad \text{and} \quad W = \sum_{i=1}^{n} b_i\{x_i, u_i\}$$

The resource available at allocation stage i is x_i, and the amount allocated at stage i is u_i. Both V and W satisfy the separability requirements of dynamic programming. The problem of maximizing V subject to the attainment of a specified level of W may be formulated as a dynamic programming problem by including W as an additional state variable as follows:

$$V_i\{x_i, W_i\} = \max_{u_i} [a_i\{x_i, u_i\} + V_{i+1}\{x_{i+1}, W_{i+1}\}] \qquad (i = n, \ldots, 1) \qquad (4.14)$$

subject to

$$x_{i+1} = x_i - u_i$$
$$W_{i+1} = W_i - b_i\{x_i, u_i\}$$

with

$$V_{n+1}\{x_{n+1}, W_{n+1}\} = 0$$
$$W_{n+1} \geq 0$$

Note that all possible states of the decision system are given by all feasible combinations of x and W. If the problem is solved numerically, the addition of W as a state variable considerably increases computation because of the increase in the number of states. However, the decision vector is unaltered. In this example there is no dependence of the stage return function or the transformation function for x on the additional variable, W. However, Tauxe *et al.* (1976*b*) give an example where there is dependence.

Just as solving the irrigation problem by backward recursion led to solutions for each possible initial dam level x_1, backward recursion of (4.14) leads to solutions for each specified value of W_1. An efficient frontier can therefore be obtained by plotting V_1 against W_1.

The way in which the Lagrange multiplier method may be used to reduce the size of this type of problem is discussed in Section 4.6.5.

4.5.2 Risk

If a resource is managed in an environment which cannot be fully controlled, the outcome of the decision sequence is uncertain. The degree of uncertainty may be an important management consideration, together with the expected present value of the outcome. In other words, the optimal decision sequence may be different if the decision maker is risk-averse or risk-loving instead of risk-neutral. A traditional approach which allows for different attitudes towards risk is to specify a utility function with expected cash return and a measure of variability of cash returns as arguments. Measures of variability may be in terms of the variance and higher moments of the distribution of cash returns, z. The expected utility of z is

$$\hat{U}\{z\} = \sum_{k=1}^{m} p\{k\}U\{z_k\}$$

where $U\{z_k\}$ is the utility from cash return z_k, and $p\{k\}$ is the probability that the k-th cash return is generated. This is based on the expected utility theorem which states that the utility of a stochastic income flow equals the expected utility of the flow. A useful exposition of the theorem and the merits of various forms of the utility function is given by Anderson *et al.* (1977).

If utility for the outcome of a decision sequence is an additive function of the utilities obtained from each decision stage, the stochastic recurrence equation (3.1) applies, with the stage return function $a_i\{x_i, u_i, k_i\}$ interpreted as utility generated, and $V_i\{x_i\}$ as the expected utility equal to the utility of following an optimal policy from stage i onwards. Utility at each stage may be discounted by a factor such as α in (3.1), though the discount rate for discounting utility is not, in general, the same as the market rate of discount.

It is often assumed in the economic analysis of optimal consumption through time that intertemporal utility is separable and subject to discounting (see, for example, Page, 1977). An alternative approach which still permits the application of dynamic programming is to work in terms of discounted money certainty equivalents. At each stage, the stage return from a decision is evaluated in terms of stage utility as before, but is then converted to a certain money equivalent $\bar{y}$ using the inverse function $\bar{y} = U^{-1}\{\hat{U}\{z\}\}$. The dynamic programming objective function to be maximized becomes the present value of stage certainty equivalents, using a discount rate suitable for discounting money flows. Hardaker (1979) has used this approach for examining the extent to which Australian farmers can reduce tax liability and stabilize post-tax income over time by buying and selling income equalisation deposits. Buying deposits reduces taxable income in the year in which they are bought, but increases taxable income in the year in which they are sold.

4.5.3 Problems involving players with conflicting objectives

Sometimes decisions are made with no knowledge of the probability distributions of random variables affecting outcomes. Various game-theory criteria have been suggested for selecting options under such uncertainty. An example is the maximin criterion, under which the option resulting in the best of all the worst outcomes is chosen. Decision making under uncertainty has been modelled as a game between two players, one player being the decision maker, the other 'player' being nature. Casting the decision problem as a game against nature and employing a maximin criterion is a very conservative or risk-averse approach because nature is a passive player. The approach is discussed by Hillier and Lieberman (1967). Dynamic programming applied to solving two-person zero-sum games with a maximin criterion is outlined by Beckmann (1968, p. 14) and Nemhauser (1966, pp. 171–4).

Game-theory frameworks are also relevant in another type of multiple-objective context. They apply if resource-management decisions are made by more than one interested party, each with a different objective. Dynamic programming models can be formulated for indicating how resources may actually be, or should be, managed by parties with conflicting objectives. As an example of how dynamic programming may be applied to such problems, a model is specified for a simple problem situation.

In game-theory terminology, the example problem is a two-person, nonzero-sum, noncooperative game in extensive form with perfect knowledge. Two parties make decisions affecting the size of a resource stock at different stages. The parties follow each other at alternate stages in making decisions. Each party knows the other's objective and transformation functions. Although one party's gain is not the other's loss, there is no cooperation between the parties to maximize joint returns. An example of this decision process would be two countries, one in the northern hemisphere and the other in the southern hemisphere, deciding on areas of wheat to plant at six-monthly intervals. Another would be two countries deciding on the harvest levels of a migrating resource such as fish crossing national boundaries at seasonal intervals.

The major difference between the usual equations and those for this problem are that each party evaluates a decision in terms of immediate return and future returns *two* stages ahead. The transformation function for each party includes information about the optimal decision vector of the other party. The solution equations are:

$$V_i\{x_i\} = \max_{u_i} [a_i\{x_i, u_i\}] \qquad (i = n, n-1) \tag{4.15}$$

with u_n implemented by party 2; and u_{n-1} implemented by party 1:

$$V_i\{x_i\} = \max_{u_i} [a_i\{x_i, u_i\} + \alpha V_{i+2}\{x_{i+2}\}] \qquad (i = n-2, \ldots, 1) \tag{4.16}$$

subject to

$$x_{i+2} = x_{i+1} + h_{i+1}\{x_{i+1}, u^*_{i+1}\{x_{i+1}\}] \tag{4.17}$$

$$x_{i+1} = x_i + h_i\{x_i, u_i\} \tag{4.18}$$

with $u_{n-2}, u_{n-4}, \ldots, u_2$ implemented by party 2 and $u_{n-3}, u_{n-5}, \ldots, u_1$ implemented by party 1.

Note that the subscript i on the stage return function a_i indicates a return to party 1 when i is odd and to party 2 when i is even. The stage transformation functions are represented by h, the subscript i again being used to denote one of two functions depending on whether i is odd or even. Solution of (4.15) for stages n and $n-1$ gives the optimal decision vectors $u^*_n\{x_n\}$ for party 2 and $u^*_{n-1}\{x_{n-1}\}$ for party 1, which are subsequently used in the solution of (4.16) to (4.18).

Although in any real-world situation the parties are unlikely to have full information on the optimal decision vector $u^*_i\{x_i\}$ of the other party, model solutions may be of interest for comparing with the cooperative outcome. The same equations could also of course be applied assuming each party has limited knowledge about $u_i^*\{x_i\}$ of the other party. A reaction function $\hat{u}_i\{x_i\}$ could be substituted for $u_i^*\{x_i\}$ which reflected a party's guesses about the other party's responses to changes in the resource stock.

Karp and McCalla (1983) explore the application of difference games with feedback control to trade models. They obtain numerical results for a model of imperfect competition in the international corn market.

4.6 Alternative Computational Methods

Important criteria for evaluating the success of a dynamic programming procedure in solving a problem are the amount of computing time and computer memory required. Given that problems often have to be simplified, another criterion is the closeness of the dynamic programming solution to the solution of the original problem. There are many practical problems for which the computation requirements of the standard algorithm would be unduly burdensome. In such cases it is natural to consider if there are alternative methods of applying the logic of dynamic programming to solve the problem. In general there are alternative methods, each entailing trade-offs between the criteria just mentioned.

As a rule of thumb, computation time for the standard algorithm rises linearly with the number of stages, but exponentially with the number of state and decision variables. High-speed memory requirements also increase exponentially with the number of state variables. The impact of increases in the number of state and decision variables on the computation burden has been termed by Bellman and Dreyfus (1962, p. 323) 'the curse of dimensionality'. It used to be thought that given the typical amount of computing power available, the curse of dimensionality limited the number of state variables to a maximum of about three. However, many suggestions for exorcising the curse have been suggested. Some of these are briefly reviewed. One method, the Lagrange multiplier method, is discussed in detail with a worked example in Section 4.6.5.

4.6.1 Approximating the value function in continuous form

The standard numerical algorithm entails storing the optimal value and/or decision for all states and all stages. The information is used to determine the optimal decision path. Memory requirements can be reduced if, for each stage, the optimal state values can be expressed approximately as a continuous function of the state variables. Instead of storing state values for each state, only the parameters of the approximating function need be stored. The function would typically be a polynomial fitted using regression or an orthogonal polynomial (see Larson and Casti, 1982, pp. 234–41).

4.6.2 *Alternative dynamic programming structures*

In some problems it may be impossible to reach some states from any state at the previous stage. If such problems are solved using backward recursion, infeasible decisions may be evaluated along with feasible decisions for many states at any stage. It may be more efficient to solve the problem using forward recursion, in which case unreachable states can be ignored with a consequent saving in computation. Alternatively, backward recursion may still be used, but with state values evaluated at each stage by what Denado (1982) terms 'reaching', instead of the more usual 'recursive fixing'. However, the inverse transformation function

$$x_i = t_i^{-1}\{x_{i+1}, u_i\}$$

must exist for these methods to be possible. Inverse transformation functions do not exist for stochastic transformation functions.

Other methods have been suggested which, whilst based on the principle of optimality, have an alternative state space to that used in the conventional approach. Cooper and Cooper (1981) have proposed a hypersurface search algorithm for integer problems, in which the state variable represents the maximum value of the value function. Dharmadhikari (1975) describes a decision-state method, in which only dominant states and associated optimal decisions are recorded at each stage. The method is similar to the imbedded state-space approach proposed by Morin and Esogbue (1974).

4.6.3 *Successive approximations around a nominal control policy*

An important parameter in the specification of a numerical solution procedure is the size of the interval between discrete values of the state and decision variables, or the coarseness of the grid. To be sure of finding the globally optimal path the grid of all possible state values across all stages has to be extensive and fine. In finding an optimal path starting from some initial state a wealth of additional information is generated on optimal decisions to take at state values far removed from the optimal path. Much of this information may be redundant. Many suggestions have been made for limiting the exploration of state space to the area through which the optimal path is expected to pass.

The basic idea is to start at the beginning of each iteration with a trial policy, and to see if there is a preferred path passing through states in the neighbourhood of the trial path. If there is, the preferred path becomes the trial path for the next iteration. If there is not, then depending on the degree of accuracy required, the grid around the trial path which is apparently optimal may be further refined. The results from running the dynamic programming algorithm on the more refined grid will indicate whether any further iterations are required to locate the optimal path.

Methods which are based on this idea have been referred to as grid refinement (Denardo, 1982, p. 75), path restriction (Norman, 1972, Chapter 4), discrete differential dynamic programming (Heidari *et al.*, 1971) and incremental dynamic programming (Turgeon, 1982). Other methods which exploit the idea but which apply to problems in which time can be treated as a continuous variable are state incremental dynamic programming (Larson and Casti, 1982, pp. 249–92) and differential dynamic programming (Yakowitz, 1982, pp. 688–98).

Memory and computation requirements can be substantially reduced with these methods, but at the risk of ending up, after several iterations, investigating paths outside the neighbourhood of the optimal path. Depending on the nature of the problem, the initial trial path may be critical. The initial trial path may be determined on the basis of *a priori* considerations, or on the basis of running the dynamic programming algorithm with a coarse grid.

If the only decisions which may be considered at each nominated state and stage are those which reach a nominated state at the next stage, a coarse grid may severely restrict the range of feasible decisions. However, this restriction is overcome if interpolation in state space is permitted. Optimal values and decisions are stored only for grid points. If a decision leads to a state intermediate between grid points, the value of the state is calculated by interpolating the state values of the nominated states immediately above and below. As pointed out by Lewis (1975), the solution system for state value interpolation is the same as for stochastic dynamic programming, with interpolation weights playing the same role as probabilities. In Chapter 10 an illustrative fisheries problem is solved using the approach.

4.6.4 Solving a sequence of problems of reduced dimension

Because the computational burden in the conventional algorithm increases exponentially with the number of state and decision variables, the solution may be reached more efficiently by making use of a series of solutions to simpler problems with fewer dimensions. At each iteration, the reduced problem is modified on the basis of the solution at the previous iteration to better approximate the original problem. Again there is a risk that the global optimum is missed.

Nemhauser (1966) describes the one-at-a-time method and its pitfalls, applicable to problems with m decision variables. The idea is to solve the problem first for decision variable u^1 with the other decision variables $u^2, \ldots, u^m$ kept fixed across all states and stages. At the next iteration the problem is solved for decision variable u^2, with $u^3, \ldots, u^m$ still fixed, and u^1 held at the optimal levels for the first iteration. Iterations continue until no further improvement in the value of the objective function is obtained.

Norman (1972) suggests six possible ways of formulating reduced problems, making use of the characteristics of the problem. For example, a problem with a large number of possible nominated state values may be reformulated as a problem with fewer state values by grouping or aggregating the state values. A one-to-many correspondence is established between state values in the reduced problem and state values in the original problem. The reduced problem is solved to give the values of the states, which are then used as approximations to the values of all of the states in the full problem. A single-stage decision vector for the full problem which is optimal or nearly optimal is obtained by solving the appropriate recursive equation such as (2.1) with $V_{i+1}\{\cdot\}$ equal to the approximately optimal values previously determined. The process is repeated sequentially for as many decision stages as required.

Burt *et al.* (1980) have used such techniques to obtain approximately optimal storage and export decisions for US wheat, a problem formulated with 15 state variables. They describe a method for dealing with state variables which are lagged variables. Taylor and Burt (1984) describe how they solved a weed-control problem with four state variables by using the solutions to a reduced problem with three state variables and fewer decision options than in the original problem.

Another method of attempting to find an optimal solution by solving a sequence of continually modified reduced problems is the method of Lagrange multipliers, and is dealt with in detail in the next section.

4.6.5 The Lagrange multiplier method

The Lagrange multiplier method may be used to eliminate state variables which record resource availability at each stage in resource allocation problems, or which represent the value of secondary objective functions in problems with several objective functions.

In the case of a problem in which stage decisions draw on two resources, the problem could be formulated with two state variables representing the availability of each resource. Alternatively, the problem could be formulated with one state variable representing the availability of one resource. Restrictions on the availability of the second resource would be implemented by charging each stage return with usage of the second resource, at a rate λ, the same for each stage. The idea is to rerun the one-state-variable dynamic programming algorithm with a larger or smaller value of λ, depending on whether the second resource was over- or under-utilized for the previous value of λ. Iterations continue until the second resource constraint is satisfied. The final value of λ is a measure of the gain in terms of the objective function which could be achieved by relaxing the constraint on the second resource by one unit. In other words, λ plays the role of a shadow price or a Lagrange multiplier.

The Lagrange multiplier method is guaranteed to find an optimal solution if the objective function is a concave function of the availability of the second resource. Unfortunately, the concavity requirement is often violated in the case of problems with integer constraints or with states represented by values at discrete intervals. For such cases the Lagrange multiplier method may fail to identify the optimal solution. However, if the method does fail it is possible to estimate a limit on the error involved.

The Lagrange multiplier method may be illustrated by applying it to the original deterministic crop-irrigation problem, this time extended to take account of an additional objective. Suppose the crop has to be processed before it is marketed, and the farmer wants to determine the best irrigation decisions for a range of processing capacities, specified in terms of total weight of crops harvested during a year. Assume that the weight of crop harvested at the end of the *i*-th season depends on the water received by the crop (w_i) and equals ($w_i - 0{\cdot}1w_i^2$). In terms of the discussion of multi-objective dynamic programming, the farmer has the twin objectives of maximizing returns and minimizing total weight of crop harvested. The farmer requires information on the efficient frontier showing trade-offs between the two objectives.

The recursive equation for the twin objective function problem becomes

$$V_i\{x_i, W_i\} = \max_{u_i} [b_i(w_i - 0{\cdot}1w_i^2) + \alpha V_{i+1}\{x_{i+1}, W_{i+1}\}] \qquad (i = 3 \text{ to } 1) \qquad (4.19)$$

subject to

$$0 \le u_i \le x_i \le 3 \qquad u_i, x_i \text{ integer}$$
$$w_i = u_i + q_i$$
$$x_{i+1} = \min(x_i - u_i + q_i, 3)$$
$$W_{i+1} = W_i - (w_i - 0{\cdot}1w_i^2)$$

with

$$V_4\{x_4, W_4\} = 0$$
$$W_4 \ge 0$$

The data are

$$x_1 = 3$$
$$b = [50 \quad 100 \quad 150]$$
$$q = [2 \quad 1 \quad 1]$$
$$\alpha = 0{\cdot}95$$

with units as before. W_i is the permissible total weight of crops in tonnes produced over the stages i to 3.

The Lagrange multiplier method may be used to eliminate W_i as a state variable in (4.19) by reformulating the problem as

$$\mathbf{V}_i\{x_i\} = \max_{u_i} [b_i(w_i - 0{\cdot}1w_i^2) - (\lambda/\alpha^{i-1})(w_i - 0{\cdot}1w_i^2) + \alpha \mathbf{V}_{i+1}\{x_{i+1}\}] \qquad (i = 3 \text{ to } 1) \qquad (4.20)$$

with

$$\mathbf{V}_4\{x_4\} = 0$$

and other constraints and data as before. The return at each stage i includes a penalty for the weight of crop produced at stage i, charged at the rate of $\$(\lambda/\alpha^{i-1})$ per tonne. Because $\mathbf{V}_1\{x_1\}$ is the sum of the discounted returns over three decision stages, λ has to be divided by α^{i-1} in (4.20) to ensure the per unit weight charge is the same for all three periods.

An iterative system of slope updating (Denardo, 1982, pp. 57–8) is used to identify at least some of the solutions which lie on the efficient W, V frontier. At each iteration j the recursive equation (4.20) is solved for $\lambda = \lambda_j$. The solutions obtained are listed in Table 4.8, and are marked by crosses for $j = 1, 2, 3, 4$ and 7 in Fig. 4.2.

TABLE 4.8
Solutions to the Crop-irrigation Problem for Alternative Values of λ

j	λ_j (\$/tonne)	W_j (000 tonnes)	V_j (\$000)	Decision vector (ha m)	Comment on decision vector
1	0·00	6·6	60·38	2, 2, 2	
2	15·00	3·4	28·73	0, 0, 0	
3	9·89	4·9	49·04	0, 0, 3	
4	6·67	5·6	55·69	0, 1, 3	
5	13·54				same as for $j=2$ or 3
6	9·50				same as for $j=3$ or 4
7	4·69	6·4	59·69	2, 1, 3	
8	3·45				same as for $j=1$ or 7
9	5·00				same as for $j=4$ or 7

The determination of the sequence of λ_j can be explained with reference to Table 4.9 and Fig. 4.2. For $j=1$, $\lambda=0{\cdot}00$ to give the optimal decision vector (2, 2, 2), the same as that for the problem without a total weight objective. At the other extreme, for $j=2$, a value of $\lambda=\$15{\cdot}00$ per tonne was found to be large enough for the minimized total weight solution with optimal decision vector (0, 0, 0) to be obtained. The two solutions are marked by crosses 1 and 2 in Fig. 4.2. For $j=3$, λ was set equal to \$9·89 per tonne, the slope of the straight line joining crosses 1 and 2 in Fig. 4.2. Any solution which lies above the straight line is thereby identified. In the example, the solution marked by cross 3 is obtained.

The process continues, with λ for $j=4$ set equal to \$6·67 per tonne, the slope of the straight line joining crosses 1 and 3 in Fig. 4.2. This

TABLE 4.9
Revision of λ Using Slope Updating

j	k	l	$\lambda_j=(V_k-V_l)/(W_k-W_l)$ (\$/tonne)
3	1	2	(60·38 − 28·73)/(6·6 − 3·4) = 9·89
4	1	3	(60·38 − 49·04)/(6·6 − 4·9) = 6·67
5	3	2	(49·04 − 28·73)/(4·9 − 3·4) = 13·54
6	4	3	(55·69 − 49·04)/(5·6 − 4·9) = 9·50
7	1	4	(60·38 − 55·69)/(6·6 − 5·6) = 4·69
8	1	7	(60·38 − 59·69)/(6·6 − 6·4) = 3·45
9	7	4	(59·69 − 55·69)/(6·4 − 5·6) = 5·00

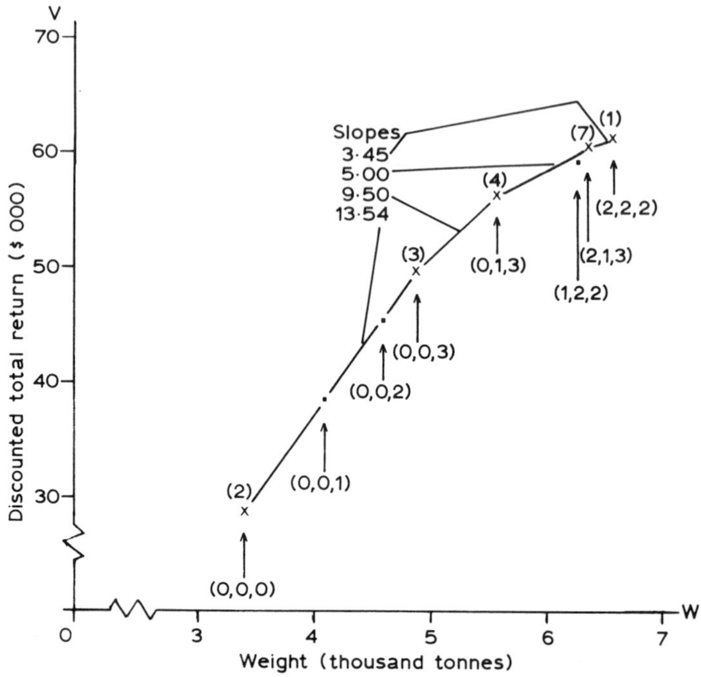

FIG. 4.2. *W–V* frontier identified using the Lagrange multiplier method. (The optimal decision vector is shown below the *j*-th solution point, with *j* above the solution point.)

results in the solution marked by cross 7. However, consider the situation when $\lambda = \$13{\cdot}54$ per tonne for $j = 5$, the slope of the straight line joining crosses 3 and 2. It is not possible to find a solution with higher **V** than that for the solutions for $j = 2$ and $j = 3$. The straight line joining crosses 2 and 3 is therefore part of the identifiable frontier.

In fact there are two other solutions ((0, 0, 1) and (0, 0, 2)) with the same value of **V** for $j = 5$, and which therefore lie on the segment joining crosses 2 and 3. Depending on how ties or close ties are dealt with in the dynamic programming routine, these solutions may also be identified for $\lambda = \$13{\cdot}54$ per tonne.

The process terminates after the nodes 2, 3, 4, 7 and 1 have been found, defining the identifiable frontier made up of segments with slopes \$13·54, 9·50, 5·00 and 3·45 per tonne, respectively. The results for $j = 5$, 6, 8 and 9 in Table 4.8 show that it is impossible to find decision vectors which are more efficient than those obtained for $j = 2$, 3, 4, 7 and 1. Subject to caveats, the slopes may be interpreted in the way Lagrange multipliers are usually interpreted. For example, it is evident from the solution for $j = 3$ in Table 4.8 that an efficient weight/value combination is \$49 040 and 4900 tonnes. The rate at which value increases as the weight constraint is relaxed is \$9·50 per tonne.

One caveat on the interpretation is necessary because the decision variable is formulated as discrete rather than continuous. The slope therefore has more the connotation of a shadow price in linear programming than a Lagrange multiplier in constrained maximization solved by calculus. A second caveat is necessary because the efficient frontier may not be concave, particularly for discrete decision variables. Any solution lying below the straight line joining two solutions fails to be identified. In the crop-irrigation problem, the solution vector (1, 2, 2) shown in Fig. 4.2 is not identified using the Lagrange multiplier method for this reason.

The problem of unidentified solutions, or gaps, is ameliorated to the extent that error bounds can be specified. For example, suppose in the crop-irrigation problem, V is to be maximized subject to $W \leq 5000$ tonnes. The solution for $j = 3$ is the identified solution with largest W below 5000 tonnes, with $W = 4900$ tonnes and $V = \$49\,040$. Because the slope of the straight line joining crosses for identified solutions 3 and 4 is \$9·50/tonne, if an unidentified solution existed for W between 4900 and 5000 tonnes, V could not be larger than

$$\$(49\,040 + 9{\cdot}50 \times (5000 - 4900)) = \$49\,990$$

A lower bound on V is \$49 040 because any solution with $V < \$49\,040$ would be dominated by the solution for $j = 3$.

4.7 Further Information on GPDP

4.7.1 The format for user-written data files

A large number of stage returns have to be calculated for some problems. It will often be more convenient to devise a special program to write the data to file in the correct format for processing by FDP and IDP, rather than to use DPD. The required format is explained in this section.

Three example programs are given which can be modified to write the data of other problems to file. The example programs are designed to write to file the data for the following three versions of the crop-irrigation problem: deterministic, finite number of stages (IRDF.BAS, Fig. 4.3); stochastic, finite number of stages (IRSF.BAS, Fig. 4.5); and stochastic,

```
100 REM----------IRRIGATION PROBLEM---IRDF----------
110 REM
120 LET N$="IRDF.DAT"
130 OPEN N$ FOR OUTPUT AS FILE #1
140 REM
150 LET N=3
160 LET D2=5.2632
170 DIM B(3),X(3)
180 MAT READ B
190 DATA 5,10,15
200 MAT READ X
210 DATA 2,1,1
220 LET S=4
230 LET C$=","
240 LET P=1
250 REM
260 REM-----WRITE BLOCK A DATA - PROBLEM PARAMETERS-----
270 PRINT #1, N
280 PRINT #1, D2
290 PRINT #1, "D"
300 PRINT #1, "N"
310 PRINT #1, "Y"
320 REM-----WRITE BLOCK B DATA - TERMINAL STATE VALUES-----
330 PRINT #1, "0,0"
340 REM-----WRITE BLOCKS CD AND ED DATA-----
350 FOR T=N TO 1 STEP -1
360    PRINT #1, " T                I    D     R                         J    P"
370    FOR I=1 TO S
380       FOR D=1 TO I
390          LET W=D-1+X(T)
400          LET R=B(T)*(W-.1*W**2)
410          LET J=I-D+1+X(T)
420          IF J<S+1 THEN 440
430          LET J=S
440          PRINT #1, T;C$,I;C$;D;C$;R;C$,J;C$;P
450       NEXT D
460    NEXT I
470    PRINT #1, T,",0,0,0,0,1"
480    IF T<N THEN 580
490 REM-----WRITE BLOCK D DATA - LABELS-----
500    PRINT #1, "LEVEL"
510    FOR I=1 TO S
520       PRINT #1, "L"+STR$(I-1)
530    NEXT I
540    PRINT #1, "RELEASE,4"
550    FOR D=1 TO 4
560       PRINT #1, "R"+STR$(D-1)
570    NEXT D
580 NEXT T
590 PRINT "DATA WRITTEN TO ";N$
600 CLOSE #1
610 END
```

FIG. 4.3. BASIC program for writing IRDF data to file.

```
 3
 5.2632
D
N
Y
0,0
 T               I   D    R                    J   P
 3 ,             1 , 1 , 13.5 ,                2 , 1
 3 ,             2 , 1 , 13.5 ,                3 , 1
 3 ,             2 , 2 , 24 ,  2 , 1
 3 ,             3 , 1 , 13.5 ,                4 , 1
 3 ,             3 , 2 , 24 ,  3 , 1
 3 ,             3 , 3 , 31.5 ,                2 , 1
 3 ,             4 , 1 , 13.5 ,                4 , 1
 3 ,             4 , 2 , 24 ,  4 , 1
 3 ,             4 , 3 , 31.5 ,                3 , 1
 3 ,             4 , 4 , 36 ,  2 , 1
 3              ,0,0,0,0,1
LEVEL
L0
L1
L2
L3
RELEASE,4
R0
R1
R2
R3
 T               I   D    R                    J   P
 2 ,             1 , 1 , 9 ,   2 , 1
 2 ,             2 , 1 , 9 ,   3 , 1
 2 ,             2 , 2 , 16 ,  2 , 1
 2 ,             3 , 1 , 9 ,   4 , 1
 2 ,             3 , 2 , 16 ,  3 , 1
 2 ,             3 , 3 , 21 ,  2 , 1
 2 ,             4 , 1 , 9 ,   4 , 1
 2 ,             4 , 2 , 16 ,  4 , 1
 2 ,             4 , 3 , 21 ,  3 , 1
 2 ,             4 , 4 , 24 ,  2 , 1
 2              ,0,0,0,0,1
 T               I   D    R                    J   P
 1 ,             1 , 1 , 8 ,   3 , 1
 1 ,             2 , 1 , 8 ,   4 , 1
 1 ,             2 , 2 , 10.5 ,                3 , 1
 1 ,             3 , 1 , 8 ,   4 , 1
 1 ,             3 , 2 , 10.5 ,                4 , 1
 1 ,             3 , 3 , 12 ,  3 , 1
 1 ,             4 , 1 , 8 ,   4 , 1
 1 ,             4 , 2 , 10.5 ,                4 , 1
 1 ,             4 , 3 , 12 ,  4 , 1
 1 ,             4 , 4 , 12.5 ,                3 , 1
 1              ,0,0,0,0,1
```

FIG. 4.4. Data file IRDF.DAT.

```
100 REM----------IRRIGATION PROBLEM---IRSF----------
110 REM
120 LET N$="IRSF.DAT"
130 OPEN N$ FOR OUTPUT AS FILE #1
140 REM
150 LET N=3
160 LET D2=5.2632
170 DIM B(3),X(3),P(3)
180 MAT READ B
190 DATA 5,10,15
200 MAT READ X
210 DATA 2,1,1
220 MAT READ P
230 DATA .25,.5,.25
240 LET S=4
250 LET C$=","
260 REM
270 REM-----WRITE BLOCK A DATA - PROBLEM PARAMETERS-----
280 PRINT #1, N
290 PRINT #1, D2
300 PRINT #1, "S"
310 PRINT #1, "N"
320 PRINT #1, "Y"
330 REM
340 REM-----WRITE BLOCK B DATA - TERMINAL STATE VALUES-----
350 PRINT #1, "0,0"
360 REM-----WRITE BLOCKS CS AND ES DATA-----
370 FOR T=N TO 1 STEP -1
380     PRINT #1, " T                  I    D     R                   J     P
390     FOR I=1 TO S
400         FOR D=1 TO I
410             LET R=0
420             FOR P9=1 TO 3
430                 LET W=D-1+X(T)+P9-2
440                 LET R=R+B(T)*(W-.1*W**2)*P(P9)
450             NEXT P9
460             FOR P9=1 TO 3
470                 LET J=I-D+1+X(T)+P9-2
480                 IF J<S+1 THEN 500
490                 LET J=S
500                 PRINT #1, T;C$,I;C$;D;C$;R;C$,J;C$;P(P9)
510             NEXT P9
520         NEXT D
530     NEXT I
540     PRINT #1, T,",0,0,0,0,1"
550     IF T<N THEN 650
560 REM-----WRITE BLOCK D DATA - LABELS-----
570     PRINT #1, "LEVEL"
580     FOR I=1 TO S
590         PRINT #1, "L"+STR$(I-1)
600     NEXT I
610     PRINT #1, "RELEASE,4"
620     FOR D=1 TO 4
630         PRINT #1, "R"+STR$(D-1)
640     NEXT D
650 NEXT T
660 PRINT "DATA WRITTEN TO ";N$
670 CLOSE #1
680 END
```

FIG. 4.5. BASIC program for writing IRSF data to file.

infinite-stage (IRSI.BAS, Fig. 4.7). As an example of the layout of files for deterministic and stochastic problems, the data files obtained by running IRDF.BAS and IRSF.BAS are shown in Figs 4.4 and 4.6, respectively. The data files are the same as those written by entering the data interactively with DPD for the problem CIDF (see Fig. 2.5) and CISF (see Fig. 3.1).

The data for a problem consist of some or all of the data blocks in Table 4.10.

TABLE 4.10
Data Structure Required in the GPDP Routines

Data block	*Description of data*
A	Problem parameters
B	Values of terminal states
C	Decisions, returns, transformations for the last decision stage—and for all decision stages if the problem is stationary
D	Labels for states and decisions (optional)
E	Decisions, returns, transformations for decision stages other than the last if the problem is non-stationary

The structure of data in blocks A, B and D is straightforward. The order of data is given in the coding sheets, and the format can be deduced from the example programs and data files.

Although there are different coding sheets for block C and block E data, depending on whether the problem is deterministic or stochastic, the format of the data on file is the same for all of the blocks CD, ED, CS and ES. Each block is headed by the line '*T IDR JP*', where *T* stands for the number of the decision stage; *I* for the state number; *D* for the decision number; *R* for the stage return given *I* and *D*; and *J* for the state number which may be reached at the next stage given *I* and *D*. Subsequent lines list values of *T*, *I*, *D*, *R*, *J* and *P* for all *I* and *D*. If the problem is deterministic, the *P* column consists of a column of ones. The block is terminated by the line '*T*, 0, 0, 0, 0, 1', where *T* is the decision-stage number.

The results for the stochastic, infinite-stage crop-irrigation problem obtained by running IDP for the problem IRSI are shown in Fig. 4.8 with discounting, and in Fig. 4.9 without discounting. The results agree with those presented in Tables 4.2 and 4.4 obtained by solving linear programming problems.

```
 3
 5.2632
S
N
Y
0,0
 T                  I   D     R                    J    P
 3 ,                1 , 1 , 12.75 ,                1 , .25
 3 ,                1 , 1 , 12.75 ,                2 , .5
 3 ,                1 , 1 , 12.75 ,                3 , .25
 3 ,                2 , 1 , 12.75 ,                2 , .25
 3 ,                2 , 1 , 12.75 ,                3 , .5
 3 ,                2 , 1 , 12.75 ,                4 , .25
 3 ,                2 , 2 , 23.25 ,                1 , .25
 3 ,                2 , 2 , 23.25 ,                2 , .5
 3 ,                2 , 2 , 23.25 ,                3 , .25
 3 ,                3 , 1 , 12.75 ,                3 , .25
 3 ,                3 , 1 , 12.75 ,                4 , .5
 3 ,                3 , 1 , 12.75 ,                4 , .25
 3 ,                3 , 2 , 23.25 ,                2 , .25
 3 ,                3 , 2 , 23.25 ,                3 , .5
 3 ,                3 , 2 , 23.25 ,                4 , .25
 3 ,                3 , 3 , 30.75 ,                1 , .25
 3 ,                3 , 3 , 30.75 ,                2 , .5
 3 ,                3 , 3 , 30.75 ,                3 , .25
 3 ,                4 , 1 , 12.75 ,                4 , .25
 3 ,                4 , 1 , 12.75 ,                4 , .5
 3 ,                4 , 1 , 12.75 ,                4 , .25
 3 ,                4 , 2 , 23.25 ,                3 , .25
 3 ,                4 , 2 , 23.25 ,                4 , .5
 3 ,                4 , 2 , 23.25 ,                4 , .25
 3 ,                4 , 3 , 30.75 ,                2 , .25
 3 ,                4 , 3 , 30.75 ,                3 , .5
 3 ,                4 , 3 , 30.75 ,                4 , .25
 3 ,                4 , 4 , 35.25 ,                1 , .25
 3 ,                4 , 4 , 35.25 ,                2 , .5
 3 ,                4 , 4 , 35.25 ,                3 , .25
 3                 ,0,0,0,0,1
LEVEL
L0
L1
L2
L3
RELEASE,4
R0
R1
R2
R3
 T                  I   D     R                    J    P
 2 ,                1 , 1 , 8.5 ,                  1 , .25
 2 ,                1 , 1 , 8.5 ,                  2 , .5
 2 ,                1 , 1 , 8.5 ,                  3 , .25
 2 ,                2 , 1 , 8.5 ,                  2 , .25
 2 ,                2 , 1 , 8.5 ,                  3 , .5
 2 ,                2 , 1 , 8.5 ,                  4 , .25
 2 ,                2 , 2 , 15.5 ,                 1 , .25
```

FIG. 4.6. Data file IRSF.DAT.

```
2 ,    2 , 2 , 15.5 ,     2 , .5
2 ,    2 , 2 , 15.5 ,     3 , .25
2 ,    3 , 1 , 8.5 ,      3 , .25
2 ,    3 , 1 , 8.5 ,      4 , .5
2 ,    3 , 1 , 8.5 ,      4 , .25
2 ,    3 , 2 , 15.5 ,     2 , .25
2 ,    3 , 2 , 15.5 ,     3 , .5
2 ,    3 , 2 , 15.5 ,     4 , .25
2 ,    3 , 3 , 20.5 ,     1 , .25
2 ,    3 , 3 , 20.5 ,     2 , .5
2 ,    3 , 3 , 20.5 ,     3 , .25
2 ,    4 , 1 , 8.5 ,      4 , .25
2 ,    4 , 1 , 8.5 ,      4 , .5
2 ,    4 , 1 , 8.5 ,      4 , .25
2 ,    4 , 2 , 15.5 ,     3 , .25
2 ,    4 , 2 , 15.5 ,     4 , .5
2 ,    4 , 2 , 15.5 ,     4 , .25
2 ,    4 , 3 , 20.5 ,     2 , .25
2 ,    4 , 3 , 20.5 ,     3 , .5
2 ,    4 , 3 , 20.5 ,     4 , .25
2 ,    4 , 4 , 23.5 ,     1 , .25
2 ,    4 , 4 , 23.5 ,     2 , .5
2 ,    4 , 4 , 23.5 ,     3 , .25
2     ,0,0,0,0,1
T      I   D     R        J    P
1 ,    1 , 1 , 7.75 ,     2 , .25
1 ,    1 , 1 , 7.75 ,     3 , .5
1 ,    1 , 1 , 7.75 ,     4 , .25
1 ,    2 , 1 , 7.75 ,     3 , .25
1 ,    2 , 1 , 7.75 ,     4 , .5
1 ,    2 , 1 , 7.75 ,     4 , .25
1 ,    2 , 2 , 10.25 ,    2 , .25
1 ,    2 , 2 , 10.25 ,    3 , .5
1 ,    2 , 2 , 10.25 ,    4 , .25
1 ,    3 , 1 , 7.75 ,     4 , .25
1 ,    3 , 1 , 7.75 ,     4 , .5
1 ,    3 , 1 , 7.75 ,     4 , .25
1 ,    3 , 2 , 10.25 ,    3 , .25
1 ,    3 , 2 , 10.25 ,    4 , .5
1 ,    3 , 2 , 10.25 ,    4 , .25
1 ,    3 , 3 , 11.75 ,    2 , .25
1 ,    3 , 3 , 11.75 ,    3 , .5
1 ,    3 , 3 , 11.75 ,    4 , .25
1 ,    4 , 1 , 7.75 ,     4 , .25
1 ,    4 , 1 , 7.75 ,     4 , .5
1 ,    4 , 1 , 7.75 ,     4 , .25
1 ,    4 , 2 , 10.25 ,    4 , .25
1 ,    4 , 2 , 10.25 ,    4 , .5
1 ,    4 , 2 , 10.25 ,    4 , .25
1 ,    4 , 3 , 11.75 ,    3 , .25
1 ,    4 , 3 , 11.75 ,    4 , .5
1 ,    4 , 3 , 11.75 ,    4 , .25
1 ,    4 , 4 , 12.25 ,    2 , .25
1 ,    4 , 4 , 12.25 ,    3 , .5
1 ,    4 , 4 , 12.25 ,    4 , .25
1     ,0,0,0,0,1
```

FIG. 4.6.—*contd.*

```
100 REM----------IRRIGATION PROBLEM---IRSI----------
110 REM
120 LET N$="IRSI.DAT"
130 OPEN N$ FOR OUTPUT AS FILE #1
140 REM
150 LET N=3
160 LET D2=5.2632
170 DIM B(3),X(3),P(3)
180 MAT READ B
190 DATA 5,10,15
200 MAT READ X
210 DATA 2,1,1
220 MAT READ P
230 DATA .25,.5,.25
240 C$=","
250 REM
260 REM-----WRITE BLOCK A DATA - PROBLEM PARAMETERS-----
270 PRINT #1, N
280 PRINT #1, D2
290 PRINT #1, "S"
300 PRINT #1, "Y"
310 PRINT #1, "Y"
320 REM
330 REM-----WRITE BLOCK B DATA - TERMINAL STATE VALUES-----
340 PRINT #1, "0,0"
350 REM-----WRITE BLOCK CS DATA-----
360 PRINT #1, " T               I    D    R                    J    P"
370 FOR I2=1 TO 3
380    FOR I3=1 TO 4
390       LET I=I3+(I2-1)*4
400       FOR D=1 TO I3
410          LET R=0
420          FOR P9=1 TO 3
430             LET W=D-1+X(I2)+P9-2
440             LET R=R+B(I2)*(W-.1*W**2)*P(P9)
450          NEXT P9
460          FOR P9=1 TO 3
470             IF I2=3 THEN 530
480             LET J=I+5-D+X(I2)+P9-2
490             IF J>(4+4*I2) THEN 510
500             GO TO 570
510             LET J=(4+4*I2)
520             GO TO 570
530             LET J=I-7-D+X(I2)+P9-2
540             IF J>4 THEN 560
550             GO TO 570
560             LET J=4
570             PRINT #1, N;C$,I;C$;D;C$;R;C$,J;C$;P(P9)
580          NEXT P9
590       NEXT D
600    NEXT I3
610 NEXT I2
620 PRINT #1, N,",0,0,0,0,1"
630 REM-----WRITE BLOCK D DATA - LABELS-----
640 PRINT #1, "S/LEVEL"
650 FOR I2 =1 TO 3
660    FOR I3=1 TO 4
670       LET S$="S"+STR$(I2)
680       LET S$=S$+"/L"
690       PRINT #1, S$+STR$(I3-1)
700    NEXT I3
710 NEXT I2
720 PRINT #1, "RELEASE,4"
730 FOR D=1 TO 4
740    PRINT #1, "R"+STR$(D-1)
750 NEXT D
760 PRINT "DATA WRITTEN TO ";N$
770 END
```

FIG. 4.7. BASIC program for writing IRSI data to file.

```
-----INFINITE STAGE DP PROBLEM-----
PROBLEM NAME ? IRSI

----------------------------------------------------------
        STOCHASTIC DYNAMIC PROGRAMMING SOLUTION
                  PROBLEM ---IRSI---

PROBLEM PARAMETERS:
INFINITE NO. OF STAGES
RATE OF DISCOUNT (PER CENT)  =  5.2632
----------------------------------------------------------

NO. OF ITERATIONS TO FIND SOLUTION =  3
```

CURRENT STATE		DECISION		NEXT STATE		PRESENT VALUE
NO.	S/LEVEL	NO.	RELEASE	NO.	S/LEVEL	
1	S1/L0	1	R0	?		366.042
2	S1/L1	1	R0	?		370.079
3	S1/L2	2	R1	?		372.579
4	S1/L3	3	R2	?		374.079
5	S2/L0	1	R0	?		363.524
6	S2/L1	1	R0	?		370.649
7	S2/L2	2	R1	?		377.649
8	S2/L3	3	R2	?		382.649
9	S3/L0	1	R0	?		363.96
10	S3/L1	2	R1	?		374.46
11	S3/L2	3	R2	?		381.96
12	S3/L3	4	R3	?		386.46

```
REQUIRE LONG-RUN STATE PROBABILITIES - Y(YES) OR N(NO)? Y
```

STATE		PROBABILITY
NO.	S/LEVEL	
1	S1/L0	.833333E-01
2	S1/L1	.166667
3	S1/L2	.833333E-01
4	S1/L3	0
5	S2/L0	0
6	S2/L1	.208333E-01
7	S2/L2	.104167
8	S2/L3	.208333
9	S3/L0	0
10	S3/L1	.833333E-01
11	S3/L2	.166667
12	S3/L3	.833333E-01

FIG. 4.8. Solution to the problem 'IRSI' using IDP with discount factor $\alpha = 0{\cdot}95$.

```
-----INFINITE STAGE DP PROBLEM-----
PROBLEM NAME ? IRSI


-------------------------------------------------------------
        STOCHASTIC DYNAMIC PROGRAMMING SOLUTION
                   PROBLEM ---IRSI---

PROBLEM PARAMETERS:
INFINITE NO. OF STAGES
RATE OF DISCOUNT (PER CENT)  =  0
-------------------------------------------------------------

NO. OF ITERATIONS TO FIND SOLUTION =  3


-------------------------------------------------------------
  CURRENT       DECISION        NEXT           RELATIVE
   STATE                        STATE           VALUE
-----------   -----------   -----------
 NO. S/LEVEL   NO. RELEASE   NO. S/LEVEL
-------------------------------------------------------------
  1  S1/L0      1  R0          ?               0
  2  S1/L1      1  R0          ?               4.25
  3  S1/L2      2  R1          ?               6.75
  4  S1/L3      3  R2          ?               8.25
  5  S2/L0      1  R0          ?              -2.89583
  6  S2/L1      1  R0          ?               4.60417
  7  S2/L2      2  R1          ?               11.6042
  8  S2/L3      3  R2          ?               16.6042
  9  S3/L0      1  R0          ?              -2.29167
  10 S3/L1      2  R1          ?               8.20833
  11 S3/L2      3  R2          ?               15.7083
  12 S3/L3      4  R3          ?               20.2083
-------------------------------------------------------------

     OPTIMAL AVERAGE STAGE RETURN =  18.8542

REQUIRE LONG-RUN STATE PROBABILITIES - Y(YES) OR N(NO)? Y


---------------------------------
   STATE         PROBABILITY
-----------
 NO. S/LEVEL
---------------------------------
  1  S1/L0        .833333E-01
  2  S1/L1        .166667
  3  S1/L2        .833333E-01
  4  S1/L3        0
  5  S2/L0        0
  6  S2/L1        .208333E-01
  7  S2/L2        .104167
  8  S2/L3        .208333
  9  S3/L0        0
  10 S3/L1        .833333E-01
  11 S3/L2        .166667
  12 S3/L3        .833333E-01
---------------------------------
```

FIG. 4.9. Solution to the problem 'IRSI' using IDP without discounting.

4.7.2 Redimensioning arrays in FDP and IDP

The program DIM.BAS can be used for redimensioning arrays in FDP and IDP if the problem is larger than can be currently accommodated. By running the program the following data are requested:

Maximum number of states at any stage (S)
Maximum number of decisions at any stage (G)
Maximum number of possible states following from any decision (C)
Maximum number of decision stages (N)

Revised dimension statements for FDP and IDP are written to files named DIMFDP.BAS and DIMIDP.BAS, respectively. Print-outs of these files may be used for manually changing the dimension statements in FDP and IDP. Alternatively, the programs FDP.BAS and IDP.BAS may be edited by placing DIMFDP.BAS and DIMIDP.BAS, respectively, at the end of the program using the BASIC command 'APPEND'.

The listings of FDP and IDP in Appendix A2 show arrays dimensioned to accommodate problems with the following size parameters: $S=20$; $G=10$; $C=10$; and $N=10$.

4.8 References

Anderson, J.R., Dillon, J.L. and Hardaker, J.B. (1977), *Agricultural Decision Analysis*, Iowa State University Press, Ames.

Beckmann, M.J. (1968), *Dynamic Programming of Economic Decisions*, Springer-Verlag, Berlin.

Bellman, R.E. and Dreyfus, S.E. (1962), *Applied Dynamic Programming*, Princeton University Press, Princeton.

Burt, O.R. (1964), Optimal resource use over time with an application to ground water, *Management Science*, **11**(1), 80–93.

Burt, O.R. (1981), Farm level economics of soil conservation in the Palouse area of the Northwest, *American Journal of Agricultural Economics*, **63**(1), 83–92.

Burt, O.R. and Cummings, R.G. (1977), Natural resource management, the steady state, and approximately optimal decision rules, *Land Economics*, **53**(1), 1–22.

Burt, O.R., Koo, W.W. and Dudley, N.J. (1980), Optimal stochastic control of US wheat stocks and exports, *American Journal of Agricultural Economics*, **62**(2), May, 172–87.

Chow, G.C. (1975), *Analysis and Control of Dynamic Economic Systems*, J. Wiley, New York.

Cooper, L. and Cooper, M.W. (1981), *Introduction to Dynamic Programming*, Pergamon Press, Oxford.

Denardo, E.V. (1982), *Dynamic Programming: Models and Applications*, Prentice-Hall, Englewood Cliffs.

Dharmadhikari, V.K. (1975), Solving discrete-variable multiple-constraint non-linear programs: The decision-state method, Institute of Technology, Southern Methodist University, Dallas.

Hadley, G. (1970), *Nonlinear and Dynamic Programming*, Addison-Wesley, Reading, Massachusetts.

Hardaker, J.B. (1979), Optimal management of income equalisation deposits, Paper presented to the Australian Agricultural Economics Society 23rd Annual Conference, Canberra, 6–8 February.

Heidari, M., Chow, V.T., Kokotovic, P.V. and Meredith, D.D. (1971), The discrete differential dynamic programming approach to water resources systems optimization, *Water Resources Research*, **7**(2), 273–82.

Hillier, F.S. and Lieberman, G.J. (1967), *Introduction to Operations Research*, Holden-Day, San Francisco.

Jacobs, O.L.R. (1974), *Introduction to Control Theory*, Clarendon Press, Oxford.

Karp, L.S. and McCalla, A.F. (1983), Dynamic games and international trade: an application to the world corn market, *American Journal of Agricultural Economics*, **65**(4), 641–50.

Keeney, R.L. and Raiffa, H. (1976), *Decisions with Multiple Objectives: Preferences and Value Trade-offs*, J. Wiley, New York.

Knapp, K.C. (1983), Steady state solutions to dynamic optimization models with inequality constraints, *Land Economics*, **59**(3), 300–4.

Larson, R.E. and Casti, J.L. (1982), *Principles of Dynamic Programming Part II, Advanced Theory and Applications*, Dekker, New York.

Lewis, T.R. (1975), Optimal resource management under conditions of uncertainty: the case of an ocean fishery, Unpublished PhD thesis, University of California, San Diego.

Morin, T.L. and Esogbue, A.M.O. (1974), The imbedded state space approach to reducing dimensionality in dynamic programs of higher dimensions, *Journal of Mathematical Analysis and Applications*, **48**, 801–10.

Nemhauser, G.L. (1966), *Introduction to Dynamic Programming*, J. Wiley, New York.

Norman, J.M. (1972), *Heuristic Procedures in Dynamic Programming*, Manchester University Press, Manchester.

Page, T. (1977), *Conservation and Economic Efficiency—An Approach to Materials Policy*, Johns Hopkins University Press, Baltimore.

Pearce, D.W. (ed.) (1978), *The Valuation of Social Cost*, George Allen and Unwin, London.

Rausser, G.C. and Hochman, E. (1979), *Dynamic Agricultural Systems: Economic Prediction and Control*, North-Holland, New York.

Sinden, J.A. and Worrell, A.C. (1979), *Unpriced Values—Decisions Without Market Prices*, J. Wiley, New York.

Tauxe, G.W., Inman, R.R. and Mades, D.M. (1979*a*), Multi-objective dynamic programming: a classic problem redressed, *Water Resources Research*, **15**(6), 1398–402.

Tauxe, G.W., Inman, R.R. and Mades, D.M. (1979*b*), Multi-objective dynamic programming with application to a reservoir, *Water Resources Research*, **15**(6), 1403–8.

Taylor, R.C. and Burt, O.R. (1984), Near-optimal management strategies for

controlling wild oats in spring wheat, *American Journal of Agricultural Economics*, **66**(1), 50–60.
Thampapillai, D. (1978), Methods of multiple objective planning: a review, *World Agricultural Economics and Rural Sociology Abstracts*, **20**(12), 803–13.
Theil, H. (1964), *Optimal Decision Rules for Government and Industry*, North-Holland, Amsterdam.
Turgeon, A. (1982), Incremental dynamic programming may yield nonoptimal solutions, *Water Resources Research*, **18**(6), 1599–604.
Yakowitz, S. (1982), Dynamic programming applications in water research, *Water Resources Research*, **18**(4), 673–96.

4.A Appendix: The Slope and Curvature of the Optimal Return Function $V_i\{x_i\}$

The basic dynamic programming equation for the general resource problem introduced in Chapter 1 is

$$\begin{aligned} V_i\{x_i\} &= \max_{u_i} [a_i\{x_i, u_i\} + \alpha V_{i+1}\{x_i + g_i\{x_i\} - h_i\{x_i, u_i\}\}] \\ &= \max_{u_i} [z_i\{x_i, u_i\}] \qquad (i = 1, \ldots, n) \end{aligned} \tag{A4.1}$$

with

$$V_{n+1}\{x_{n+1}\} = F\{x_{n+1}\}$$

It was stated in Section 1.4 that a necessary condition for finding an interior solution was that $\partial z_i/\partial u_i = 0$. The necessary and sufficient conditions are that $\partial z_i/\partial u_i = 0$ and z_i be concave in u_i. Some of the circumstances under which z_i is concave in u_i are investigated in this appendix. It will be seen that this requires consideration of the sign of the slope of the final value function $F\{x_{n+1}\}$, and of $V_{i+1}\{x_{i+1}\}$ for $i < n$.

It might be supposed that $V_i\{x_i\}$ would have a positive slope, but declining as x_i is increased, reflecting the idea that resources have value but are subject to diminishing returns. This was assumed in the example in Section 4.4, but another question examined in this appendix is under what circumstances does the assumption hold.

The slope and curvature of $V_i\{x_i\}$ depend on the final value function $F\{x_{n+1}\}$, and on $a_j\{x_j, u_j\}$, $g_j\{x_j\}$ and $h_j\{x_j, u_j\}$ for all $j \geq i$. They can be derived by following the backward process of dynamic programming, starting with stage n.

It is assumed that all functions are continuously differentiable. For the last decision stage

$$\begin{aligned} V_n\{x_n\} &= \max_{u_n} [a_n\{x_n, u_n\} + \alpha F\{x_n + g_n\{x_n\} - h_n\{x_n, u_n\}\}] \\ &= a_n\{x_n, u_n^*\} + \alpha F\{x_n + g_n\{x_n\} - h_n\{x_n, u_n^*\}\} \end{aligned} \tag{A4.2}$$

Sufficient conditions for u_n^* to be an interior solution are that

$$\partial a_n/\partial u_n - \alpha(\mathrm{d}F/\mathrm{d}x_{n+1})\,\partial h_n/\partial u_n = 0 \tag{A4.3}$$

and that $\partial^2 a_n/\partial u_n^2 \leq 0$, $\partial^2 h_n/\partial u_n^2 \geq 0$ and $\mathrm{d}F/\mathrm{d}x_{n+1} \geq 0$ for concavity.

Totally differentiating (A4.2) with respect to x_n gives

$$\begin{aligned}\frac{\mathrm{d}V_n}{\mathrm{d}x_n} &= \frac{\partial a_n}{\partial x_n} + \frac{\partial a_n}{\partial u_n}\frac{\mathrm{d}u_n^*}{\mathrm{d}x_n} + \alpha\frac{\mathrm{d}F}{\mathrm{d}x_{n+1}} \\ &\quad \times \left(1 + \frac{\mathrm{d}g_n}{\mathrm{d}x_n} - \frac{\partial h_n}{\partial x_n} - \frac{\partial h_n}{\partial u_n}\frac{\mathrm{d}u_n^*}{\mathrm{d}x_n}\right) \\ &= \frac{\partial a_n}{\partial x_n} + \alpha\frac{\mathrm{d}F}{\mathrm{d}x_{n+1}}\left(1 + \frac{\mathrm{d}g_n}{\mathrm{d}x_n} - \frac{\partial h_n}{\partial x_n}\right) + \frac{\mathrm{d}u_n^*}{\mathrm{d}x_n} \\ &\quad \times \left(\frac{\partial a_n}{\partial u_n} - \alpha\frac{\mathrm{d}F}{\mathrm{d}x_{n+1}}\frac{\partial h_n}{\partial u_n}\right)\end{aligned}$$

Given (A4.3), the last term disappears so that

$$\frac{\mathrm{d}V_n}{\mathrm{d}x_n} = \frac{\partial a_n}{\partial x_n} + \alpha\frac{\mathrm{d}F}{\mathrm{d}x_{n+1}}\left(1 + \frac{\mathrm{d}g_n}{\mathrm{d}x_n} - \frac{\partial h_n}{\partial x_n}\right) \tag{A4.4}$$

Thus even though u_n^* is a function of x_n, $\mathrm{d}V_n/\mathrm{d}x_n$ can be specified without partial derivatives taken with respect to u_n. Alternatively, this result could have been stated by appealing to the envelope theorem. The theorem is used implicitly below.

From (A4.4) sufficient conditions for $\mathrm{d}V_n/\mathrm{d}x_n > 0$ are $\partial a_n/\partial x_n > 0$, $\mathrm{d}F/\mathrm{d}x_{n+1} \geq 0$ and $(1 + \mathrm{d}g_n/\mathrm{d}x_n - \partial h_n/\partial x_n) \geq 0$. Differentiating (A4.4) gives

$$\frac{\mathrm{d}^2 V_n}{\mathrm{d}x_n^2} = \frac{\partial^2 a_n}{\partial x_n^2} + \alpha\left[\frac{\mathrm{d}F}{\mathrm{d}x_{n+1}}\left(\frac{\mathrm{d}^2 g_n}{\mathrm{d}x_n^2} - \frac{\partial^2 h_n}{\partial x_n^2}\right) + \frac{\mathrm{d}^2 F}{\mathrm{d}x_{n+1}^2}\left(1 + \frac{\mathrm{d}g_n}{\mathrm{d}x_n} - \frac{\partial h_n}{\partial x_n}\right)^2\right]$$

Sufficient conditions for $\mathrm{d}^2 V_n/\mathrm{d}x_n^2 \leq 0$ are therefore $\partial^2 a_n/\partial x_n^2 < 0$, $(\mathrm{d}^2 g_n/\mathrm{d}x_n^2 - \partial^2 h_n/\partial x_n^2) \leq 0$, $\mathrm{d}F/\mathrm{d}x_{n+1} > 0$ and $\mathrm{d}^2 F/\mathrm{d}x_{n+1}^2 \leq 0$.

Sufficient conditions for $z_i\{x_i, u_i\}$ to be concave in u_i, and for $V_i\{x_i\}$ to be positive and concave have been identified for $i = n$. By induction, sufficient conditions can be identified for $i = n-1, \ldots, 1$. Sufficient conditions for all i which must hold over the feasible ranges of x_i and u_i are summarized in Table A4.1. In the case of the fishery, it would be plausible

TABLE A4.1
Sufficient Conditions on the Signs of Derivatives, $j \geq i$, for Slope and Curvature Requirements at Stage i

Requirement ($i = n, \ldots, 1$)	*Order of derivative*	$a_j\{x_j, u_j\}$		$x_j + g_j\{x_j\} - h_j\{x_j, u_j\}$		$V_{j+1}\{x_{j+1}\}$	$F\{x_{n+1}\}$
		x_j	u_j	x_j	u_j	x_{j+1}	x_{n+1}
$\partial^2 z_i/\partial u_i^2 \leq 0$	first					>0	>0
	second		≤ 0		≤ 0		
$\mathrm{d}V_i/\mathrm{d}x_i > 0$	first	>0		≥ 0		>0	>0
$\mathrm{d}^2 V_i/\mathrm{d}x_i^2 \leq 0$	first					>0	>0
	second	≤ 0		≤ 0		≤ 0	≤ 0

to assume that stage return and harvest functions are positive and concave in x_i and u_i, and that $g_i\{x_i\}$ is concave. It is clear, however, that harvest and growth functions of this type do not guarantee that $z_i\{x_i, u_i\}$ is concave in u_i, or that $V_i\{x_i\}$ is concave. If the problem is simplified by changing the control variable from harvesting effort to harvest level, so that $h_i\{x_i, u_i\}$ in (A4.1) is replaced by u_i, the state transformation is linear in u_i. Retaining the other assumptions, the concavity of $z_i\{x_i, u_i\}$ is assured, provided also that $\mathrm{d}g_j/\mathrm{d}x_j \geq -1$ for all $j \geq i$.

PART III

Dynamic Programming Applications to Agriculture

Chapter 5

SCHEDULING, REPLACEMENT AND INVENTORY MANAGEMENT

In this chapter agricultural applications of dynamic programming are considered in which state transformation functions can be specified on the basis of logic rather than on the basis of empirical biological relationships. An example is the replacement problem with age as a state variable. A decision at stage i to replace or keep an asset determines directly the age of the asset at stage $i+1$. Another example is the inventory problem with stock as a state variable. The stock at stage $i+1$ equals the stock at stage i, minus net outgoings between stage i and stage $i+1$. Two of the earliest agricultural applications of dynamic programming were to replacement (White, 1959), and inventory (Gustafson, 1958) problems. As mentioned in Chapter 1, many problems which involve biological relationships, such as the management of crops, livestock and trees, are in part replacement problems.

The first two sections deal with farm level problems. The optimal scheduling of farm activities using critical path analysis is examined in Section 5.1. A replacement problem involving the selection of tractor capacity in a whole-farm context is studied in Section 5.2. In the third section a number of dynamic programming formulations are presented of the problem of determining optimal levels of food and feed-grain stocks to hold at national and international levels.

5.1 Critical Path Analysis

A large number of activities may have to be carried out before the completion of a job. It may also be the case that one or more activities cannot be started until one or more other activities have been finished. If the job has to be completed in minimum time, how should the activities be scheduled?

The problem can be conceptualized as a network problem consisting of nodes connected by arrows. Nodes represent points in time by which certain activities must finish before others may begin. Arrows represent activities. A sequence of arrows connecting the start-of-job node to the completion-of-job node via intermediate nodes is a path through the network. The time taken to traverse the path is the sum of the durations of each activity on the path. The shortest time in which a job can be completed is the time taken to traverse the path of greatest duration. The path is referred to as the critical path, because of its key role in determining the duration of the job. There can be no slack time between activities on the critical path. The overall job time can only be shortened by reducing the duration of activities on the critical path.

The identification of the longest path through a network is a simple dynamic programming problem, akin to the least-cost path problem introduced in Chapter 2, and is easily solved by hand in the case of a small network. However, there are two aspects of dynamic programming for critical path analysis which still require discussion. One is the numbering of nodes, and its relation to the states and stages normally used in dynamic programming. The other is the calculation of activity float times, which indicate the time available for an activity relative to the duration of the activity. In order to calculate float times, the critical path is identified twice by dynamic programming, once using forward recursion, and again using backward recursion. Both of these aspects are discussed after introducing a simple farm example.

5.1.1 A farm example

The job is the ploughing, planting, harvesting, carting and storing associated with the production of two crops, A and B. Crop A takes 60 days to mature after planting, whereas crop B takes only 40 days. Activities essential to the planting of crop A are therefore planned so that they precede those for crop B. However, crop B is harvested first, and only after this is it possible to harvest crop A. There are ten real activities, *A* to *J*, which must be performed to complete the job and two dummy activities of zero duration, *K* and *L*. Dummy activities are necessary where, by the nature of the activities, one activity cannot start before the completion of another. These are listed in Table 5.1, together with their durations and necessary precedent activities.

TABLE 5.1
Activities and Floats

Activity			Precedent activities	Nodes (i,j)	Float (days)	
Code	*Description*	*Duration* $d\{i,j\}$ *(days)*			*Total* TF_{ij}	*Free* FF_{ij}
A	Plough land for crop A	10	—	(1, 2)	0	0
B	Plant crop A	6	*A*	(2, 3)	0	0
C	Crop A matures	60	*B*	(3, 8)	0	0
D	Harvest crop A	4	*C, L*	(8, 9)	0	0
E	Cart and store crop A	2	*D*	(9, 10)	0	0
F	Plough land for crop B	5	*A*	(2, 4)	16	1
G	Plant crop B	3	*F, K*	(4, 5)	15	0
H	Crop B matures	40	*G*	(5, 6)	15	0
I	Harvest crop B	2	*H*	(6, 7)	15	0
J	Cart and store crop B	1	*I*	(7, 10)	20	20
K	Dummy (crop A planted)	0	*B*	(3, 4)	15	0
L	Dummy (crop B planted)	0	*I*	(7, 8)	15	15

A network for these nodes is given in Fig. 5.1. Activity durations are shown in brackets alongside each activity arrow. Note that the system of numbering nodes, whilst not unique, is not arbitrary. The numbering system is progressive in the sense that for any activity defined by start node i and finish node j, $j > i$. Progressive numbering of an n-node network can be effected using the following rules:

(1) Label the node marking the start of the job '1', and the node marking the end of job 'n'.
(2) For $i = 2, \ldots, n-1$, after label 'i' has been allocated to a node, label any node '$i+1$' which has all input arrows coming from nodes already labelled.

Progressive numbering facilitates the process of identifying the critical path using the recursive logic of dynamic programming. The longest path through the network shown in Fig. 5.1 can be found by forward recursion using the following equation:

$$V^{f}\{j\} = \max_{i<j} [d\{i,j\} + V^{f}\{i\}] \qquad (j = 2, \ldots, 10) \qquad (5.1)$$

with

$$V^{f}\{1\} = 0$$

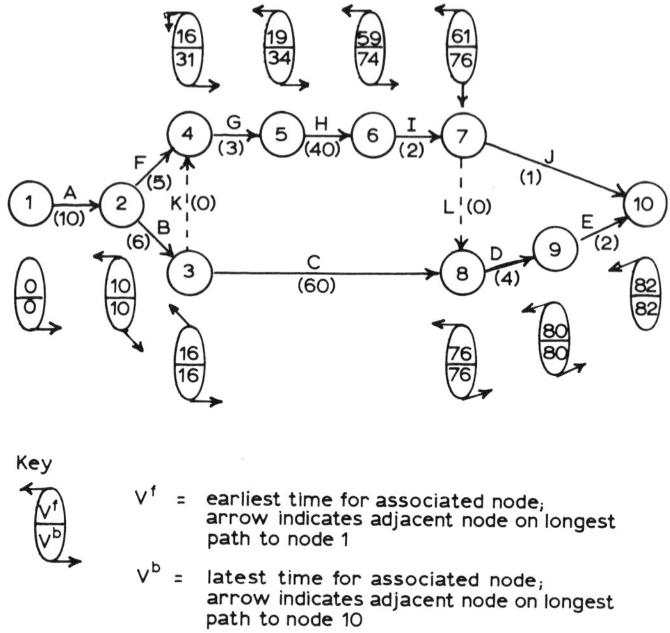

FIG. 5.1 Activity network for farm example.

where $V^f\{j\}$ is the earliest time for the occurrence of node j relative to the start time $V^f\{1\}$, and $d\{i,j\}$ is the duration of the activity linking nodes i and j. The process of forward recursion involves identifying at each successive node j the longest path to that node from node 1. Maximization is with respect to i, the adjacent node on the longest path. Note that for the first j considered in (5.1), $j = 2$, optimal i must equal 1. The values of $V^f\{j\}, j = 1, \ldots, 10$, for the farm problem are given in Table 5.2 and also in the upper half of each capsule associated with each node

TABLE 5.2
Earliest ($V^f\{i\}$) and Latest ($V^b\{i\}$) Node Times

	Node i									
	1	*2*	*3*	*4*	*5*	*6*	*7*	*8*	*9*	*10*
$V^f\{i\}$	0	10	16	16	19	59	61	76	80	82
$V^b\{i\}$	0	10	16	31	34	74	76	76	80	82
i on critical path	√	√	√					√	√	√

in Fig. 5.1. The optimal i at each j node is also shown by an arrow emanating from the upper half of each capsule. The longest path through the network can be found from Fig. 5.1 by following the arrows back from node 10 to node 1. The nodes on this critical path are 1, 2, 3, 8, 9, 10 and the associated duration, which is the minimum completion time for the project, is 82 days.

It is also possible to determine the slack or float associated with each activity. To calculate these, the latest time of occurrence of each node must be calculated. These can be determined by backward recursion using the equation:

$$V^{b}\{i\} = \min_{j>i} [V^{b}\{j\} - d\{i,j\}] \qquad (i = 9, \ldots, 1) \qquad (5.2)$$

with

$$V^{b}\{n\} = V^{f}\{n\}$$

where $V^{b}\{i\}$ is the latest time of occurrence of node i relative to the start time $V^{f}\{1\}$. Essentially the longest path is again being found, but this time by minimizing $\sum -d\{i,j\}$, and by starting at the end node and working backwards. In working through (5.2) $\sum d\{i,j\}$ at any stage i is subtracted from $V^{f}\{10\} = V^{b}\{10\} = 82$, so that $V^{b}\{i\}$ refers to the maximum number of days following node 1 instead of the minimum number of days preceding node 10. The values of $V^{b}\{i\}$, $i = 1, \ldots, 10$ are given in Table 5.2, and also in the lower half of each capsule in Fig. 5.1.

Because the earliest and latest times are the same for each node on the critical path, Table 5.2 provides confirmation that the critical path connects nodes 1, 2, 3, 8, 9, 10.

Two frequently calculated types of float are total float (TF) and free float (FF). The total float associated with an activity is the maximum time within which the activity must be completed in order not to extend the minimum overall duration of the job. For the activity with start node i and finish node j it is the latest time for node j less the earliest time for node i, minus the duration of the activity, or

$$\text{TF}_{ij} = V^{b}\{j\} - V^{f}\{i\} - d\{i,j\}$$

The free float associated with an activity is the free time available in carrying out the activity if nodes i and j occur at their earliest times, or

$$\text{FF}_{ij} = V^{f}\{j\} - V^{f}\{i\} - d\{i,j\}$$

The total and free floats for each activity in the farm example are shown in Table 5.1. It is apparent that it is possible for there to be

$31 - 10 - 5 = 16$ days of slack time associated with activity F (plough land for crop B) if the succeeding activity G (plant crop B) is postponed to the last moment. This is the total float. However, if G is started at the earliest possible time, the free time associated with activity F, or the free float, is only 1 day.

5.1.2 Solution using GPDP

Most critical path problems in farming are readily solved by hand using the method explained in the previous section. If the network is large and complex, it may be useful to identify the critical path using a computer algorithm such as GPDP. However, to use GPDP for this purpose, the nodes have to be renumbered, and extra dummy nodes may have to be defined.

Problems can only be solved with GPDP if there are the same number of decision stages on all paths from start to finish. In network problems this means that every possible path through the network must consist of the same number of nodes. Therefore, to use GPDP, the path with the largest number of nodes must be identified, and then the number of nodes on all other paths must be expanded to the maximum number by adding dummy activities and nodes. The path with most nodes can be found by backward recursion using the equation:

$$m\{i\} = \max_j [1 + m\{j\}] \qquad (i = n-1, \ldots, 1) \qquad (5.3)$$

subject to

$$j > i$$

with

$$m\{n\} = 1$$

where $m\{i\}$ is the maximum inclusive number of nodes which can be traversed from node i to the final node, n, and j is a node adjacent to i. The network for the farm problem is redrawn in Fig. 5.2, showing the

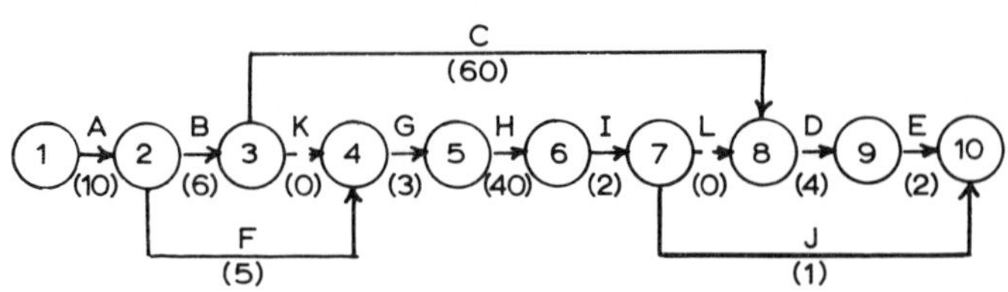

FIG. 5.2. Network redrawn to show path with maximum nodes.

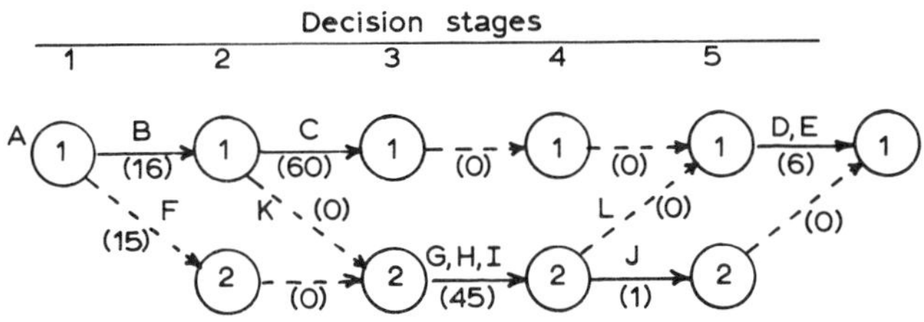

FIG. 5.3. Network redrawn so that all paths have an equal number of nodes.

nodes on the path with most nodes laid out on a straight line. For convenience a node with only one activity input and one activity output may be eliminated by combining the two activities and corresponding durations. Dummy activities and nodes are then added to the network to ensure an equal number of nodes on all possible paths. The modified network for the farm problem is shown in Fig. 5.3. The network is now in a form suitable for analysis using GPDP.

5.1.3 Selected applications

Cooke-Yarborough (1964) describes the application of critical path analysis to farm operations in an early expository article, but without reference to dynamic programming. Most modern texts on operations research techniques devote chapters to both critical path analysis and dynamic programming, but the linkage between the techniques is often not made explicit. Interest in farm applications of critical path analysis may be revived with the publication of the book by Smith (1984).

5.2 Farm Investment Decisions

Linear programming is often used to find optimal farm production decisions given a planning horizon of one year. The planning horizon of one year may be satisfactory if the production activities for one year do not affect the optimal selection of activities for a subsequent year. Examples of such activities may be cropping or hiring labour.

Decisions such as buying in livestock in one year which lead to returns in subsequent years are investment decisions. Multiperiod linear programming may be an appropriate technique for determining simultaneously optimal investment and annual production decisions. A requirement which is more likely to arise with investment than with production decisions is that the activity levels be integer. This is the case for example with livestock decisions, where livestock units are integer. The number of cows in an optimal farm plan obtained from running a multiperiod linear program without integer constraints may be 27·3. The loss of net revenue through implementing the non-integer plan but with 27 cows compared with implementing the optimal integer-constrained plan is likely to be low.

It is usually only important to constrain what is actually an integer activity to integer levels if its optimal level is relatively low, such as 0, 1, 2 or 3. Examples of such activities are the buying of complex pieces of machinery, such as tractors, harvesters and milking parlours. Optimal farm plans incorporating integer investment decisions can be obtained by using mixed-integer multiperiod linear programming. However, mixed-integer linear programming computer codes are not readily available, although Benders' (1962) decomposition algorithm has been used successfully for solving farm machinery selection problems (Danok *et al.*, 1978, 1980), and grain subterminal location problems (Hilger *et al.*, 1977). The algorithm involves the iterative solution of integer and linear programming problems. An alternative approach for smaller problems employing linear and dynamic programming is explained in the next section.

5.2.1 Optimal tractor replacement

Suppose the farm investment decision concerns the optimal replacement of a tractor. The tractor, when it is replaced, need not be replaced with one of the same capacity. It may be optimal to change tractor capacity at the time of replacement because of changes in tractor technology, changes in crop prices relative to livestock prices, or changes in fuel prices. Assume that the farmer aims to choose annual production and tractor replacement plans over the next n years so as to maximize the present value of net revenue. All costs, prices, resource availabilities and technical coefficients are known with certainty.

Linear programming is used to find efficient production plans for given tractor capacity available. Net revenue is maximized over one year, and is used in calculating stage returns in the dynamic programming routine. If tractor capacity only affects the available number of tractor hours in the resource vector of the linear program, farm plans for different tractor capacities may be obtained by parameterizing tractor hours. However, tractor capacity may affect the cultivation techniques employed, necessitating other changes in the matrix of technical coefficients or in the vector of gross margins. The linear program is run for each of the n years with appropriate data on prices, costs, technology and resource availability. In this way net returns from production are obtained for all n years for all tractor capacities. Dynamic programming is then used to determine the optimal scheduling of tractor capacity through time.

Both the production and investment decisions determine taxable income each year, and hence the rate of tax and total tax paid each year. Because the formulation for maximizing the present value of after-tax net revenues is more complicated, the problem without tax is considered first.

5.2.2 Formulation of the problem without tax

The combined farm production and tractor replacement problem is[1]

$$\max_{u_1,\ldots,u_n;\mathbf{y}_1,\ldots,\mathbf{y}_n} \sum_{i=1}^{n} \alpha^{i-1}[\mathbf{z}_i'\mathbf{y}_i - k_i\{a_i, c_i, u_i\}] + \alpha^n P_{n+1}\{a_{n+1}, c_{n+1}\} \quad (5.4)$$

subject to

$$A_i\mathbf{y}_i \leq \mathbf{b}_i\{c_i, u_i\}$$

$$\left.\begin{array}{lll} \text{if } u_i = 0 \text{ (keep):} & a_{i+1} = a_i + 1; & c_{i+1} = c_i \\ \text{if } u_i > 0 \text{ (replace):} & a_{i+1} = 0; & c_{i+1} = u_i \end{array}\right| \quad (i = 1, \ldots, n)$$

where a_1 and c_1 are given. Here:

u_i is the tractor replacement decision implemented at the beginning of the i-th year: $u_i = 0$ for keep; $u_i = j$ for replace with capacity number j;
$\mathbf{y}_i$ is the vector of production activity levels;
$\mathbf{z}_i$ is the vector of gross margins for each production activity;
a_i is tractor age at the beginning of the i-th year;
c_i is tractor capacity at the beginning of the i-th year;
A_i is the matrix of technical coefficients;
$\mathbf{b}_i\{c_i, u_i\}$ is the vector of resource levels available, with available tractor hours dependent on c_i and u_i;
$k_i\{a_i, c_i, u_i\}$ is the cash outlay for the tractor replacement decision;
$P_{n+1}\{a_{n+1}, c_{n+1}\}$ is the tractor salvage price at the end of the planning horizon;
α is the discount factor; and
n is the number of years in the planning horizon.

[1] In this section capital letters denote matrices, lower-case bold letters are vectors, and remaining symbols are scalars.

The cash outlay function is determined as follows. If the decision is to keep the tractor

$$k_i\{a_i, c_i, 0\} = g_i\{a_i, c_i\}$$

where $g_i\{a_i, c_i\}$ is the outlay on repairs, maintenance, insurance and shelter paid at the beginning of year i for a tractor of age a_i and capacity c_i.

If the decision is to replace the tractor, $u_i > 0$ and

$$k_i\{a_i, c_i, u_i\} = g_i\{0, u_i\} - P_i\{0, u_i\} + P_i\{a_i, c_i\}$$

where $P_i\{a_i, c_i\}$ is the price of a tractor of age a_i and capacity c_i at the beginning of year i. Buying and selling prices are equal for simplicity.

It is assumed that a tractor on hand at the end of the n-th year must be sold to give a return $P_{n+1}\{a_{n+1}, c_{n+1}\}$.

Problem (5.4) can be decomposed into the following two-stage problem

$$\max_{u_1,\ldots,u_n} \sum_{i=1}^{n} \alpha^{i-1}[\max_{\mathbf{y}_i} [\mathbf{z}_i'\mathbf{y}_i] - k_i\{a_i, c_i, u_i\}] + \alpha^n P_{n+1}\{a_{n+1}, c_{n+1}\} \qquad (5.5)$$

subject to the same constraints as before. The inner maximization with respect to $\mathbf{y}_i$ can be performed using linear programming, and the outer maximization with respect to u_i using dynamic programming. The recursive equation for solving problem (5.5) is

$$V_i\{a_i, c_i\} = \max_{u_i} [\max_{\mathbf{y}_i} [\mathbf{z}_i'\mathbf{y}_i] - k_i\{a_i, c_i, u_i\} + \alpha V_{i+1}\{a_{i+1}, c_{i+1}\}]$$
$$(i = n, \ldots, 1) \qquad (5.6)$$

with

$$V_{n+1}\{a_{n+1}, c_{n+1}\} = P_{n+1}\{a_{n+1}, c_{n+1}\}$$

subject to the same constraints as before.

5.2.3 Formulation of the problem with tax

The problem is more complex if the present value of net revenue after tax is to be maximized and the tax rate in year i depends on the net revenue after tax in year i. Also, the way in which tax payments are affected by rules for calculating depreciation and profit or loss on sale of an asset must be taken into account. Suppose in year i for a tractor of age a_i and capacity c_i depreciation $d\{a_i, c_i\}$ can be charged against tax. Suppose as well that the book value of a tractor sold in year i is the purchase price, i.e. $P_{i-a_i}\{0, c_i\}$, minus depreciation charged against tax over the years the tractor has been held, i.e.

$$\sum_{l=1}^{a_i} d\{a_i - l + 1, c_i\}$$

Tax is payable (refunded) on the profit (loss) on sale of the tractor, calculated as the salvage value ($P_i\{a_i, c_i\}$) less the book value of the tractor. The amount of tax payable is $t\{\Pi\}$, where Π is taxable income.

The constraints are the same as in (5.4), but the objective function becomes

$$\max_{u_1,\ldots,u_n;\mathbf{y}_1,\ldots,\mathbf{y}_n} \sum_{i=1}^{n} \alpha^{i-1} f_i\{a_i, c_i, u_i, \mathbf{y}_i\} \tag{5.7}$$

where the stage return function for $u_i = 0$ (keep) is

$$f_i\{a_i, c_i, 0, \mathbf{y}_i\} = \mathbf{z}_i'\mathbf{y}_i - g_i\{a_i, c_i\} - t\{\mathbf{z}_i'\mathbf{y}_i - g_i\{a_i, c_i\} - d\{a_i, c_i\}\}$$

and for $u_i > 0$ (replace with capacity u_i) is

$$\begin{aligned} f_i\{a_i, c_i, u_i, \mathbf{y}_i\} &= \mathbf{z}_i'\mathbf{y}_i - g_i\{0, u_i\} - P_i\{0, u_i\} + P_i\{a_i, c_i\} \\ &\quad - t\Bigg\{\mathbf{z}_i'\mathbf{y}_i - g_i\{0, u_i\} - d\{0, u_i\} \\ &\quad + P_i\{a_i, c_i\} - P_{i-a_i}\{0, c_i\} - \sum_{l=1}^{a_i} d\{a_i - l + 1, c_i\}\Bigg\} \end{aligned}$$

The recursive equation for solving problem (5.7) is

$$V_i\{a_i, c_i\} = \max_{u_i, \mathbf{y}_i} [f_i\{a_i, c_i, u_i, \mathbf{y}_i\} + \alpha V_{i+1}\{a_{i+1}, c_{i+1}\}] \quad (i = n, \ldots, 1) \tag{5.8}$$

with

$$\begin{aligned} V_{n+1}\{a_{n+1}, c_{n+1}\} &= P_{n+1}\{a_{n+1}, c_{n+1}\} \\ &\quad - t\Bigg\{P_{n+1}\{a_{n+1}, c_{n+1}\} - P_{n+1-a_{n+1}}\{0, c_{n+1}\} \\ &\quad - \sum_{l=1}^{a_{n+1}} d\{a_{n+1} - l + 1, c_{n+1}\}\Bigg\} \end{aligned}$$

Net revenue from production, $\mathbf{z}_i'\mathbf{y}_i$, still must be maximized at each stage for all c_i and u_i. To see this, consider the following simple scalar problem where y = output. Net revenue after tax is

$$R = \Pi\{y\} - t\{\Pi\{y\}\}$$

A necessary condition for R to be a maximum is that

$$\mathrm{d}R/\mathrm{d}y = \mathrm{d}\Pi/\mathrm{d}y(1 - \mathrm{d}t/\mathrm{d}\Pi) = 0$$

For $(1 - \mathrm{d}t/\mathrm{d}\Pi) \neq 0$ the equation is satisfied by

$$\mathrm{d}\Pi/\mathrm{d}y = 0$$

which is a requirement that net revenue before tax be maximized.

5.2.4 Discussion

A change of tractor capacity may have other effects on production possibilities besides altering the number of tractor hours available. Tractors are used with other implements such as ploughing equipment which may perform less efficiently with a change of tractor capacity due to mismatching. Some of the coefficients in the technical matrix A_i may therefore be dependent on c_i and u_i. It may also be appropriate to consider the replacement of ancillary equipment. This would involve including the age and capacity of tractor implements as state variables in the dynamic programming formulation, an extension of the problem which would be demanding in terms of both information and computing. Conceptually, other investment decisions such as the purchase of additional land could also be handled by increasing the number of state and decision variables in the dynamic programming problem.

The formulation of the problem in (5.4) shows the vector of resources available $\mathbf{b}_i$ to depend on just c_i and u_i, and not also on tractor age, a_i. The decline in tractor productivity with increasing age is assumed to result in an increase in the repairs and maintenance component of the fixed costs $g_i\{a_i, c_i\}$ incorporated in the stage return. However, a replacement decision affects the farm's cash flow. Receipts from selling a tractor are a function of the age of the tractor. Therefore, dependent on how liquidity constraints are specified, it may be necessary to solve the linear programming problem for all a_i for replacement decisions $(u_i > 0)$. However, this is not necessary if a perfect capital market applies, so that any amount of funds can be borrowed or saved at the same rate of interest (or if the borrowing rate is only very slightly higher than the saving rate). As long as the discount factor α in (5.4) is based on the same rate of interest as that used for the borrowing and saving elements in the gross margin vector $\mathbf{z}_i$ in the linear programming problem, the opportunity cost of funds is constant. It is not affected by the opening cash available to the farm, or by a requirement to raise finance for a new tractor. This means that the optimal production plan is not affected by the liquidity position of the farm.

If there are limits to the amount that can be borrowed at one rate of interest then cash on hand at the end of the year joins tractor status as a linkage between years, and has to be added as a state variable in the dynamic programming problem.

5.2.5 Selected applications

Sowell (1967) uses dynamic programming to study the optimal replacement decision for self-propelled cotton pickers. The objective is the minimization of holding costs for planning horizons of 25 years and infinity. The state variable is machine age. Ambrosius (1970) develops a model for determining replacement of tractor complements and combines on US corn farms so as to maximize the present value of net revenue over ten years. The state variables are the ages and sizes of tractor and combine. The model is run for different farm sizes and farm growth rates. Hesselbach (1974) also uses dynamic programming for machinery selection on US corn farms with the aim of maximizing the present value of net revenue. The decision is whether to change the set of harvesting equipment, the set being described by the number, age and capacity of harvesters.

The need to optimize the present value of farm net revenue with respect to both all production plans and all machinery investment decisions in the planning horizon has been little discussed in the literature. It is recognized that returns from farm machinery affect the replacement decision in the dynamic programming studies of Ambrosius (1970) and Hesselbach (1974), and in conceptual analytical models such as that outlined by Baquet (1981). However, there is no suggestion of optimizing net revenue from production in these models. Mixed-integer programming models which integrate production and machinery decisions have been suggested by Danok *et al.* (1980), but the models are static. Kelly (1981, 1985) has proposed the integrated use of linear programming and dynamic programming for the tractor replacement decision along the lines described in the previous section. Given the uncertainty of future tractor, commodity and input prices, and technology, a key question that such a model can be used to examine is the sensitivity of the first-year replacement decision to forecasts of these parameters.

5.3 Buffer Stock Policies

The economic welfare generated by the production and consumption of grain in year i can be identified in Fig. 5.4 which shows a downward-sloping linear demand curve for grain consumption and an upward-sloping linear supply curve for grain production. The demand and supply equations are

$$q_i^d = f + gP_i \tag{5.9}$$

$$q_i^s = r + tP_i \tag{5.10}$$

where q_i^d and q_i^s are quantities of grain demanded and supplied, P_i is the price of grain and f, g, r and t are parameters, the same for all i. If production is perfectly competitive, and producers set production so that the marginal cost of production (MCP$\{q^s\}$) equals price, (5.10) implies the marginal cost of production equation

$$\text{MCP}_i = (q_i^s - r)/t \tag{5.11}$$

The value to consumers of q_i^d tonnes of grain is the area under the demand curve from $q = 0$ to $q = q_i^d$, or

$$C\{q_i^d\} = (-fq_i^d + (q_i^d)^2/2)/g \tag{5.12}$$

The total cost of producing q_i^s tonnes of grain is the area under the supply curve from $q = 0$ to $q = q_i^s$, or

$$c\{q_i^s\} = (-rq_i^s + (q_i^s)^2/2)/t \tag{5.13}$$

The economic welfare generated by producing and consuming q' is $C\{q'\} - c\{q'\}$, equal to the hatched area in Fig. 5.4. It can be seen from Fig. 5.4 that in order to maximize economic welfare with respect to q_i^d and q_i^s subject to $q_i^d \leq q_i^s$, the optimal levels are

$$q_i^d = q_i^s = q_i^* \tag{5.14}$$

If the demand and supply equations (5.9) and (5.10) hold for all years, and the objective is to maximize the present value of the stream of annual economic welfare, (5.14) specifies the optimal q_i^d and q_i^s for all i. The present value of economic welfare could not be increased by foregoing current consumption and transferring stocks of $q_i^s - q_i^d$ to year $i + 1$.

However, it may be optimal to carry buffer stocks if the supply curve is stochastic, even though the same stochastic supply curve applies over all years. Actual harvest may differ from planned harvest on account of

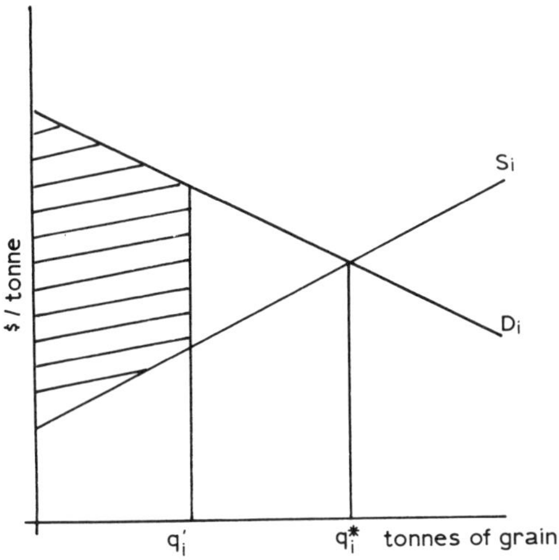

FIG. 5.4. Economic welfare generated by production and consumption of grain.

yield variability. In this case, the determination of optimal annual changes in buffer stocks, $q_i^s - q_i^d$, can be formulated as a stochastic dynamic programming problem. The problem can be expanded to allow for stochastic demand. Two dynamic programming approaches to these problems are examined in the following sections. One is the value-of-consumption approach, and the other is the price approach. The latter offers substantial computational advantages when both supply and demand are stochastic. Solution by the price approach does not follow the usual dynamic programming process, but still relies on recursive equations. Problems can be solved by the price method using the FDP program with minor changes.

5.3.1 Stochastic yields: planned production and demand constant

To start with a simple problem suppose that the demand curve is given by (5.9), but that the supply is stochastic and not responsive to price, given by

$$q_i^s = \bar{q}^s + e_{s,i} \tag{5.15}$$

where $\bar{q}^s$ is fixed expected production before harvest, and $e_{s,i}$ is a random production deviation taking on the value $e_{s,i}^k$ with probability $p_i^s\{k_i\}$. The marginal cost of storing one tonne of grain for one year is assumed to be constant and equal to s_i. In the value-of-consumption approach, the objective is to maximize the expected present value of grain consumption less the expected present value of storage costs. Because production costs are fixed, they do not affect storage decisions and are dropped for the moment. The decision variable, u_i, denotes the stocks made available for carryover before the harvest q_i^s is realized. Consumption in year i is therefore

$$q_i^d = u_{i-1} + q_i^s - u_i \tag{5.16}$$

The stage return is $C\{q_i^d\} - s_i u_i$.

This leaves the state variable to be defined. A first thought might be to use two state variables. These would be, for year i, the stock after consumption but before harvest in year $i-1$, and the stochastic component of supply in year $i-1$. This leads to the following dynamic programming formulation:

$$V_i\{u_{i-1}, e_{s,i-1}\} = \max_{u_i} [C\{u_{i-1} + \bar{q}^s + e_{s,i-1} - u_i\} - s_i u_i$$

$$+ \alpha \sum_{k_i} p_i^s\{k_i\} V_{i+1}\{u_i, e_{s,i}^k\}] \qquad (i = n, \ldots, 1) \tag{5.17}$$

subject to

$$0 \leq u_i \leq u_{i-1} + \bar{q}^s + e_{s,i-1}$$

with

$$V_{n+1}\{u_n, e_{s,n}\} = 0$$

The constraints on u_i ensure that the optimal u_i is non-negative and less than the total amount of grain available. Storage capacity may set a lower upper limit.

Although this is a valid formulation, the number of discrete values of the combinations of the two state variables is larger than necessary. This can be illustrated even for a problem with relatively few values of u_i and $e_{s,i}$. Suppose that u_i takes one of the four values 0, 10, 20, 30 thousand tonnes, and $e_{s,i}$ one of the five values -20, -10, 0, 10, 20 thousand tonnes. The number of state values required in (5.17) is $4 \times 5 = 20$. However, it is clear from the stage return function and the upper constraint on u_i that only the sum of u_{i-1} and $e_{s,i-1}$ and not their individual values are required on the right-hand side of (5.17). Consequently, a computationally more efficient formulation is

$$V_i\{x_i\} = \max_{0 \leq u_i \leq x_i} \left[C\{x_i - u_i\} - s_i u_i + \alpha \sum_{k_i} p_i^s\{k_i\} V_{i+1}\{u_i + \bar{q}^s + e_{s,i}^k\} \right] \quad (i = n, \ldots, 1) \qquad (5.18)$$

with

$$V_{n+1}\{x_{n+1}\} = 0$$

where $x_i = u_{i-1} + \bar{q}^s + e_{s,i-1}$, referred to as total supply. Year i is defined as starting at that point in the calendar year when the decision on carryover is made, some time after the harvest (and hence $e_{s,i-1}$) for year $i-1$ is known. If $\bar{q}^s$ is 50 thousand tonnes, the minimum value of x_i is $\min(u_i) + \min(e_{s,i}) + 50 = 30$ thousand tonnes, and the maximum value of x_i is $\max(u_i) + \max(e_{s,i}) + 50 = 100$ thousand tonnes. The range of x_i is represented by only eight discrete values if the interval between values is 10 thousand tonnes.

The infinite-stage solution to this problem was obtained by running the BASIC program shown in Fig. 5.5 to write the problem data to file, and then running IDP. The assumed probability distribution $p_i^s\{k_i\}$ is given in Table 5.3.

TABLE 5.3
Probability Distribution for the Yield Deviations

k_i	*1*	*2*	*3*	*4*	*5*
e_i^k	−20	−10	0	+10	+20
$p_i^s\{k_i\}$	0·1	0·2	0·4	0·2	0·1

Storage cost $s_i = \$10$/tonne and α is the discount factor for a rate of discount of 5 per cent per annum. The parameters of the demand equation (5.9) are $f = 100$ and $g = -0{\cdot}5$ where quantity is in thousands of tonnes and price is in dollars per tonne.

The infinite-stage solution is presented in Table 5.4. There is no carryover for supply of 60 thousand tonnes or less. The optimal storage policy ensures that consumption (supply less carryover) is never greater than 70 thousand tonnes. The long-run probabilities of supply levels obtained from the output of IDP are also shown, and are used for determining the long-run probabilities of different consumption levels if

```
100 REM----------STORE----------
110 REM
120 LET N$="STORE.DAT"
130 OPEN N$ FOR OUTPUT AS FILE #1
140 REM
150 LET N=5
160 LET D1=5
170 DIM X(8),U(4),P(5)
180 MAT READ X
190 DATA 30,40,50,60,70,80,90,100
200 MAT READ U
210 DATA 0,10,20,30
220 MAT READ P
230 DATA .1,.2,.4,.2,.1
240 LET F=100
250 LET G=-.5
260 LET S=10
270 LET C$=","
280 REM
290 REM-----WRITE BLOCK A DATA - PROBLEM PARAMETERS-----
300 PRINT #1, N
310 PRINT #1, D1
320 PRINT #1, "S"
330 PRINT #1, "Y"
340 PRINT #1, "Y"
350 REM
360 REM-----WRITE BLOCK B DATA - TERMINAL STATE VALUES-----
370 PRINT #1, "0,0"
380 REM-----WRITE BLOCKS CS AND ES DATA-----
390 PRINT #1, " T                    I    D     R                  J     P"
400 FOR I=1 TO 8
410    FOR D=1 TO 4
420       IF U(D)>X(I) THEN 510
430       Q=X(I)-U(D)
440       LET R=Q*(-F/G+.5*Q/G)-S*U(D)
450       FOR P9=1 TO 5
460          LET J=(D-1)+P9
470          LET P2=P(P9)
480          PRINT #1, T;C$,I;C$;D;C$;R;C$,J;C$;P2
490       NEXT P9
500    NEXT D
510 NEXT I
520 PRINT #1, T,",0,0,0,0,1"
530 REM-----WRITE BLOCK D DATA - LABELS-----
540 PRINT #1, "SUPPLY"
550 FOR I=1 TO 8
560    PRINT #1, STR$(X(I))
570 NEXT I
580 PRINT #1, "STOCKS,4"
590 FOR D=1 TO 4
600    PRINT #1,STR$(U(D))
610 NEXT D
620 PRINT "DATA WRITTEN TO ";N$
630 CLOSE #1
640 END
```

FIG. 5.5. Program for writing data for the storage problem with stochastic yields to the file STORE.DAT.

TABLE 5.4
Infinite-Stage Solution to the Storage Problem with Stochastic Yields

Supply x *(thousand tonnes)*	*Carryover* u *(thousand tonnes)*	*Long-run probability of* x	*Present value (million dollars)*
30	0	0·087 5	153
40	0	0·187 5	154
50	0	0·375 0	155
60	0	0·225 0	156
70	10	0·112 5	157
80	10	0·012 5	158
90	20	0·000 0	158
100	30	0·000 0	159

the optimal policy were pursued indefinitely. These are compared with the consumption probabilities in the absence of a storage policy in Table 5.5. Whilst the range of consumption is 30 to 70 thousand tonnes and the mean consumption is 50 thousand tonnes in both cases, under the optimal storage policy the probability of consuming 60 thousand tonnes is much greater, whilst the probabilities of consuming all other levels are reduced.

It is useful at this stage to introduce the alternative set of recursive equations which are the basis of the price approach. The approach does not offer any computational advantage for the problem considered in this section, but it does provide an economic rationale for the optimal storage policy in terms of marginal analysis. Rules for optimal storage are derived

TABLE 5.5
The Probability Distribution of Grain Consumption With and Without Storage

Consumption (thousand tonnes)	*Probability of consumption*	
	Without storage	*With storage*
30	0·100 0	0·087 5
40	0·200 0	0·187 5
50	0·400 0	0·375 0
60	0·200 0	0·337 5
70	0·100 0	0·012 5

by differentiating the recursive equation (5.18), written more compactly as

$$V_i\{x_i\} = \max_{0 \leqslant u_i \leqslant x_i} [C\{x_i - u_i\} - s_i u_i + \alpha \mathrm{E} V_{i+1}\{x_{i+1}^k\}] \qquad (i = n, \ldots, 1) \qquad (5.19)$$

with

$$V_{n+1}\{x_{n+1}\} = 0$$

where $x_{i+1}^k = u_i + \bar{q}^s + e_{s,i}^k$ and E is the expectation operator. Provided the demand curve given by (5.9) is downward sloping ($g < 0$), and ignoring the upper limit on u_i (which it can be shown is not limiting for a stationary demand curve), the conditions for the expression in square brackets to be a maximum are

$$-\mathrm{d}C/\mathrm{d}u_i - s_i + \alpha E(\mathrm{d}V_{i+1}/\mathrm{d}x_{i+1})(\partial x_{i+1}/\partial u_i) \leq 0 \quad [=0, \quad \text{if } u_i > 0] \qquad (i = 1, \ldots, n) \qquad (5.20)$$

The term $C\{x_i - u_i\}$ in (5.19) is the area under the demand curve to $q_i^d = x_i - u_i$, so the value of consumption lost by a one unit increase in storage $\mathrm{d}C/\mathrm{d}u_i$ is the demand price of grain defined as $P_i\{x_i - u_i\}$.

Denoting optimal u by u_i^*, differentiating (5.19) with respect to x_i gives

$$\mathrm{d}V_i/\mathrm{d}x_i = P_i\{x_i - u_i^*\} \qquad (i = 1, \ldots, n) \qquad (5.21)$$

which after substitution in (5.20) yields

$$-P_i\{x_i - u_i^*\} - s_i + \alpha E P_{i+1}\{x_{i+1}^k - u_{i+1}^* \mid k\} \begin{vmatrix} = 0 & \text{if} \quad u_i^* > 0 \\ < 0 & \text{if} \quad u_i^* = 0 \end{vmatrix} \qquad (i = n, \ldots, 1) \qquad (5.22)$$

Condition (5.22) has the following economic interpretation: if for zero carryover to year $i + 1$ the price of grain in year i would be higher than the discounted expected price of grain in year $i + 1$ minus the marginal cost of storage, the optimal level of carryover is zero; otherwise, the optimal level of carryover to year $i + 1$ is the level which equates price in year i with discounted expected net unit return in year $i + 1$.

The optimal storage policy may be found by backward induction using (5.22). Under the price approach, $V_i\{x_i\}$ is the price of grain at stage i given total supply x_i and that the optimal policy $u_i^*, \ldots, u_n^*$ is followed. Recursion starts with $V_n\{x_n\} = P_n\{x_n\}$ because zero storage is optimal for the final stage. Optimal prices for all levels of supply at stage $n-1$ are then considered. Values of u_{n-1}^* and $P_{n-1}\{x_{n-1} - u_{n-1}^*\}$ are found which satisfy (5.22) and recorded for each x_{n-1}. The process continues for all i, after which the optimal sequence of storage levels and prices for an initial supply of $\bar{x}_1$, and for a particular sequence of yield deviations $k_1, \ldots, k_n$, is determined by forward tracking. Note that unlike with the value-of-consumption approach, the expected value of economic welfare from following the optimal policy is not obtained directly.

Solution of the following sequence of recursive optimization problems gives $u_i^*\{x_i\}$ for all i which satisfy (5.22):

$$\left.\begin{array}{l} \max\limits_{u_i \geq 0} [-|-P_i\{x_i - u_i\} - s_i + \alpha E V_{i+1}\{x_{i+1}^k\}|] \\ V_i\{x_i\} = P_i\{x_i - u_i^*\} \end{array}\right\} \quad (i = n-1, \ldots, 1) \qquad (5.23)$$

with

$$V_n\{x_n\} = P_n\{x_n\}$$

The formulation can be checked as follows. If $P_i\{x_i\} > \alpha E V_{i+1}\{x_{i+1}^k\} - s_i$, the expression in square brackets is made smaller by making u_i positive, so u_i^* must equal zero. Otherwise $u_i^* > 0$ can be found to bring the expression as close as possible to zero, which ensures the optimal balance between $P_i\{x_i - u_i^*\}$ and $\alpha E V_{i+1}\{x_{i+1}^k\} - s_i$.

The optimization in (5.23) is similar to the usual procedure used for solving dynamic programming problems numerically. The procedure can be performed using FDP by suitably altering the objective function and the recording of $V_i\{x_i\}$. The stage return is the negative of price plus marginal storage cost. The changes required to the data generation program (STORE.BAS) presented in Fig. 5.5 and to FDP listed in Appendix A2 to obtain a solution by the price approach are:

In STORE.BAS; insert '445 LET R = −(Q − F)/G − S'
In FDP.BAS; insert '1025 LET E(L) = −ABS(E(L))'
'1175 LET F(I2, N2) = −A(I2, N2) − 10'

The optimal policy for all stages with more than one decision stage remaining obtained from running the modified FDP program is the same as that shown in Table 5.4. Optimal expected prices are printed in place of expected present values in the output.

5.3.2 Stochastic yields and demand: planned production constant

The demand curve (5.9) may be subjected to random shifts. For example, demand may include export demand which increases due to production shortfalls elsewhere. The demand equation may be respecified as

$$q_i^{\mathrm{d}} = f + gP_i + e_{\mathrm{d},i} \tag{5.24}$$

where $e_{\mathrm{d},i}$ is a random demand variable. Suppose for concreteness $e_{\mathrm{d},i} = -10,\ 0,\ +10$ with probabilities 0.25, 0.5, 0.25, respectively. The recursive equation for solving the problem using the value-of-consumption approach is:

$$V_i\{x_i, e_{\mathrm{d},i-1}\} = \max_{u_i \geq 0}\Big[C\{x_i - u_i, e_{\mathrm{d},i-1}\} - s_i u_i + \alpha \sum_{k_i} p_i^{\mathrm{s}}\{k_i\} \times \sum_{l_i} p_i^{\mathrm{d}}\{l_i\} V_{i+1}\{u_i + \bar{q}^{\mathrm{s}} + e_{\mathrm{s},i}^{k}, e_{\mathrm{d},i}^{l}\}\Big] \quad (i = n, \ldots, 1) \tag{5.25}$$

with

$$V_{n+1}\{x_{n+1}, e_{d,n}\} = 0$$

The set of state values consists of all possible combinations of x_i and $e_{\mathrm{d},i-1}$, which means that for three values of $e_{\mathrm{d},i-1}$ there is a three-fold increase in the number of state values over the problem with deterministic demand. The problem can be solved for a much lower increase in the number of state values if the price approach is taken. Suppose, for a given level of supply with $e_{\mathrm{s},i} = 0$, and for $e_{\mathrm{d},i} = 0$, grain prices are calculated for the feasible range of u_i. The prices would be exactly the same if $e_{\mathrm{s},i} = e_{\mathrm{d},i} = 10$ thousand tonnes, (i.e.) if the demand and supply curves shifted to the right by the same amount. Values of consumption are of course different in the two cases.

It is this insight which permits a reduction in the dimension of the state variable using the price approach. Prices are the same for all combinations of $e_{\mathrm{s},i}$ and $e_{\mathrm{d},i}$ which give a particular value of $(e_{\mathrm{s},i} - e_{\mathrm{d},i}) \equiv \varepsilon_i$. The recursive equation is

$$\left.\begin{aligned} &\max_{u_i \geq 0}\Big[-\Big|-P_i\{x_i - u_i\} - s_i + \alpha \sum_{j_i} P_i^{\varepsilon}\{j_i\} V_{i+1}\{u_i + \bar{q}^{\mathrm{s}} + \varepsilon_i^{j}\}\Big|\Big] \\ &V_i\{x_i\} = P_i\{x_i - u_i^*\} \end{aligned}\right\} \quad (i = n-1, \ldots, 1) \tag{5.26}$$

with

$$V_n\{x_n\} = P_n\{x_n\}$$

where x_i is redefined as $u_{i-1} + \bar{q}^s + \varepsilon_{i-1}$, and

$$p_i^t\{j\} = \sum_{k_i} (p_i^s\{k_i\} p_i^d\{l_i \mid e_{s,i}^k - e_{d,i}^l = \varepsilon_i^j\})$$

assuming $e_{s,i}$ and $e_{d,i}$ are independent with probability distributions $p_i^s\{k_i\}$ and $p_i^d\{l_i\}$. For the example problem the number of state values is $8 + 2 = 10$ using the price approach, as against $8 \times 3 = 24$ using the value-of-consumption approach. The optimal carryover vector is the same as that given in Table 5.4 for the problem with deterministic demand, after allowing for the changed interpretation of x. Additionally, for $x = 20$, $u^* = 0$ and for $x = 110$, $u^* = 30$.

5.3.3 *Planned production a decision variable*

Planned production is in practice another decision variable. It will be some function of the expected future price of grain and the costs of production. Suppose that the decision on planned production q_i^s is made in year i, and actual production turns out to be $q_i^s + e_{s,i}$ in year $i + 1$.

The recursive equation for maximizing the present value of economic welfare is

$$V_i\{x_i\} = \max_{\substack{0 \leq u_i \leq x_i \\ 0 \leq q_i^s}} \left[C\{x_i - u_i\} - s_i u_i - c\{q_i^s\} + \alpha \sum_{k_i} p_i\{k_i\} V_{i+1}\{u_i + q_i^s + e_{s,i}^k\} \right] \quad (i = n, \ldots, 1) \quad (5.27)$$

with

$$V_{n+1}\{x_{n+1}\} = 0$$

where $x_i = u_{i-1} + q_{i-1}^s + e_{s,i-1}$ and $c\{q_i^s\}$ is the cost of planned production given by (5.13). To keep notation simple, demand is deterministic in (5.27). There is no increase in the number of state values as a result of including planned production as a decision variable, but maximization is carried out over all combinations of u_i and q_i^s.

The first steps towards deriving the recursive equation for the price approach analogous to (5.23) are to:

(i) differentiate the right-hand side of (5.27) with respect to u_i and q_i^s and to apply the Kuhn–Tucker conditions for a maximum; and
(ii) differentiate (5.27) with respect to x_i.

This leads as before to (5.22) as one of the conditions for optimality, and to

$$-\partial c_i/\partial q_i^s + \alpha E P_{i+1}\{x_{i+1}^k\} = 0 \tag{5.28}$$

as the other, on the assumption that optimal q_i without restriction is positive. The economic interpretation of (5.28) is that planned production should be set so that the marginal cost of planned production in year i equals the discounted expected price in year $i+1$. Note that expectation in year i of price in year $i+1$ is not P_i, as would be the case for a cobweb-type process, but is formed on the assumption of rational expectations.

If the demand curve is subject to a stochastic shift factor, the price approach is again more economical on computation. However, further changes would have to be made to FDP to be able to use it for solving the problem with variable planned production and rational expectations.

5.3.4 Selected applications

Stochastic dynamic programming is an appropriate approach for determining optimal stockholding policies. As has been pointed out by Gardner (1979), the certainty-equivalence theorem does not apply to the problems formulated above. Although the objective function is quadratic and the state transformation equation is linear, the theorem does not hold because of the restrictions on carryover such as that $u_i \geq 0$. For this reason applications of optimal control theory assuming certainty equivalence, such as that by Taylor and Talpaz (1979), may be inaccurate to some degree.

The simple framework described above can be extended in various ways to allow for further real-world factors. The random production-deviation term, $e_{s,i}$, might be better specified as multiplicative rather than additive to reflect production the result of the product of acreage planted and stochastic yield. The deviation terms $e_{s,i}$ and $e_{d,i}$ may be serially correlated, in which case the problem must be formulated with additional state variables. Producers may be risk averse in determining their production decisions. Food and feed grains are often substitutes in both production and consumption, so it may be important to consider optimal policies for several grains simultaneously.

Gustafson (1958) pioneered the dynamic programming approach. He considered many extensions, such as non-linear demand curves, and multiregional production and consumption. A critical and wideranging exposition of the dynamic programming approach which builds on Gustafson's work is presented in the book by Gardner (1979). He also emphasizes that the interaction between public and private stockholders should be modelled, and describes an application of the dynamic programming approach to US wheat stockpiling.

Plato and Gordon (1983) in a very useful article explain the value-of-consumption and price approaches, and describe and compare algorithms developed by Gardner (1979) and Ippolito (1979) which allow planned production as a decision variable. Plato and Gordon (1984) use Ippolito's algorithm for determining optimal stockpiling of US soybeans.

Burt *et al.* (1980) develop a model for determining the optimal level of US wheat exports, carryover being a residual after allowance for carryin, domestic consumption and production, and exports. Recognition of lagged variables in consumption and production leads to a dynamic programming problem with a formidable number of state variables. Methods of obtaining approximately optimal export decisions are described. Koo and Burt (1980) report results for a similar model, but with a penalty added to the objective function for price variability.

Cochrane and Danin (1976) experiment with various models under the objective of minimizing price variability. A simplified world model is described with different categories of aggregated grain. A second model of US wheat is also developed. State variables in the models are price of grain and reserve stocks.

Other grain applications include those by: Johnson and Sumner (1976) to optimal carryover in various developing countries and regions for three insurance programmes guaranteeing deliveries in the event of production shortfalls; Alaouze *et al.* (1978, 1979) to the situation where a country's export demand is perfectly elastic, which results in a bang-bang control policy of either zero or maximum stock-holding; and Kennedy (1979) to carryover of world wheat using assumptions made by Reutlinger (1976) in a simulation model.

The general topic of stockholding policies is dealt with comprehensively in the book by Newbery and Stiglitz (1981). They develop a stochastic dynamic programming approach, and derive the price approach (Chapters 29 and 30). This is also elaborated in their 1982 article.

Another expository article based on the price approach and using dynamic programming is that by Knapp (1982). Optimal storage, trade and foreign exchange dealing are analysed graphically using marginal cost-of-supply and demand-for-grain schedules.

Applications to agricultural commodities other than grains include those by Townsley (1964) to New Zealand butter; by Kenyon and Carman (1971) to Californian apples; and by Dalton (1976) and Hinchy and Simmons (1983) to Australian wool.

5.4 References

Alaouze, C.M., Sturgess, N.H. and Watson, A.S. (1978), Australian wheat storage: a dynamic programming approach, *Australian Journal of Agricultural Economics*, **22**(3), 158–74.

Alaouze, C.M., Sturgess, N.H. and Watson, A.S. (1979), Australian wheat storage: a dynamic programming approach—a correction, *Australian Journal of Agricultural Economics*, **23**(3), 231–2.

Ambrosius, J.E. (1970), Optimal tractor and combine replacement policies on corn farms under alternative firm growth patterns, unpublished PhD thesis, Purdue University.

Baquet, A.E. (1981), A theory of production, investment and disinvestment, in *The Rural Challenge, Contributed Papers Read at the 17th International Conference of Agricultural Economists*, M.A. Bellamy and B.L. Greenshields (eds), Gower, Aldershot.

Benders, J.F. (1962), Partitioning procedures for solving mixed-variables programming problems, *Numerische Mathematick*, **4**, 238–52.

Burt, O.R., Koo, W.W. and Dudley, N.J. (1980), Optimal stochastic control of US wheat stocks and exports, *American Journal of Agricultural Economics*, **62**(2), 172–87.

Cochrane, W.W. and Danin, Y. (1976), Reserve stock grain models, the world and the United States, 1975–1985, University of Minnesota, Agricultural Experiment Station, Technical Bulletin 305.

Cooke-Yarborough, R.E. (1964), Critical path planning and scheduling, *Review of Marketing and Agricultural Economics*, March, 36–48.

Dalton, M.E. (1976), Dynamic stockholding policies for stabilizing the wool market, *Quarterly Review of Agricultural Economics*, **29**(3), 179–92.

Danok, A.B., McCarl, B.A. and White, T.K. (1978), Machinery selection and crop planning on a state farm in Iraq, *American Journal of Agricultural Economics*, **60**(3), 544–9.

Danok, A.B., McCarl, B.A. and White, T.K. (1980), Machinery selection modelling: Incorporation of weather variability, *American Journal of Agricultural Economics*, **62**(4), 700–8.

Gardner, B.L. (1979), *Optimal Stockpiling of Grain*, D.C. Heath, Lexington, Massachusetts.

Gustafson, R.L. (1958), *Carryover levels for grains: a method for determining amounts that are optimal under specified conditions*, United States Department of Agriculture, Technical Bulletin No. 1178, Washington.

Hesselbach, J. (1974), The solution of agricultural machine investment problem by dynamic optimization (in German), *Landarbeit*, **25**(4), 31–5.

Hilger, D.A., McCarl, B.A. and Uhrig, J.W. (1977), Facilities location: The case of grain subterminals, *American Journal of Agricultural Economics*, **59**(4), 674–82.

Hinchy, M. and Simmons P. (1983), An optimal-control approach to stabilizing Australian wool prices, *Australian Journal of Agricultural Economics*, **27**(1), 44–72.

Ippolito, P. (1979), Private speculative storage and government stabilization programs, unpublished working paper, United States Department of Agriculture, Economic Research Services, June.

Johnson, D.G. and Sumner, D. (1976), An optimization approach to grain reserves for developing countries, *Analysis of Grain Reserves, A Proceedings*, United States Department of Agriculture, Economic Research Services Report No. 634, August.

Kelly, P.D. (1981), Incorporation of long-term investment decisions into recursive programming models: a case of farm tractors. Paper presented to Australian Agricultural Economics Society 25th Annual Conference, Christchurch, New Zealand, 10–12 February.

Kelly, P.D. (1985), Simulating investment behaviour in recursive programming models, with particular reference to the farm tractor replacement decision, unpublished PhD thesis, La Trobe University, Melbourne.

Kennedy, J.O.S. (1979), Optimal buffer stock policies for wheat at the world level, *Australian Journal of Agricultural Economics*, **23**(3), 163–75.

Kenyon, D.E. and Carman, H.F. (1971), Marketing rules for California fresh newtown apple sales: An application of dynamic programming, Department of Agricultural Economics, University of California, Davies.

Knapp, K. (1982), Optimal grain carryovers in open economies: a graphical analysis, *American Journal of Agricultural Economics*, **64**(2), 197–204.

Koo, W.W. and Burt, O.R. (1980), US Wheat storage control under joint criteria of mean benefits and joint variation, Staff Papers in Economics 80–14, Montana State University, Bozeman.

Newbery, D.M.G. and Stiglitz, J.E. (1981), *The Theory of Commodity Price Stabilization: A Study in the Economics of Risk*, Clarendon Press, Oxford.

Newbery, D.M.G. and Stiglitz, J.E. (1982), Optimal commodity stock-piling rules, *Oxford Economic Papers*, **34**(3), 403–27.

Plato, G. and Gordon, D. (1983), Dynamic programming and the economics of optimal grain storage, *Agricultural Economics Research*, **35**(1), 10–22.

Plato, G. and Gordon, D. (1984), Stockpiling US agricultural commodities with volatile world markets: the case of soybeans, *Agricultural Economics Research*, **36**(2), 1–9.

Reutlinger, S. (1976), A simulation model for evaluating worldwide buffer stocks of wheat, *American Journal of Agricultural Economics*, **58**(1), 1–12.

Smith, P. (1984), *Agricultural Project Management: Monitoring and Control of Implementation*, Elsevier Applied Science, Barking, Essex.

Sowell, R.S. (1967), A farm machinery replacement study with application to the replacement of self-propelled cotton pickers, Unpublished PhD thesis, North Carolina State University, Raleigh.

Taylor, C.R. and Talpaz, H. (1979), Approximately optimal carryover levels for wheat in the United States, *American Journal of Agricultural Economics*, **61**(1), 32–40.

Townsley, R. (1964), The maximization of revenue from New Zealand sales of butter on the United Kingdom market—a dynamic programming problem, *Australian Journal of Agricultural Economics*, **9**(2), 169–80.

White, W.C. (1959), The determination of an optimal replacement policy for a continually operating egg production enterprise, *Journal of Farm Economics*, **41**(5), 1535–42.

Chapter 6

CROP MANAGEMENT

6.1 The Crop Decision Problem

The season's return from annual cropping depends on certain uncontrollable events such as the timing and amount of rainfall and exposure to sunlight. It also depends on many controllable events such as the timing and density of planting, the timing of harvesting and the timing and amount of maintenance inputs. Maintenance inputs include water for irrigation and soil leaching, pesticides, herbicides and fertilizers.

The treatment of states, stages, decisions and returns in dynamic programming formulations of crop problems in general is considered in this section. Subsequent sections deal with dynamic programming applications to particular types of cropping decision: the management of water and pesticides, and the determination of crop rotation. The optimal application of fertilizer when some fertilizer carries over from one growing season to another is examined in detail in Section 6.5. Rules are derived which have received relatively little attention to date. The derivation illustrates the way in which dynamic programming may be used to obtain analytical instead of numerical results.

6.1.1 States

The returns from cropping decisions depend on state variables such as:

(i) Crop growth or maturity
(ii) Soil-salinity, moisture, nutrients, depth, weed/pest infestation
(iii) Pest and predator (beneficiary) population by size and stage of development (egg, larvae, pupae, adult)
(iv) Water stock—surface and below ground.

Transformation functions are required which describe how the state at stage $i+1$ is determined by the state and the decision made at stage i. They may be complex hydrological and/or biological functions, and may be subject to uncertainty. In many studies, particularly in the area of water management, state transition probabilities for stochastic transformation functions have been obtained from simulation models.

6.1.2 Stages

For some input decisions, only one application per season need be considered. Such decision problems may be simple static problems which can be solved by marginal analysis, or, if the input decision affects a relevant state variable in the subsequent season, they may be treated as interseasonal problems. The decision on fertilizer application when carryover occurs is one such problem. Early formulations of the pesticide decision problem were of this type (e.g. Headley, 1972; Hall and Norgaard, 1973), but more recent formulations allow for multiple applications throughout the crop season.

The typical intraseasonal cropping problem is to find the optimal timing and levels of successive applications of managed inputs over the crop season. All possible application times are designated as stages in the dynamic programming formulation. The interval between stages over the crop season may be variable. They may be set according to stages in crop development. The first decision stage may be the time for inputs prior to planting, and the next stage the time when the crop is first mature enough to accept an input such as irrigation water. The intervals between subsequent stages may be smaller and regular up to the stage of harvesting (see Dudley *et al.*, 1971*b*). It may be useful in some cases to have stages equally spaced in a time dimension which is a transformation of normal time. For example, if the input is pesticide, the problem may be specified in physiological time such as degree days.The number of degree days in a given period is obtained by integrating the excess of the temperature over a minimum threshold temperature, depending on the pest (Talpaz and Borosh, 1974).

In many dynamic programming applications an intraseasonal decision sequence is a subproblem of an interseasonal decision sequence. If state transformation and stage return functions are non-stationary intraseasonally but stationary interseasonally, policy solutions for an infinite planning horizon may be obtained by the method illustrated in Section 3.4.

6.1.3 Returns

In market economies cropping decisions are typically made by farmers according to their own criteria of say expected net revenue, or utility a function of expected net revenue and variability of net revenue. However, the cropping decisions made by one farmer may affect the returns of other farmers and also of the non-farming community. The farm decisions judged optimal for society may differ from those optimal for the farmer himself. Instances of the external effects of cropping decisions are the run-off of fertilizers to pollute streams and lakes, the raising of watertables through irrigation resulting in salt intrusion in soils, and the resistance of pests to pesticides. Dynamic programming may be used to find socially optimal management solutions across farms in a region and across years. If the decisions of farmers maximizing private objective functions are also modelled, it is possible to determine Pigouvian taxes or other regulations which will bring about the socially optimal solution through the market system.

Net revenue consists of returns from the sale of the harvested crop, less the costs of input application, and less other ancillary costs. The costs of input application are the material costs plus any application or set-up costs, such as aerial spraying costs in the case of pesticides. Set-up costs are easily handled in numerical dynamic programming formulations (in the case of pesticide applications see e.g. Raphael, 1979, although they have been handled by other solution methods: see e.g. Talpaz and Borosh, 1974). The returns from sale of the harvested crop may be calculated in one of two ways. One way is to make crop yield or size a state variable. This method has the advantage that change in crop size from stage i to $i+1$ can be specified quite flexibly as dependent on size and input decision at stage i. Despite the flexibility permitted the assumed Markov property of state transition may not apply strictly unless additional state variables are specified. For example, the response of crop size to irrigation at stage i may also depend on the extent of water deficit in some period before stage i. The effect on change in crop size may be only partially allowed for if crop size at stage i is the only state variable. With this method stage returns for stages before harvest are just the negative costs of input applications. Harvest proceeds are terminal values.

The alternative, and more usual, method is to assume that returns from changes in yield can be calculated for each decision stage. Crop size is not treated as a state variable. Instead, stage return depends on the input decision and some other state variable, such as soil moisture or pest population. The stage return also depends on the stage number, which can be viewed as a state variable measuring the age and hence the maturity of the crop. An example of an additive multistage crop-yield function is

$$Y = A - \sum_{i=1}^{n} d_i\{x_i, u_i\} \tag{6.1}$$

where A is the crop yield which would be achieved if there were no need for input control u_i (e.g. if the crop were not subject to lack of water, or to pests); and d_i is the reduction in yield damage from stage i to $i+1$ given u_i and state variable x_i (e.g. soil moisture or pest population). Stage return is $-p^y d_i$ less costs of input application, u_i, where p^y is the price obtained for the harvested crop.

There are limited alternative formulations of the crop yield function if it is used to calculate stage returns, rather than to specify the state of the system at each stage. This is because of the requirement of separability of stage returns discussed in Section 2.3. It is generally desirable to make the objective function additive rather than say multiplicative in stage returns. Two problems arise with a multiplicative objective function. One is that it would not be possible to use it directly in a stochastic dynamic programming formulation, because expected returns are probability weighted sums of random returns. The second is that it would not be possible to take account of the costs of input application, because costs are additive.

However, it is possible to overcome these problems by making a multiplicative objective function sequentially additive. Suppose the objective function V is the product of three stage returns, and equals $a_1 \times a_2 \times a_3$. It is simple to check that this equals $a_1 + a_1(a_2 - 1) + a_1a_2(a_3 - 1)$. In general,

$$V = \prod_{i=1}^{n} a_i = a_1 + \sum_{j=2}^{n} \left[\left(\prod_{k=1}^{j-1} a_k \right)(a_j - 1) \right] \tag{6.2}$$

Moreover, if stage returns are random and independent,

$$E(V) = E\left(\prod_{i=1}^{n} a_i \right) = E(a_1) + \sum_{j=2}^{n} \left[\left(\prod_{k=1}^{j-1} E(a_k) \right)(E(a_j - 1)) \right]$$

This means that the contribution of stage return a_i to total return can be calculated provided $\prod_{k=1}^{i-1} a_k$ is known. Dynamic programming solutions for a multiplicative objective function can therefore be obtained using an additive objective function in the dynamic programming routine provided a record is kept of the product of the stage returns of the original multiplicative objective function. If backward recursion is used, the order of calculation in (6.2) is reversed and products of all stage returns to the last one in the planning horizon have to be stored for all states and stages.

This permits the use of a multiplicative yield function

$$Y = A \prod_{i=1}^{n} b_i^{\delta_i} \tag{6.3}$$

in dynamic programming problems, where b_i is a crop reduction coefficient and δ_i is a measure of crop damage, a function of x_i and u_i. Again $Y = A$ if $\delta_i = 0$ for all i. Yaron *et al.* (1980) have used (6.3) in finding the optimal irrigation of sorghum in Israel using saline water.

Dudley and Burt (1973) provide a useful discussion of the two methods of calculating returns from crop yield.

6.1.4 Decisions

Besides decisions on the frequency and level of input applications, decisions have to be made on the timing and density of planting, and the timing and intensity of harvesting. Often the latter decisions are not dealt with in dynamic programming formulations of the crop problem, possibly because they are largely determined by the weather and by physiological characteristics of the plant. If the timing of planting or harvesting affects parameters in the multistage crop-yield functions such as A, d_i or δ_i in (6.1) and (6.3) it may be best not to use these functions and instead to formulate the problem with crop yield as a state variable.

It is usually assumed in dynamic programming formulations that all of the crop available for harvesting is harvested. In some formulations the crop available for harvesting is reduced as a result of deciding to stop injecting inputs at some decision stage before harvesting. For example, Dudley *et al.* (1971*b*) allow for the abandonment of irrigating some land during the growth period before harvest. Corrie and Boyce (1972) present a dynamic programming model for determining rules for the selective and successive harvesting of cauliflowers over the maturing period.

6.2 Applications to Water Management

An example problem of how best to apply water over time to a crop was developed in Chapters 2 and 3 in order to demonstrate some of the solution techniques of dynamic programming. A typical problem would be specified as follows: the optimal quantity of water to apply at each of 10 stages over a growing season of 100 to 200 days is to be determined; the state variable is reservoir level; rainfall is a random variable, affecting reservoir level and the water received by the crop; the stage return is a function of stage reached in the growing season and water received by the crop; and the objective is to maximize expected net revenue over the growing season.

Some of the studies in which dynamic programming has been applied to water management for cropping are summarized in Table 6.1. The early studies tend to be of the type just outlined. Note, however, that Asopa (1971) and Burt and Stauber (1971) used crop condition as a state variable.

An important extension is the consideration of the interseasonal as well as the intraseasonal problem. Total water usage over one growing season affects the amount of water available for the subsequent season. Most of the studies listed in Table 6.1 consider both problems. The objective function becomes the maximization of the present value of expected net revenue. The formulation of the interseasonal problem with intraseasonal decisions to be made over an infinite planning horizon as a stationary problem has been described in Section 3.4. This type of model has been used to analyze strategic investment decisions, such as the optimal size of reservoir, and the area to devote to irrigated crops (Burt, 1964, 1966; O'Loughlin, 1971; Dudley, 1972).

The stage return function can often be made more realistic by making crop growth depend on soil moisture and/or soil salinity as well as water applied. In some of the studies this is accomplished by adding state variables. If soil salinity is a problem, the decision options are extended to include the quality of water to use at different times of the year for both leaching and irrigation (Yaron and Olian, 1973; McFarland, 1975; Matanga and Miguel, 1979; Yaron *et al.*, 1980; Yaron and Voet, 1983; Knapp, 1984).

Other state variables that have been considered are groundwater storage (Burt, 1964, 1966; McFarland, 1975), and soil salinity by soil layer (Knapp, 1984).

In some studies linear programming is used to determine the optimal mix of crops and intraseasonal irrigation strategies, within an interseasonal dynamic programming framework (McFarland, 1975; Dudley *et al.*, 1976; Yaron and Voet, 1983). A linear programming problem may be formulated to find the cropping and irrigation plan which maximizes net revenue

TABLE 6.1
Summary of Dynamic Programming Applications to Water Management

Reference	*Interval between stages*	*Number of stages*	*State(s)*[a]	*Decision(s)*	*Objective function maximized*[b]	*Comments*
Hall *et al.* (1968)	1 month	84	Water storage	(i) Water released (ii) Peak electricity generated (iii) Off-peak electricity generated	NR	Describes the use of LP for generating shadow prices for water released for different purposes, including electricity generation and irrigation
Burt (1964, 1966)						
(a) Surface water	1 month	∞	Surface water storage (stoch.)	Surface water released	PV of ENR	Considers the integration of a range of models dealing with investment decisions, land allocation and water management policy
(b) Groundwater	5 years	∞	Groundwater storage (stoch.)	Groundwater released	PV of ENR	
Flinn and Musgrave (1967)	30 days	8	Water storage	Number of irrigations (0 to 6)	NR	Model's use for pricing irrigation water also considered
Hall and Butcher (1968)	Unspecified	10	(i) Water storage (ii) Soil moisture	Irrigation water applied	NR	Total return from the crop a multiplicative function of period returns
de Lucia (1969)	1 week	22	(i) Surface water storage (stoch). (ii) Previous stream flow (stoch.) (iii) Soil moisture (stoch.)	(i) Water applied from surface (ii) Water applied from aquifer	ENR	Random variables are net stream inflow into surface storage and effective rainfall
Asopa (1971)	9 days	5	Crop condition (stoch.)	Irrigation water applied	ENR	No constraint on water available; there is a charge for water use
Burt and Stauber (1971)	10 days	5	(i) Crop condition (stoch.) (ii) Water storage	Irrigation water applied	ENR	Rainfall stochastic; state variable (ii) eliminated by the use of a Lagrange multiplier

Dudley *et al.* (1971*a*)	15 days	6	(i) Soil moisture (stoch.) (ii) Water storage (stoch.)	Terminal soil moisture	ENR	State transition probabilities based on the result of a plant water simulation model; the way in which results could be used for the intermediate-run problem of area of crop to plant is discussed in Dudley *et al.* (1971*b*); see also Dudley *et al.* (1972)
O'Loughlin (1971)	1 month	∞	(i) Reservoir level (stoch.) (ii) Area irrigated	Irrigation water applied	PV of ENR	Optimal area of crop to be planted at the start of each growing season is determined; three models run end-to-end to model a changing crop sequence throughout the year
Dudley (1972)	1 year	20	Reservoir level (stoch.)	Irrigation water applied	PV of ENR	Used in conjunction with other models (Dudley *et al.* 1971*a* and *b*, 1972) to find PV of ENR a function of acreage developed for irrigation
Biere and Lee (1972)	4 months	∞	Surface reservoir level	Reservoir water released to recharge valley aquifers	PV of reduction in expected pumping costs	Solved assuming a stationary process with respect to years, non-stationary with respect to seasons
Dudley and Burt (1973)	Irregular	7 within year; ∞ number of years	(i) Soil moisture (stoch.) (ii) Reservoir level (stoch.) (iii) Area available for irrigation	(i) Within-year irrigation (ii) Area irrigated	PV of ENR	State variable transition probabilities are calculated by simulation; variance of net benefits treated as a criterion for trade-off against expected value of net benefits
Yaron and Olian (1973)	1 year	50	Soil salinity (stoch.)	Water applied for leaching before summer irrigation	PV of ENR	Estimates the value of water quality in irrigation
McFarland (1975)	1 year	50	(i) Soil salinity (ii) Groundwater storage	(i) Water for irrigation (ii) Water applied for leaching during the dormant season	PV of NR	Yearly cropping plan determined using LP

(*continued*)

TABLE 6.1—*contd.*

Reference	*Interval between stages*	*Number of stages*	*State(s)*[a]	*Decision(s)*	*Objective function maximized*[b]	*Comments*
Dudley *et al.* (1976)	6 months	∞	(i) Reservoir level (stoch.) (ii) Area of winter wheat in spring	(i) Area of winter wheat (ii) Irrigation water applied	PV of ENR	Used in conjunction with LP for finding the optimal crop mix and intraseasonal irrigation, and simulation for finding transition probabilities; the effect of area irrigated on the mean and variance of returns examined
Matanga and Miguel (1979)	1 year	∞	Soil salinity (stoch.)	(i) Water applied for leaching prior to the irrigation season (ii) Water applied for irrigation	PV of ENR	Results from running the model for three crops on a per acre basis used in an LP model to determine the allocation of land to each crop (Matanga and Marino, 1979)
Yaron *et al.* (1980)	16 days	18	(i) Soil moisture (ii) Soil salinity	Irrigation water applied	NR	Season divided into 4 growing subperiods, yield function multiplicative; salinity of irrigation water exogenous
Yaron and Dinar (1982)	1 day	81	(i) Soil moisture (ii) Water available per unit of new LP crop activity	Water input per unit of new LP crop activity	Gross margin of new LP crop activity	DP used for finding new LP crop activities efficiently by an iterative process; yield function additive
Yaron and Voet (1983)	1 year	25	Soil salinity in two fruit groves (stoch.)	Water applied for leaching and irrigation to each grove	PV of ENR	LP used for finding returns from irrigating field crops with water not used for leaching and irrigating fruit groves
Knapp (1984)	1 year	∞	Vector of soil salinity by soil layer	Water applied	PV of NR	Up to 20 soil layers considered; solution obtained by differentiating the recursive DP equation and following an iterative procedure

[a] Stochastic state variables are denoted by '(stoch.)'.
[b] 'PV of ENR' stands for 'present value of expected net revenue'.

over the growing season, subject to parametric variation of constraints on water availability. The maximum net revenue obtained by linear programming may form part of the stage return in the dynamic programming routine. There have been other methods of integrating linear programming and dynamic programming. Hall *et al.* (1968) run a linear program to generate shadow prices of water from streams used over each month for various purposes including electricity generation and irrigation. The shadow prices are fed into a dynamic programming model for allocating water to stream subsystems. An iterative process is followed until there is no further improvement in the overall objective function. Shadow prices are revised by rerunning the linear programs for each allocation of water to stream subsystems.

Another integration has been reported by Matanga and Miguel (1979) and Matanga and Marino (1979). The idea is to obtain optimal allocations of water for leaching and irrigation of a crop on a per acre basis across years. This is done for several crops and an infinite planning horizon. The results for each crop are used in a linear programming model to determine the acreages to devote to each crop, and the irrigation strategy, with and without limitations on total water available. However, it is not clear whether the price of water used in the dynamic programming model described by Matanga and Miguel (1979) is consistent with the water availability constraints in the linear programming model described by Matanga and Marino (1979).

Yaron and Dinar (1982) integrate the running of dynamic and linear programming models in a quite different and novel way. Linear programming is used to find the plan which maximizes the income of a farm producing cotton. Dynamic programming is used for finding optimal linear programming activities given shadow prices of water obtained from a previous run of the linear programming model. New linear programming activities are generated via an iterative process until farm income cannot be further improved. The dynamic programming model takes account of the non-linear relationship between soil moisture and crop yield.

6.3 Applications to Pesticide Management

The question of how best to apply pesticides can be formulated as a very complex problem indeed. The complexity arises for various reasons. Knowledge is required about the dynamics of the pest population, and how these are affected by weather patterns, pesticide applications, and by parasite or beneficiary populations which attack the pests. The adverse effect of the pesticides on the beneficiary population needs to be considered. Pest and beneficiary populations are difficult to estimate accurately. A function relating crop damage to pest population is necessary. The external effects of applying pesticides need consideration. These are: the human health risks involved in application; the benefits to neighbouring farms of one farmer applying pesticide; and the increase in resistance to pesticides in the general pest population. Because of the adverse side effects of pesticide applications, the question of how much pesticide to apply cannot be divorced from the question of the use of alternative measures, such as biological controls or crop rotation.

Up until the early 1970's, entomologists viewed the pesticide problem as one of deciding economic pest population thresholds. The threshold was the population level at which pesticides should be applied to virtually eliminate the pest. Hall and Norgaard (1973) criticized this position and developed a model which incorporated the timing and level of pesticide application as decision variables. Talpaz and Borosh (1974) emphasized that in many cases multiple applications and the set-up costs of application had to be considered. They and Talpaz *et al.* (1978) developed models for the control of pests in cotton production which were solved using Newton–Raphson and Fletcher–Powell–Davidson methods of non-linear optimization respectively.

Watt (1963) was the first to suggest dynamic programming as a technique for determining pesticide usage. The objective function to be minimized was total defoliation over a span of 60 years. Mann (1971) developed simple conceptual multiperiod models using dynamic programming for finding threshold pest populations. Pest population transitions followed a Markov process. The objective function to be maximized was discounted expected net revenue. Shoemaker (1973*a*, *b* and *c*) in a wideranging series of articles considered how dynamic programming could be applied to pesticide management taking account of many economic and biological realities. Taylor and Headley (1975) formulated a dynamic programming problem in which optimal pesticide application to each generation of pest is determined taking account of the fact that applications lead to increased pesticide resistance in subsequent generations of the pest. Three state variables were specified, being the number of pests in each of three categories defined by degree of resistance to the pesticide. Offspring survival was formulated as stochastic, conditional on the total population after pesticide application. Hueth and Regev (1974) also formulated a problem taking account of pesticide resistance in an optimal control theory framework, with potential plant product, pest population and pest susceptibility as state variables. Regev *et al.* (1979) and Regev *et al.* (1983) subsequently applied models allowing for pesticide resistance to weevil control in crops of alfalfa. Lazarus and Dixon (1984) applied a non-linear programming model to the problem of controlling corn rootworm in the presence of pesticide resistance. The decision options include a switch of crop from corn to soybeans, as well as corn production with and without pesticide application.

Some dynamic programming applications to pesticide management are summarized in Table 6.2. The usual model structure is as follows: at least one state variable is the pest population; the stage return depends on crop damage, which is proportional to the pest population; the pest population grows at a natural rate less a kill rate dependent on pesticide application. Shoemaker (1973*c*), Conway *et al.* (1975) and Raphael (1979) address the problem of multiple applications within a season, whilst Watt (1963) and Shoemaker (1979a) determine annual application policies across years.

Regev *et al.* (1979) study both the intraseasonal and interseasonal problems. They use a non-linear optimizing technique (generalized reduced gradient algorithm) for the intraseasonal problem of how to shift from one level of pesticide resistance at the beginning of year i to another at the beginning of year $i+1$ with least costs of crop loss and of pesticide applications. The results are used in a dynamic programming model which has pesticide resistance as the state variable. The model determines the optimal sequence of resistance across years.

The dynamic programming model developed by Regev *et al.* (1979) is used to find the application decisions over time which a regional planner might regard as optimal. It is assumed that the planner controls a sealed region which has no exchange of pests with other regions. The results are contrasted with the application decisions over time made by individual farmers unable to appropriate for themselves the benefits of reduced application. In this case farmers can be assumed to apply pesticide so as to sequentially minimize losses each year. The Pigouvian taxes on pesticides which would be necessary to reduce applications to socially optimal levels can be calculated.

An important area that does not appear to have been tackled so far in either dynamic or other non-linear programming models of pesticide management is uncertainty and decision making attitudes towards uncertainty. Non-linear programming applications have been deterministic, as have most of the dynamic programming applications except that by Shoemaker (1979*a*) which incorporates weather pattern as a stochastic

TABLE 6.2
Summary of Dynamic Programming Applications to Pesticide Management

Reference	*Interval between stages*	*Number of stages*	*State(s)*[a]	*Decision(s)*	*Objective function maximized*[b]	*Comments*
Watt (1963, 1968)	1 year	60	Pest density	Kill rate once pest density exceeds specified threshold	Minus total defoliation	A pioneering non-economic application
Shoemaker (1973*a*, *b*, *c*)	Growing season divided by 9	9	(i) Pest density (ii) Parasite density	Pesticide applied	Minus sum of costs of pesticide application and pest damage	A hypothetical numerical application; considers extensions for dealing with many real world additional complexities
Conway *et al.* (1975)	Interval between brood emergence	4	Size of brood	1 out of 10 spraying strategies	NR	Application to control of the sugar cane froghopper in Trinidad; planning horizon is one season
Raphael (1979)	Half-week	24	(i) Population density of 3 sizes of pest larvae (ii) Population density of pest's predator	Pesticide applied	Minus sum of costs of pesticide application and pest damage	Argues that further refinements must await results of more biological research

Regev *et al.* (1979)	1 year	6	Resistance to pesticide	(i) Increase in resistance to pesticide (ii) Stage at which to switch to alternative control technology	PV of NR	Intraseasonal pesticide applications determined using a non-linear programming code; solutions under centralized decision making and under competition compared
Shoemaker (1979*a*)	1 year	3	(i) Adult weevil density (ii) Adult parasite density (iii) Weather pattern (stoch.)	(i) Time of harvest (ii) Pesticide applied after harvest	PV of ENR	Application to control of weevils in alfalfa; population model developed in detail; see Shoemaker (1982) for development of the model to include four state variables
Shoemaker (1979*b*)	1 week	28	(i) Time of last pesticide application (ii) Time of second-last pesticide application	Pesticide applied or not applied	NR	Claimed that with this formulation many more age classes of pest can be considered—11 in this application to Egyptian alfalfa weevils

[a] Stochastic state variables are denoted by '(stoch.)'.
[b] 'PV of ENR' stands for 'present value of expected net revenue'.

TABLE 6.3
Summary of Dynamic Programming Applications to Crop Selection

Reference	*Interval between stages*	*Number of stages*	*State(s)*[a]	*Decision(s)*	*Objective function maximized*[b]	*Comments*
Burt and Allison (1973)	1 year	∞	Initial soil moisture (1 of 5 levels) (stoch.)	Plant land to wheat or leave fallow	PV of ENR	Initial soil moisture depends on whether wheat was planted the previous year, and precipitation in the previous year
Throsby (1964*a*)	1 year	6	Land available to different activities	Land allocated to different activities with different production periods	NR	A simple numerical example in an expository article
Throsby (1964*b*)	1 year	3	Land available for pasture improvement	Land allocated to pasture improvement	NR	A simple numerical example in an article which examines in detail how the qualitative nature of solutions depends on the period return functions
Burt and Johnson (1967)	1 year	∞	Initial soil moisture (1 of 6 levels) (stoch.)	(i) Plant land to wheat or leave fallow (ii) Mix of wheat/fallow strategies	(i) PV of ENR (ii) Weighted sum of ENR and VNR	Extension of Burt and Allison (1963)—considers more empirical data, and trade-off possibilities between mean returns and variance of returns when mixed strategies are allowed
Burt (1971)	Irregular	∞	Length of previous renewal cycle (stoch.)	Length of current cycle	PV of ENR	Describes various models for determining the optimal cycle length for removing scrub from pasture, dependent on previous cycle length and time since last renewal

Iskakov and Finkelstejn (1973)	1 year	3	Crop at current stage	Crop at next stage	PV of ENR	A simple numerical example of crop rotation in an expository article
Kennedy *et al.* (1973)	4 months	∞	(i) Land fallow or crop age (ii) Season	Continue, plant or replace	PV of NR	Optimal fertilizer application when there is fertilizer carry-over is deduced by induction: investigates a cropping cycle for sorghum in the Northern Territory
Burt (1981)	1 year	∞	Percentage organic matter in the upper six inches of soil	Percentage of land in wheat	PV of NR	DP results shown to be closely approximated by results from the method proposed by Burt and Cummings (1977)
Fisher and Lee (1981)	6 months	20	(i) Soil moisture (ii) Soil crown rot-disease (iii) Wild oats infestation	Crop selection	PV of NR	Optimal wheat/fallow rotations derived for north-western New South Wales
Taylor and Burt (1984)	1 year	∞	(i) Soil seed reserve (stoch.) (ii) Price of spring wheat (stoch.) (iii) Previous land use	Crop/herbicide	PV of ENR	Optimal wheat/fallow rotations derived for north central Montana; results used in two other DP models, one accounting for soil moisture, and the other for possible emergent herbicide applications

[a] Stochastic state variables are denoted by '(stoch.)'.
[b] 'PV of ENR' stands for 'Present value of expected net revenue'; 'VNR' stands for 'Variance of net revenue'.

state variable. Work to date which formulates the stochastic pesticide problem in terms of maximizing utility has been couched in a static framework (see e.g. Carlson, 1979; Feder, 1979; Lazarus and Swanson, 1983).

Shoemaker (1981) reviews dynamic programming applications to pest management, and compares the dynamic programming approach with others which make use of simulation and gradient methods.

6.4 Applications to Crop Selection

In crop selection problems a decision has typically to be made at the beginning of the growing season on what type of crop, if any, to produce. The decision variable is often discrete or qualitative rather than quantitative as in the problems of water, pesticide and fertilizer application. Such problems are particularly amenable to numerical solution using dynamic programming. However, in some crop selection problems the decision variable is continuous, as is the case for example if the amount of land to devote to a crop is to be determined.

Crop selection affects the states of the crop system such as: land available for other crops; soil moisture, organic matter in the soil; pest, weed or scrub infestation; and the age of a plant of a perennial crop. These crop-system states at the beginning of the growing season together with the crop selection decision jointly determine net revenue for the current period, and the state at the beginning of the next growing season. These states are therefore the state variables in dynamic programming formulations of a crop selection problem. A summary of some applications is presented in Table 6.3.

There is scope for further applications of dynamic programming in which the timing and level of inputs such as water, pesticide or fertilizer have to be determined together with the type of crop.

6.5 Applications to Fertilizer Management

Rules for optimal fertilizer application when there is carryover of various types have been derived by Kennedy *et al.* (1973), Dillon (1977), Kennedy (1981) and Taylor (1983) by a process of induction based on dynamic programming. In this section a simpler and more direct method of deriving the rules is explained. The rules are obtained by differentiating the recursive equation used in dynamic programming.

6.5.1 Optimal rules for single-period carryover functions

Consider the simplest possible problem situation. The way in which crop yield responds in period i to available fertilizer r_i is given by the response function $y_i\{r_i\}$. Available fertilizer (r_i) is the sum of applied fertilizer (u_i) and carryover (x_i). Carryover is a proportion (b_i) of available fertilizer in period $i-1$. The price of the crop, p_i^y, and of the fertilizer, p_i^f, are known for all periods. Fertilizer costs are paid at the beginning of each period, and crop returns are received at the end of each period. The objective is to maximize the present value of net returns from fertilizer applications over n periods, given a discount factor α and initial fertilizer carryover, x_1. Residual fertilizer at the end of period n has zero value. The problem can be formulated as

$$\max_{u_1,\ldots,u_n} \sum_{i=1}^{n} \alpha^{i-1}(\alpha p_i^y y_i\{x_i + u_i\} - p_i^f u_i)$$

subject to

$$\left.\begin{aligned} u_i &\geq 0 \\ x_{i+1} &= b_i(x_i + u_i) \end{aligned}\right\} \quad (i = 1, \ldots, n)$$

$$x_1 \text{ given}$$

This problem can be solved by using the recursive equation

$$V_i\{x_i\} = \max_{u_i} [\alpha p_i^y y_i\{r_i\} - p_i^f u_i + \alpha V_{i+1}\{b_i r_i\}] \qquad (i = n, \ldots, 1) \quad (6.4)$$

subject to

$$u_i \geq 0$$

$$r_i = x_i + u_i$$

$$x_1 \text{ given}$$

with

$$V_{n+1}\{x_{n+1}\} = 0$$

where $V_i\{x_i\}$ is the present value of net returns from carryover x_i obtained by following the optimal policy of fertilizer application over periods i to n.

If the response function is concave, differentiating the term in square brackets with respect to u_i gives the first-order condition for an interior maximum

$$\alpha p_i^y \, \partial y_i/\partial u_i - p_i^f + \alpha b_i \, \mathrm{d}V_{i+1}/\mathrm{d}x_{i+1} = 0 \qquad (i = 1, \ldots, n) \quad (6.5)$$

Note that, because $\partial r_i/\partial u_i = \partial r_i/\partial x_i = 1$, $\partial y_i/\partial u_i = \partial y_i/\partial r_i$; also that, because $x_{i+1} = b_i(x_i + u_i)$, $\partial x_{i+1}/\partial u_i = b_i$.

If $\mathrm{d}V_{i+1}/\mathrm{d}x_{i+1}$ were known, then (6.5) could be solved for optimal u_i^*, provided $u_i^* > 0$. In fact it is easy to show that $\mathrm{d}V_{i+1}/\mathrm{d}x_{i+1}$ (evaluated at $x_{i+1} = b_i(x_i + u_i^*)$) is equal to the price of fertilizer in period $i+1$ for all $i \leq n-1$. Rewriting (6.4) as

$$V_i\{x_i\} = \alpha p_i^y y_i\{x_i + u_i^*\} - p_i^f u_i^* + \alpha V_{i+1}\{b_i(x_i + u_i^*)\}$$

and differentiating with respect to x_i gives

$$\mathrm{d}V_i/\mathrm{d}x_i = \alpha p_i^y\, \partial y_i/\partial u_i + \alpha b_i\, \mathrm{d}V_{i+1}/\mathrm{d}x_{i+1} \qquad (i = 1, \ldots, n)$$

But from (6.5),

$$\alpha p_i^y\, \partial y_i/\partial u_i + \alpha b_i\, \mathrm{d}V_{i+1}/\mathrm{d}x_{i+1} = p_i^f \qquad (i = 1, \ldots, n)$$

and therefore

$$\mathrm{d}V_i/\mathrm{d}x_i = p_i^f \qquad (i = 1, \ldots, n) \tag{6.6}$$

This result is not surprising. Equation (6.6) merely states that the value of an additional unit of fertilizer carried over from period $i-1$ to period i should equal the price of fertilizer in period i. After substitution in (6.5), the condition for optimality becomes

$$\alpha p_i^y\, \partial y_i/\partial u_i = p_i^f - \alpha b_i p_{i+1}^f \qquad (i = 1, \ldots, n-1) \tag{6.7}$$

Terminal values equal zero for all x_{n+1}, so $\mathrm{d}V_{n+1}/\mathrm{d}x_{n+1} = 0$. Thus the rule for the final decision period is found directly from (6.5) to be

$$\alpha p_n^y\, \partial y_n/\partial u_n = p_n^f$$

Condition (6.7) states that for u_i to be optimal, the present value of the marginal product of fertilizer must equal the opportunity cost of the marginal unit of fertilizer. The latter equals the price of fertilizer in period i, less the present value of savings of b_i units of applied fertilizer in period $i+1$ because of carryover. The condition can be compared with that for maximizing net revenue for the static case without carryover:

$$\alpha p^y\, \mathrm{d}y/\mathrm{d}u = p^f$$

The only additional information required to solve the dynamic problem with carryover is the values of b_i and p_{i+1}^f. Thus in order to find the optimal application rate u_i^*, there is no need to know $y_j\{r_j\}$, b_j or p_j^y for $j > i$, nor p_j^f for $j > i+1$.

Note that the condition for optimality (6.7) is the same regardless of how small or how large n may be. The reason for the separability of u_i^* from parameters beyond period $i+1$ is that, whatever the level $x_i + u_i^*$ and hence carryover $b_i(x_i + u_i^*)$ to period $i+1$, the total amount of fertilizer, r_{i+1}^*, to be made available in period $i+1$ is the same.[1] Hence carryover to period $i+2$, equal to $b_{i+1}r_{i+1}^*$, is independent of u_i^*.

So far it has been assumed that carryover to period i is directly proportional to fertilizer available in $i-1$. More complex single-period carryover functions have been suggested. For example, Fuller (1965) has used

$$x_{i+1} = r_i(a + c\exp(hr_i))$$

and Stauber *et al.* (1975)

$$x_{i+1} = qr_i^s$$

where a, c, h, q and s are constants. It is simple to show that for the general case with a concave single-period carryover function

$$x_{i+1} = g_i\{r_i\} \tag{6.8}$$

the optimality condition is (6.7) after replacing b_i with $\partial g_i/\partial u_i$. Because $\partial g_i/\partial u_i$ may be a function of u_i, it may be necessary to find u_i^* numerically.

6.5.2 *Optimal rules for a multiperiod carryover function*

In using a single-period carryover function such as (6.8), it is assumed that carryover x_{i+1} is dependent only on r_i, regardless of the periods in which the fertilizer making up r_i was applied. Bowden and Bennett (1974) and Godden and Helyar (1980) have proposed more complex multiperiod functions in which the proportion of applied fertilizer which is carried over to another period is dependent on the number of periods of carryover, j. If the proportion carried over is denoted by w_j, the available fertilizer in period i is

$$r_i = u_i + \sum_{j=1}^{i-1} w_{i-j}u_j \tag{6.9}$$

and fertilizer carryover to period $i+1$ is

$$x_{i+1} = \sum_{j=1}^{i} w_{i+1-j}u_j \tag{6.10}$$

[1] It must be emphasized that (6.7) holds only for an interior solution, that is, for $u_i^* > 0$. In other words, carryover x_i must be less than r_i^*.

Bowden and Bennett (1974) suggested $w_j = 1/(1+j)$ and Godden and Helyar (1980) $w_j = z/(z+j)$ where z is a constant. In order to incorporate (6.9) and (6.10) in a dynamic programming framework, fertilizer applications in all periods prior to i have to be state variables. Suppose m is large enough that, for practical purposes, w_j for $j > m$ can be taken to be zero. If the objective is to maximize the present value of net returns the recursive functional equation is

$$V_i\{u_{i-m}, \ldots, u_{i-1}\} = \max_{u_i} [\alpha p_i^y y_i\{r_i\} - p_i^f u_i + \alpha V_{i+1}\{u_{i-m+1}, \ldots, u_i\}] \quad (i = n, \ldots, 1) \qquad (6.11)$$

subject to

$$u_i \geq 0$$

$$r_i = u_i + \sum_{j=1}^{m} w_{m-j+1} u_{i+j-m-1}$$

$$x_1 \text{ given}$$

with

$$V_{n+1}\{u_{n-m+1}, \ldots, u_n\} = 0$$

Noting from (6.9) that $\partial r_i/\partial u_i = 1$, the first-order condition for the term in square brackets in (6.11) to be an interior maximum is

$$\alpha p_i^y \, \partial y_i/\partial u_i - p_i^f + \alpha \, \partial V_{i+1}/\partial u_i = 0 \qquad (i = 1, \ldots, n) \qquad (6.12)$$

Again the unknown is $\partial V_{i+1}/\partial u_i$. Differentiating (6.11) with respect to the state variables $u_{i-m}, \ldots, u_{i-1}$, and after some substitution and manipulation, it is found that

$$\partial V_{i+1}/\partial u_i = \sum_{j=1}^{n-i} \alpha^j w_j p_{i+j}^y \, \partial y_{i+j}/\partial u_{i+j} \qquad (i = 1, \ldots, n-1) \quad (6.13)$$

Substituting (6.13) in (6.12) leads to one way of expressing the condition for optimality as

$$\alpha p_i^y \, \partial y_i/\partial u_i + \sum_{j=1}^{n-i} \alpha^{j+1} w_j p_{i+j}^y \, \partial y_{i+j}/\partial u_{i+j} = p_i^f \qquad (i = 1, \ldots, n-1) \qquad (6.14)$$

The condition states that for optimality the price of fertilizer applied in period i must equal the present value of the marginal product of the fertilizer, obtained in periods i through to n. Equation (6.14) is a set of n m-or-lower-order difference equations. Optimal values of $\partial y_i/\partial u_i$ for $i = 1, \ldots, n$ may be found in terms of $w_1, w_2, \ldots, p_i^y, p_i^f, \ldots, p_n^f$ and α by using the matrix method of solving simultaneous equations. The solution is the same as that obtained by Kennedy (1981) through a process of induction, and is more easily expressed as

$$\alpha p_i^y \, \partial y_i/\partial u_i = p_i^f - \sum_{j=1}^{n-i} \alpha^j \beta_j p_{i+j}^f \qquad (i = 1, \ldots, n-1) \qquad (6.15)$$

where $\beta_1 = w_1$ and $\beta_k = w_k - \sum_{l=1}^{k-1} \beta_{k-l} w_l$ for $k > 1$. Because terminal values are zero for all u_n, $\partial f_{n+1}/\partial u_n = 0$, and it follows directly from (6.12) that for the final decision period

$$\alpha p_n^y \, \partial y_n/\partial u_n = p_n^f$$

In this way of expressing the optimality condition u_i^* does not depend on future product prices or response functions. Although the right-hand side of (6.15) looks a lot more complicated than that of (6.7), it can still be rationalized as the opportunity cost of the marginal unit of applied fertilizer. It is first necessary to see that β_k is the fertilizer saved k periods into the future through applying an additional unit now. It equals something less than w_k, because by adding a unit now, less will be added over each of the next $k - 1$ periods if the optimal rule is followed. That is, β_1 is saved next year, β_2 the following year, and so on. The reduction in fertilizer available after k carryover periods is thus $\beta_1 w_{k-1} + \beta_2 w_{k-2} + \cdots$. Hence $\beta_k = w_k - \beta_1 w_{k-1} - \beta_2 w_{k-2} - \cdots$. The right-hand side of (6.15) is therefore the price of fertilizer in period i, less the present value of savings in applied fertilizer over all subsequent periods in the planning horizon resulting from the application of an additional unit in period i.

If crop and fertilizer prices and response functions remain unchanged over all future years, it is simple to find the optimal $\partial y/\partial u$ for an infinite planning horizon. In this stationary case optimal $\partial y/\partial u$ will be the same for all years. It follows from (6.14) that

$$\alpha p^y \, \partial y/\partial u = p^f \Bigg/ \left(1 + \sum_{j=1}^{m} \alpha^j w_j\right)$$

6.5.3 A numerical example

The implications for the levels of applied fertilizer and for net revenue of using different application rules are brought out by considering an example. The optimal rules refer to the situations where: (i) there is zero carryover (R0); (ii) there is carryover which can be described by a single-period carryover function (R1); and (iii) there is carryover which can be described by a multiperiod carryover function (R2). The example is based on one selected by Godden and Helyar (1980). They illustrated the determination of fertilizer applications in the case of a multiperiod carryover function, but used a heuristic algorithm instead of the R2 rule of (6.15).

The problem is to determine optimal levels of phosphate fertilizer to apply each year over six years for growing sorghum in the Northern Territory, Australia. No fertilizer has been applied prior to year 1. Parameters are the same for each year. The sorghum response function is the following Mitscherlich function

$$y_i = 2129(1 - \exp(-0{\cdot}04(u_i + x_i + 8{\cdot}8)))$$

where

y_i is the yield of sorghum in kg/ha;
u_i is the rate of phosphate application per year in kg/ha; and
x_i is phosphate carried over from year $i - 1$ in kg/ha.

It is assumed that 8.8 kg of phosphate are always accessible in the soil whether fertilizer is applied or not.

Available fertilizer from current or past applications is defined as $r_i \equiv u_i + x_i$. The marginal product of available fertilizer is

$$\mathrm{d}y/\mathrm{d}r = (2129 \times 0{\cdot}04)\exp[-0{\cdot}04(r + 8{\cdot}8)] \qquad (6.16)$$

It follows that the r required for any particular $\mathrm{d}y/\mathrm{d}r$ is

$$r = -[\ln(\mathrm{d}y/\mathrm{d}r \div (2129 \times 0{\cdot}04))]/0{\cdot}04 - 8{\cdot}8 \qquad (6.17)$$

The values of the other parameters are:

p^y, the price of sorghum, \$0·08 per kg;
p^f, the price of phosphate fertilizer, \$0·83 per kg;
α, the discount factor, 0·9091, based on a discount rate of 10 per cent per annum; and
w_j, the phosphate available j years after applying 1 kg of phosphate, $0{\cdot}8/(j + 0{\cdot}8)$ kg.

In order to illustrate the application of *R*1, a single-period carryover function

$$x_{i+1} = b(u_i + x_i)$$

was selected with $b = 0{\cdot}570$. The value of b was set such that

$$\sum_{j=1}^{5} b^j = \sum_{k=1}^{5} w_k \tag{6.18}$$

in an attempt to keep the total carryover for the single-period carryover function the same as that for the multiperiod carryover function.

The calculations for optimal fertilizer applications following rules R0, R1 and R2 for the respective carryover processes of (i) zero carryover (C0); (ii) carryover described by a single-period carryover function (C1); and carryover described by a multiperiod carryover function (C2) are dealt with in turn. Table 6.4 shows the results, as well as the outcomes of applying rules inappropriate for the actual carryover process.

(i) Fertilizer applications under rule R0 and carryover process C0
Because none of the parameters change from year to year, the same rate of fertilizer is applied in each of the six years. The required marginal product under R0 is

$$\partial y_i/\partial u_i = \mathrm{d}y_i/\mathrm{d}r_i = p^f/(p^y\alpha) = (0{\cdot}83)(1{\cdot}1)/0{\cdot}08 = 11{\cdot}41$$

Using (6.17) the required rate of fertilizer availability is found to be 41·4 kg of phosphate per ha. Because fertilizer carryover is assumed to be zero, this is also the optimal rate of application in each year.

(ii) Fertilizer applications under rule R1 and carryover process C1
By (6.7), the required marginal product under R1 for $i = 1$ to 5 is

$$\partial y_i/\partial u_i = \mathrm{d}y_i/\mathrm{d}r_i = p^f(1 - \alpha b)/\alpha p^y = (0{\cdot}83)(1{\cdot}1)(1 - 0{\cdot}570/1{\cdot}1)/0{\cdot}08 = 5{\cdot}50$$

From (6.17) the corresponding optimal fertilizer availability for $i = 1$ to 5 is 59·7 kg of phosphate per ha. As there is no fertilizer carryover from year 0, this is the optimal fertilizer application rate for year 1. In years 2 to 6 there is carryover from the previous year of $br = 0{\cdot}570\ (59{\cdot}7) = 34{\cdot}0$. The optimal application rate for years 2 to 5 is therefore $(59{\cdot}7 - 34{\cdot}0) = 25{\cdot}7$ kg of phosphate per ha.

The R1 rule for the final year is the same as the R0 rule for all years.

TABLE 6.4
Application Rates (u), Carryover (x) and Yields (y) (kg/ha) for Alternative Application Rules and Carryover Processes

Year i	u_i^*	No carryover (C0)		Carryover 1 (C1)		Carryover 2 (C2)	
		$x_{0,i}$	$y_i^*\{u_i^* + x_{0,i}\}$	$x_{1,i}$	$y_i^*\{u_i^* + x_{1,i}\}$	$x_{2,i}$	$y_i^*\{u_i^* + x_{2,i}\}$
For u_i^* based on rule R0							
1	41	0	1 844	0	1 844	0	1 844
2	41	0	1 844	24	2 018	18	1 992
3	41	0	1 844	37	2 064	30	2 044
4	41	0	1 844	45	2 081	39	2 069
5	41	0	1 844	49	2 089	46	2 084
6	41	0	1 844	52	2 092	52	2 093
PV of NR[a] ($/ha)		478*		539		535	
For u_i^* based on rule R1							
1	60	0	1 992	0	1 992	0	1 992
2	26	0	1 593	34	1 992	27	1 943
3	26	0	1 593	34	1 992	28	1 957
4	26	0	1 593	34	1 992	31	1 976
5	26	0	1 593	34	1 992	34	1 992
6	7	0	1 016	34	1 884	37	1 872
PV of NR[a] ($/ha)		437		566*		562	
For u_i^* based on rule R2							
1	61	0	1 998	0	1 998	0	1 998
2	33	0	1 732	35	2 030	27	1 995
3	27	0	1 623	39	2 021	32	1 989
4	23	0	1 539	38	1 997	34	1 980
5	19	0	1 433	35	1 955	35	1 959
6	7	0	981	31	1 792	35	1 844
PV of NR[a] ($/ha)		434		565		562*	

[a] Present value of net revenue.

The optimal fertilizer availability for year 6 is therefore 41·4 kg. Because carryover to year 6 is 34·0 kg, the optimal application rate is (41·4 − 34·0) = 7·4 kg of phosphate per ha.

(iii) Fertilizer applications under rule R2 and carryover process C2

By (6.15), for optimality under R2 the marginal product for $i = 1$ to 5 is

$$\partial y_i/\partial u_i = \mathrm{d}y_i/\mathrm{d}r_i = p^f \gamma_i/\alpha p^y \tag{6.19}$$

where $\gamma_i = 1 - \sum_{j=1}^{6-i} \alpha^j \beta_j$. Because γ_i changes from year to year, the optimal

TABLE 6.5
Values of γ_i Given w_j

Number of carryover years, j	w_j	β_j	*Year number, i*	γ_i
5	0·138	0·020	1	0·458
4	0·167	0·028	2	0·471
3	0·211	0·044	3	0·490
2	0·286	0·088	4	0·523
1	0·444	0·444	5	0·596

fertilizer availability also varies. Values of γ_i, and of w_j and β_j on which they depend, are shown in Table 6.5.

The optimal fertilizer availability for year 1, calculated given $\gamma_1 = 0{\cdot}458$ and using (6.19), (6.16) and (6.17), is 61·0 kg of phosphate per ha. Because there is no carryover to year 1, this is also the optimal application rate. Carryover to year 2 is $61w_1 = 61(0{\cdot}444) = 27{\cdot}1$ kg per ha. Optimal fertilizer availability for year 2 is 60·3 kg per ha, which implies an optimal application rate equal to $(60{\cdot}3 - 27{\cdot}1) = 33{\cdot}2$ kg per ha. Optimal application rates for years 3, 4 and 5 are 27·1, 23·3 and 19·1 kg per ha, respectively. Carryover to year 6 is 34·8 kg per ha. Given that optimal availability in year 6 is 41·4 kg per ha, the optimal application rate in year 6 is $(41{\cdot}4 - 34{\cdot}8) = 6{\cdot}6$ kg per ha.

It is clear from Table 6.4 that, in this example, if the carryover process is C1 or C2, the use of rules R1 or R2 leads to *less* fertilizer being applied over six years than would be incorrectly applied under R0 ignoring carryover. Crop yields are consequently lower also.

Two interesting questions arise: (a) how much net revenue is foregone if an incorrect rule is implemented?; and (b) what is the loss if R1 and R2 are implemented when C2 and C1 respectively operate? Answers to these questions may be found in Table 6.4. The present value of net revenue is reduced by about 5 per cent if R0 instead of R1 or R2 is applied when C1 or C2 operates. If R1 or R2 instead of R0 is applied when there is no carryover, the loss in net revenue is about 9 per cent.

Foregone net revenue is insignificant if R1 or R2 is implemented when C2 or C1 operate. Given that R1 is a simpler rule to apply, this finding suggests that if the carryover process is thought to be C2 little will be lost if R1 is implemented instead of R2, using (6.18) to estimate b from w_k.

6.5.4 Extensions

The problems considered so far have been deterministic, but in practice the response and carryover functions are stochastic and prices are uncertain. Suppose the objective is to maximize the present value of expected net returns. Taylor (1983) considers the case of a stochastic response function, a carryover function which is a stochastic version of (6.8), and crop and fertilizer prices which have a first-or-higher-order Markovian structure. He shows that the optimality condition is the same as (6.7), but with stochastic variables and functions set at their expected values. Taylor uses the method originally suggested by Kennedy *et al.* (1973) of first finding the optimality condition for the final period in the planning horizon, then for the penultimate period, and so on. The optimality condition for all periods but the last is found by inductive logic. The result for the stochastic case can be obtained more directly by the method used above for the deterministic case.

For the problem with prices following a first-order Markovian process, the recursive functional equation is

$$V_i\{x_i, p_i^y, p_i^f\} = \max_{u_i} [E(\alpha p_i^y y_i\{r_i, \varepsilon_i^y\} - p_i^f u_i + \alpha V_{i+1}\{g_i\{r_i, \varepsilon_i^x\}, p_{i+1}^y, p_{i+1}^f\})] \qquad (i = n, \ldots, 1) \qquad (6.20)$$

subject to

$$u_i \geq 0$$

$$r_i = x_i + u_i$$

with

$$V_{n+1}\{x_{n+1}, p_{n+1}^y, p_{n+1}^f\} = 0, \quad x_1 \text{ given}$$

where ε_i^y and ε_i^x are random variables affecting yield and carryover, respectively.

Because next period's crop and fertilizer prices depend on the current period's prices, current prices (assumed known) are included in the optimal return function. It is useful to refer to $E(y_i\{r_i, \varepsilon_i^y\})$ as $\bar{y}_i\{r_i\}$, and to $E(g_i\{r_i, \varepsilon_i^x\})$ as $\bar{g}_i\{r_i\}$. Note that $E(\partial y_i/\partial r_i) = \mathrm{d}\bar{y}_i/\mathrm{d}r_i$ and that $\partial r_i/\partial u_i = \partial r_i/\partial x_i = 1$. Again making the usual assumptions about concavity and that $u_i^* > 0$ for all i, the first order condition for a maximum is

$$\alpha p_i^y \, \partial \bar{y}_i/\partial u_i - p_i^f + \alpha E((\partial V_{i+1}/\partial g_i)(\partial g_i/\partial r_i)) = 0 \qquad (i = 1, \ldots, n) \qquad (6.21)$$

Partially differentiating (6.20) with respect to x_i gives

$$\partial V_i/\partial x_i = \alpha p_i^y \, \partial \bar{y}_i/\partial u_i + \alpha E((\partial V_{i+1}/\partial g_i)(\partial g_i/\partial r_i)) \qquad (i = 1, \ldots, \mathrm{n}) \qquad (6.22)$$

From (6.21) and (6.22) it follows that

$$\partial V_i/\partial x_i \equiv \partial V_i/\partial g_{i-1} = p_i^f \qquad (i = 1, \ldots, n) \qquad (6.23)$$

Substituting (6.23) in (6.21) gives

$$\alpha p_i^y \, \partial \bar{y}_i/\partial u_i - p_i^f + \alpha E(p_{i+1}^f(\partial g_i/\partial r_i)) = 0 \qquad (i = 1, \ldots, n-1)$$

Assuming that fertilizer price and carryover are independent, the condition for optimality simplifies to

$$\alpha p_i^y \, \partial \bar{y}_i/\partial u_i = p_i^f - \alpha(E(p_{i+1}^f \mid p_i^f)) \, \partial \bar{g}_i/\partial r_i \qquad (i = 1, \ldots, n-1)$$

which states that the discounted value of the expected marginal product of applied fertilizer should equal the expected opportunity cost of the marginal unit of applied fertilizer. Because the terminal value of carried-over fertilizer is zero, the optimal rate for the final decision period is $\alpha p_n^y \, \partial \bar{y}_n/\partial u_n = p_n^f$. It can be shown that a similar 'certainty-equivalent' version of (6.15) holds if the carryover function is multiperiod.

The fertilizer decision problem may not be just how much fertilizer to apply each year, but also, whether to apply any. Application costs may be high enough to make it more economic to apply fertilizer infrequently at high rates. Stauber *et al.* (1975) have shown that this problem is equivalent to the inventory problem of when and how much to reorder, and therefore has the same type of solution. They use stochastic dynamic programming numerically to find optimal nitrogen fertilization policies for seeded grasses in semiarid regions in the US.

An important study of the fertilizer carryover problem which uses separable linear programming instead of dynamic programming as a solution technique is that by Lanzer and Paris (1981). A deterministic problem of fertilizer application to wheat and soybeans production in Southern Brazil is examined. Agronomic considerations are discussed in detail.

6.6 References

Asopa, V.N., (1971), An application of mathematical decision theory to the economic evaluation of water use in crop production, Unpublished PhD thesis, University of Illinois.

Biere, A.W. and Lee, I.M. (1972), A model for managing reservoir water releases, *American Journal of Agricultural Economics*, **54**(3), 411–21.

Bowden, J.W. and Bennett, D. (1974), The decide model for predicting superphosphate requirements, Proceedings of a symposium, *Phosphate in Agriculture*, Australian Institute of Agricultural Science (Victorian Branch).

Burt, O.R. (1964), The economics of conjunctive use of ground and surface water, *Hilgardia*, **36**(2), 31–111.

Burt, O.R. (1966), Economic control of groundwater reserves, *Journal of Farm Economics*, **48**(3), 632–47.

Burt, O.R. (1971), A dynamic economic model of pasture and range investments, *American Journal of Agricultural Economics*, **53**(2), 197–205.

Burt, O.R. (1981), Farm level economics of soil conservation in the Palouse area of the North west, *American Journal of Agricultural Economics*, **63**(1), 83–92.

Burt, O.R. and Allison, J.R. (1963), Farm management decisions with dynamic programming, *Journal of Farm Economics*, **45**(1), 121–36.

Burt, O.R. and Cummings, R.G. (1977), Natural resource management, the steady state, and approximately optimal decision rules, *Land Economics,* **53**(1), 1–22.

Burt, O.R. and Johnson, R.D. (1967), Strategies for wheat production in the Great Plains, *Journal of Farm Economics*, **49**(4), 881–99.

Burt, O.R. and Stauber, M.S. (1971), Economic analysis of irrigation in subhumid climate, *American Journal of Agricultural Economics*, **53**(1), 33–46.

Carlson, G.A. (1979), Pest control risks in agriculture, in *Risk, Uncertainty and Agricultural Development*, J.A. Roumasset, J.M. Boussard and I. Singh (eds), SEARCA and ADC, Laguna, Philippines and New York.

Conway, G.R., Norton, G.A., Small, N.J. and King, A.B.S. (1975). A systems approach to the control of the sugar cane froghopper, in *Study of Agricultural Systems*, G.E. Dalton (ed.), Applied Science, London.

Corrie, W.J. and Boyce, D.S. (1972), A dynamic programming method to optimize policies for the multistage harvest of crops with an extended maturity period, *Journal of Agricultural Engineering Research*, **17**, 348–54.

de Lucia, R.J. (1969), Operating policies for irrigation systems under stochastic regimes, Unpublished PhD thesis, Harvard University.

Dillon, J.L. (1977), *The Analysis of Response in Crop and Livestock Production*, 2nd edition, Pergamon Press, Oxford.

Dudley, N.J. (1972), Irrigation planning 4: optimal interseasonal water allocation, *Water Resources Research*, **8**(3), 586–94.

Dudley, N.J. and Burt, O.R. (1973), Stochastic reservoir management and system design for irrigation, *Water Resources Research*, **9**(3), 507–22.

Dudley, N.J., Howell, D.T. and Musgrave, W.F. (1971*a*), Optimal intraseasonal irrigation water allocation, *Water Resources Research*, **7**(4), 770–88.

Dudley, N.J., Howell, D.T. and Musgrave, W.F. (1971*b*), Irrigation planning 2: choosing optimal acreages within a season, *Water Resources Research*, **7**(5), 1051–63.

Dudley, N.J., Howell, D.T. and Musgrave, W.F. (1972), Irrigation planning 3: the best size of irrigation area for a reservoir, *Water Resources Research*, **8**(1), 7–17.

Dudley, N.J., Reklis, D.M. and Burt, O.R. (1976), Reliability, trade-offs, and water resources development modeling with multiple crops, *Water Resources Research*, **12**(6), 1101–8.

Feder, G. (1979), Pesticides, information, and pest management under uncertainty, *American Journal of Agricultural Economics*, **61**(1), 97–103.

Fisher, B.S. and Lee, R.R. (1981), A dynamic programming approach to the economic control of weed and disease infestations in wheat, *Review of Marketing and Agricultural Economics*, **49**(3), 175–87.

Flinn, J.C. and Musgrave, W.F. (1967), Development and analysis of input–output relations for irrigation water, *Australian Journal of Agricultural Economics*, **11**(1), 1–19.

Fuller, W.A. (1965), Stochastic fertilizer production functions for continuous corn, *Journal of Farm Economics*, **47**, 105–19.

Godden, D. P. and Helyar, K. R. (1980), An alternative method for deriving optimal fertilizer rates, *Review of Marketing and Agricultural Economics*, **48**(2), 83–97.

Hall, D.C. and Norgaard, R.B. (1973), On the timing and application of pesticides, *American Journal of Agricultural Economics*, **55**(2), 198–201.

Hall, W.A. and Butcher, W.S. (1968), Optimal timing of irrigation, *Journal of Irrigation and Drainage Division of the American Society of Civil Engineers*, **94**, 267–75.

Hall, W.A., Butcher, W.S. and Esogbue, A. (1968), Optimization of the operation of a multiple-purpose reservoir by dynamic programming, *Water Resources Research*, **4**(3), 471–7.

Headley, J.C. (1972), Defining the economic threshold, in *Pest Control Strategies for the Future*, National Academy of Sciences, Washington DC.

Hueth, D. and Regev, U. (1974), Optimal agricultural pest management with increasing pest resistance, *American Journal of Agricultural Economics*, **56**(3), 543–52.

Iskakov, B.I. and Finkelstejn, J.J. (1973), Application of dynamic programming planning to crop rotation, *Ekonomika i Matematiceskie Metody*, **9**(4), 734–8.

Kennedy, J.O.S. (1981), An alternative method for deriving optimal fertilizer rates: comment and extension, *Review of Marketing and Agricultural Economics*, **49**(3), 203–9.

Kennedy, J.O.S., Whan, I.F., Jackson, R. and Dillon, J.L. (1973), Optimal fertilizer carryover and crop recycling policies for a tropical grain crop, *Australian Journal of Agricultural Economics*, **17**(2), 104–13.

Knapp, K.C. (1984), Steady-state solutions to soil salinity optimization problems, *American Journal of Agricultural Economics*, **66**(3), 279–85.

Lanzer, E.A. and Paris, Q. (1981), A new analytical framework for the fertilization problem, *American Journal of Agricultural Economics*, **63**(1), 93–103.

Lazarus, W.F. and Dixon, B.L. (1984), Agricultural pests as common property: control of the corn rootworm, *American Journal of Agricultural Economics*, **66**(4), 456–65.

Lazarus, W.F. and Swanson, E.F. (1983), Insecticide use and crop rotation under risk: rootworm control in corn, *American Journal of Agricultural Economics*, **65**(4), 738–47.

Mann, S.H. (1971), Mathematical models for the control of pest populations, *Biometrics*, **27**(2), 357–68.

Matanga, G.B. and Marino, M.A. (1979), Irrigation planning 1: cropping pattern, *Water Resources Research*, **15**(3), 672–8.

Matanga, G.B. and Miguel, A.M. (1979), Irrigation planning 2: water allocation for leaching and irrigation purposes, *Water Resources Research*, **15**(3), 679–83.

McFarland, J.W. (1975), Groundwater management and salinity control: a case study in Northwest Mexico, *American Journal of Agricultural Economics*, **57**(3), 456–62.

O'Loughlin, G.G. (1971), Optimal reservoir operation, Unpublished PhD thesis, University of New South Wales.

Raphael, B.C. (1979), Economics of cotton–pest interactions: a dynamic programming approach, *Proceedings of the Fourth Annual Conference of the Australian Society for Operations Research*, Melbourne, August, 188–203.

Regev, U., Shalit, H. and Gutierrez, A.P. (1979), Economic conflicts in plant protection: the problems of pesticide resistance, theory and application to the Egyptian Alfalfa weevil, in *Pest Management*, G.A. Norton and C.S. Holling (eds), Pergamon Press, Oxford.

Regev, U., Shalit, H. and Gutierrez, A.P. (1983), On the optimal allocation of pesticides with increasing resistance: the case of Alfalfa Weevil, *Journal of Environmental Economics and Management*, **10**, 86–100.

Shoemaker, C. (1973*a*), Optimization of agricultural pest management I: biological and mathematical background, *Mathematical Biosciences*, **17**, 143–75.

Shoemaker, C. (1973*b*), Optimization of agricultural pest management II: formulation of a control model, *Mathematical Biosciences*, **17**, 357–65.

Shoemaker, C. (1973*c*), Optimization of agricultural pest management III: results and extensions of a model, *Mathematical Biosciences*, **18**, 1–22.

Shoemaker, C.A. (1979*a*), Optimal management of an alfalfa ecosystem, in *Pest Management*, G.A. Norton and C.S. Holling (eds), Pergamon Press, Oxford.

Shoemaker, C.A. (1979*b*), Optimal timing of multiple applications of pesticide with residual toxicity, *Biometrics*, **35**, 803–12.

Shoemaker, C.A. (1981), Applications of dynamic programming and other optimization methods in pest management, *Institute of Electrical and Electronic Engineers Transactions on Automatic Control*, **26**(5), 1125–32.

Shoemaker, C.A. (1982), Optimal integrated control of univoltine pest populations with age structure, *Operations Research*, **30**(1), 40–61.

Stauber, M.S., Burt, O.R. and Linse, F. (1975), An economic evaluation of nitrogen fertilization of grass where carry-over is significant, *American Journal of Agricultural Economics*, **57**(3), 463–71.

Talpaz, H. and Borosh, I. (1974), Strategy for pesticide use: frequency and applications, *American Journal of Agricultural Economics*, **56**(4), 769–75.

Talpaz, H., Curry, G.L., Sharpe, P.J., De-Michele, D.W. and Frisbie, R.E. (1978), Optimal pesticide application for controlling the boll weevil on cotton, *American Journal of Agricultural Economics*, **60**(3), 469–75.

Taylor, C. R. (1983), Certainty equivalence for determination of optimal fertilizer application rates with carry-over, *Western Journal of Agricultural Economics*, **8**(1), 64–7.

Taylor, C.R. and Burt, O.R. (1984), Near-optimal management strategies for controlling wild oats in spring wheat, *American Journal of Agricultural Economics*, **66**(1), 50–60.

Taylor, R.C. and Headley, J.C. (1975), Insecticide resistance and the evaluation of control strategies for an insect population, *Canadian Entomologist*, **107**(3), 237–42.

Throsby, C.D. (1964*a*), Some dynamic programming models for farm management research, *Journal of Agricultural Economics*, **16**(1), 98–110.

Throsby, C.D. (1964*b*), Theoretical aspects of a dynamic programming model for studying the allocation of land to pasture improvement, *Review of Marketing and Agricultural Economics*, **32**(3), 149–181.

Watt, K.E.F. (1963), Dynamic programming, 'look ahead programming', and the strategy of insect pest control, *Canadian Entomologist*, **95**, 525–636.

Watt, K.E.F. (1968), *Ecology and Resource Management: A Quantitative Approach*, McGraw-Hill, New York.

Yaron, D. and Dinar, A. (1982), Optimal allocation of farm irrigation water during peak seasons, *American Journal of Agricultural Economics*, **64**(4), 681–89.

Yaron, D. and Olian, A. (1973), Application of dynamic programming in Markov chains to the evaluation of water quality in irrigation, *American Journal of Agricultural Economics*, **55**(3), 467–71.

Yaron, D. and Voet, H. (1983), Application of an integrated dynamic and linear programming model to the analysis of optimal irrigation on a farm with dual quality (salinity) water supply, in *Planning and Decision in Agribusiness: Principles and Experiences*, C-H. Hanf and G.W. Schiefer (eds), Elsevier Scientific, Amsterdam.

Yaron, D., Bresler, E., Bielorai, H. and Harpinist, B. (1980), A model for optimal irrigation scheduling with saline water, *Water Resources Research*, **16**(2), 257–62.

Chapter 7

LIVESTOCK MANAGEMENT

7.1 Livestock Decision Problems

Livestock units are managed either for the produce extracted from them over their lives (such as milk, eggs and wool) or for the carcase meat harvested at the end of their lives (such as beef, mutton, pork and chicken), or both. A necessary maintenance input that has to be managed is feed. Decisions have to be made on the quality, quantity and timing of the feed input. The product return to feed input changes continuously over the life of the unit. Typically, productivity first increases and then declines with livestock age. If the livestock enterprise is to be a continuing one, a decision must be made on when to replace the aging unit with a younger unit.

It follows that livestock producers face the twin problems of finding optimal feeding regimes and optimal replacement schedules. The optimal replacement decision has been a major field of application of dynamic programming. Many applications have been industrial, but the same scope for dynamic programming applications exists in livestock management. Applications to the replacement of livestock are discussed in the next section. The importance of the other major decision, the feeding decision, depends on the type of livestock system being managed. If the livestock obtain their feed by grazing pasture, the feed decisions are more strategic than tactical. Decisions have to be made on stocking densities and pasture maintenance, but thereafter much is left to the natural interaction of weather, plant-growth and animal systems. If the interaction is variable and unpredictable, management intervention in the form of supplementary feeding or livestock disposal may be required in times of stress, such as during a drought.

If livestock are managed intensively as in a feedlot, the uncertainty of production may be reduced, but the feed input decision assumes much greater importance for two reasons. In an intensive system feed inputs can be used to control the quality and quantity of livestock output, and hence gross returns. The other reason is that feed costs are a major item of expenditure in an intensive system. In the case of intensive feeding for fattening, the techniques of marginal analysis and linear programming have been used to determine least-cost rations for livestock of particular liveweight and for particular rates of liveweight gain. However, the problem of finding the optimal sequence of feed input over the life of the livestock unit cannot be separated from those of finding the optimal sequence of liveweight gains, and the optimal replacement policy. In Sections 7.3 and 7.4 it is shown how dynamic programming can be used to find simultaneously both the optimal replacement pattern and the optimal sequence of liveweight gains, making use of least-cost rations.

7.2 Livestock Replacement Problems

One of the first published suggestions for applying dynamic programming to solving on-farm decision problems was made by White (1959). He illustrated the potential with an application to decisions on the replacement of flocks of laying hens, and drew attention to the scope of the technique for application to other types of livestock. Throsby (1964) refers to White's work when he describes how dynamic programming can be applied to the decision to cull livestock in an early expository article. He states: 'A promising avenue of application of dynamic programming in agricultural economics is its use in determining optimal replacement policies for continually operating livestock enterprises such as are found in beef, milk, egg and broiler processes.' Some of the published work done in the area since this statement was made is documented in this section. Summaries of applications to laying hens, grazing animals and dairy cows appear in Tables 7.1 to 7.3. Some of the different types of replacement problem are discussed in Section 7.2.1.

White's work inspired many applications of dynamic programming to the dairy cow replacement problem. Some of these are considered in Section 7.2.2. An adaptive application by McArthur (1973) to milk production, which could also be applied to other livestock enterprises, is discussed in Section 7.2.3.

TABLE 7.1

Dynamic Programming Applications to the Management of Laying Hens

Reference	*Interval between stages*	*Number of stages*	*State(s)*	*Decision(s)*	*Objective function maximized*[a]	*Comments*
White (1959); Halter and White (1962)	1 month	120	Age of flock	Keep or replace	NR	One of the earliest applications of DP to agriculture
Brookhouse and Low (1967)	4 weeks	13 and 130	Age of flock (0–64 weeks)	Keep or replace	PV of NR	Extension considered to deal with one rearing unit and more than one flock
Pouliquen (1970)	Depends on decision	∞	Time from start to arrival of current batch	Time from start date to arrival of previous batch	NR	Solved by forward recursion; egg prices seasonal and cyclical; profitability of forced moulting investigated

[a] 'PV of NR' stands for 'Present value of net revenue'.

TABLE 7.2
Summary of Dynamic Programming Applications to the Management of Grazing Animals

Reference	*Interval between stages*	*Number of stages*	*State(s)*[a]	*Decision(s)*	*Objective function maximized*[b]	*Comments*
Scobie (1967)	1 year	∞	Age of sheep	Keep or replace with choice of age	PV of NR	Applied to wether flocks
Fisher (1974)	1 month	∞	(i) Vegetation per paddock (stoch.) (ii) Number of sheep per paddock (stoch.)	Buy or sell sheep	ENR	State increment DP used for extending states to two types of vegetation and three age classes of sheep; transition probabilities obtained from simulation of grazing
Toft and O'Hanlon (1979)	1 month	18	(i) Drought continues or finishes (ii) Number of stock previously sold (iii) Number of stock previously agisted	(i) Number agisted (ii) Number sold	Minus EC	A drought tactic model for sheep or cattle; subjective probability distribution of drought length elicited from the grazier; allowance for an upper limit on cash outlay
Karp and Pope (1984)	1 year	∞	Range condition (stoch.)	(i) Range treated or not (ii) Stocking rate	PV of ENR	DP solution obtained by LP; utility maximization discussed; experiments with alternative stochastic specifications reported

[a] Stochastic variables are denoted by '(stoch.)'.
[b] 'PV' stands for 'Present value'; 'ENR' stands for 'Expected net revenue'; 'EC' stands for 'Expected cost'.

TABLE 7.3
Summary of Dynamic Programming Applications to Dairy Production

Reference	Interval between stages	Number of stages	State(s)[a]	Decision(s)	Objective function maximized[b]	Comments
Jenkins and Halter (1963)	1 year	∞ and 12	Lactation number (12) (stoch.)	Keep or replace with cows in selected lactation number	ENR	Results given for (i) constant prices of cows, feed and milk, and (ii) 1950–1961 actual prices
Giaever (1966)	Variable; zero to 21 months (stoch.)	∞	(i) Lactation number (5) (stoch.) (ii) Production history (7) (stoch.) (iii) Expected length of current calving interval (3) (stoch.)	Keep or replace	PV of ENR	Stage length is stochastic and depends on state and transition
Redman and Kuo (1969)	1 year	10	Lactation number (7) (stoch.)	Keep or replace with cows in selected lactation number	PV of ENR	Applied to Holstein cows of three possible production levels; extensive sensitivity analysis
Smith (1971, 1973)	1 year	15	(i) Lactation number (6) (stoch.) (ii) Previous calving intervals (3) (iii) Production levels for each of the two immediately prior lactations (29)	Keep or replace	PV of ENR	Applied to Jersey cows; extensive sensitivity analysis
McArthur (1973)	1 year	∞	(i) Number of lactation records (7) (stoch.) (ii) Production records averaged over all previous lactations	Keep or replace	PV of ENR	An example of adaptive DP; DP policy evaluated against a Monte Carlo generation of lactation records with normal benchmark culling rules; DP policy showed no appreciable benefit over benchmark rules
Stewart *et al.* (1977)	1 year	10	(i) Lactation number (7) (stoch.) (ii) 305-day milk yield (11) (iii) 305-day milk fat % (7) (iv) Body weight (5)	Keep or replace	PV of ENR	Applied to Holstein cows; extensive sensitivity analysis

[b] 'PV of ENR' stands for 'Present value of expected net revenue'.

7.2.1 Types of problem

One type of problem arises in grazing situations in which the amount of feed available is subject to uncertainty and cannot be carried over to subsequent periods. The decision is the change in livestock numbers. A second type of problem arises when any amount of feed is available but at a cost. In this case, the optimal scale of the enterprise in terms of livestock numbers is not so much a decision-making concern as the optimal age or weight of replacement.

Examples of applications to the first type of problem are those by Fisher (1974), Toft and O'Hanlon (1979), and Karp and Pope (1984) to grazing situations (see Table 7.2). Fisher, and Karp and Pope, include a measure of feed availability as a stochastic state variable. They consider the problem of determining optimal stocking rates over an infinite time horizon. Toft and O'Hanlon develop a model for determining the optimal number of livestock to sell, agist or feed, over the limited period of a drought. The maximum length of a drought is taken to be 18 months. At each decision stage, which is the start of each month, an estimate of the probability of the drought continuing beyond the end of the month is required. Toft and O'Hanlon derive the probability from the grazier's subjective probability distribution of drought length. At the start of the i-th month of a drought it equals

$$\sum_{j=i+1}^{18} p\{j\} \Big/ \sum_{k=i}^{18} p\{k\}$$

where $p\{l\}$ is the grazier's estimate of the probability of a drought lasting between l and $l+1$ months. Henderson and Toft (1979) report the results of field testing the model.

Most dynamic programming applications to livestock replacement have been of the second type where the problem is one of how long to keep a batch of producing animals before replacing them. Although in most of the studies summarized in Tables 7.1 to 7.3 livestock age is a state variable, and a decision to keep or replace is made at frequent regular intervals, this is not the only way of formulating this type of problem. An alternative is to regard the time of introduction of a new batch of animals as a decision stage. The decision variable is the length of time for which the new batch is to be kept. The state at stage i is represented by the time available from the introduction of the i-th batch to the completion of the n-th batch if backward recursion is used. If forward recursion is used the state is the time span from the start of the first batch to the completion of the i-th batch. The approach requires less computation if the 'replace' option is restricted to a subset of livestock ages, such as those ages by which declining productivity has set in. Pouliquen (1970) has applied this approach using forward recursion to egg production.

Many of the applications in which the decision to be made at each stage is either to keep or replace have been to dairy cows. Applications to dairy cows are considered next.

7.2.2 Applications to dairy cows

A dairy cow is managed for up to about eight lactations for the calves and milk it produces. Important management inputs are feed, labour and veterinary supplies. Given that, typically, milk production in each lactation initially increases from the first lactation, but at a decreasing rate and eventually declines, the question arises as to what is the optimal number of lactations to keep a cow before replacing it. The answer depends on the current and future prices of feeds, milk, culled cows and replacement heifers. Another management problem which should ideally be solved with the replacement problem is the optimal feeding regime. To date dynamic programming applications have been concentrated on only the replacement problem, and it is this problem that is discussed in this subsection. However, the next logical step in the development of dynamic programming to problems in dairy production is to the integrated feed and replacement problem with liveweight a state variable. A start has been made with an application to a beef cow–calf feeding system by Topham (1979), mentioned in Section 7.4.4.

The simplest formulation of the replacement problem has one state variable, lactation number (l_i) at stage i, and the decision option to keep the cow for at least one more lactation, or replace the cow with a heifer about to start its first lactation. The decision stage is the start of each lactation. Net returns over the lactation $a_i\{l_i\}$ depend on the cost of feeds, the price of milk and the price of calves. If the decision is to keep the cow, and the lactation is successful, the state at stage $i+1$ is (l_i+1). If the decision is to replace the cow, and the lactation of the replacement cow is successful, the state at stage $i+1$ is 2. The return from the sale of the culled cow is denoted by $b_i\{l_i\}$, and the cost of the replacement heifer by c_i.

The lactation of the cow may be unsuccessful either because of failure through low yield or a disease problem, or because of the death of the cow. If the lactation is unsuccessful, replacement is forced. In the case of forced replacement through failure, which has a probability of $p^f\{l_i\}$, it is assumed that the stage net return is still $a_i\{l_i\}$. In the case of forced replacement through death, which has a probability of $p^d\{l_i\}$, it is assumed that the stage net return is also $a_i\{l_i\}$ but that there is no return from the carcase of the cow. The probability of a successful transition from lactation l_i to l_i+1 is therefore $(1-p^f\{l_i\}-p^d\{l_i\})$ denoted by $p^s\{l_i\}$. The recursive equation for maximization of the present value of expected net revenue is

$$V_i\{l_i\}=\max\,[K_i\{l_i\},R_i\{l_i\}] \qquad (i=n,\ldots,1) \tag{7.1a}$$

with

$$V_{n+1}\{l_{n+1}\}=b_{n+1}\{l_{n+1}\} \tag{7.1b}$$

where

$$\begin{aligned}K_i\{l_i\}=a_i\{l_i\}&+\alpha[p^s\{l_i\}V_{i+1}\{l_i+1\}+(p^f\{l_i\}+p^d\{l_i\})\\&\times(V_{i+1}\{1\}-c_{i+1})+p^f\{l_i\}b_{i+1}\{l_i+1\}]\end{aligned} \tag{7.1c}$$

for the decision to keep the cow for at least one more lactation; and

$$\begin{aligned}R_i\{l_i\}=b_i\{l_i\}-c_i+a_i\{1\}&+\alpha[p^s(1)V_{i+1}\{2\}+(p^f\{1\}+p^d\{1\})\\&\times(V_{i+1}\{1\}-c_{i+1})+p^f\{1\}b_{i+1}\{2\}]\end{aligned} \tag{7.1d}$$

for the decision to replace the cow with another about to start its first lactation.

The problem can be reformulated to allow the lactation number of the replacement cow to be chosen. The cost of the replacement cow now depends on its lactation, and is denoted by $c_i\{l_i\}$. The problem is formulated with (7.1a and b) as before but with (7.1c and d) replaced by

$$\begin{aligned}K_i\{l_i\}=a_i\{l_i\}&+\alpha[p^s\{l_i\}V_{i+1}\{l_i+1\}+p^f\{l_i\}b_{i+1}\{l_i+1\}\\&+\max_{j_{i+1}}(V_{i+1}\{j_{i+1}\}-c_{i+1}\{j_{i+1}\})(p^f\{l_i\}+p^d\{l_i\})]\end{aligned} \tag{7.1c$'$}$$

where j_{i+1} is the lactation number of a forced replacement at stage $i+1$; and

$$R_i\{l_i\}=\max_{m_i}\,[b_i\{l_i\}-c_i\{m_i\}+K_i\{m_i\}] \tag{7.1d$'$}$$

where m_i is the lactation number of a voluntary replacement at stage i. Because $K_i\{m_i\}$ appears in (7.1d′), $K_i\{l_i\}$ in (7.1c′) must be found for all l_i before selecting the optimal action according to (7.1a).

Some reported applications of dynamic programming to the replacement of dairy cows are summarized in Table 7.3. Problems allowing for the choice of lactation number of the replacement cow have been solved by Jenkins and Halter (1963), and Redman and Kuo (1969). State variables additional to lactation number which have been included in some studies are length of calving interval, body weight, and perhaps most importantly, previous production performance.

Whilst the milk yield of an individual cow changes from lactation to lactation, there is considerable variation in yield between cows with the same lactation number. Individual yield potential should be a determinant of whether or not to keep a cow for continued milk production. To allow for variations in butterfat content of milk and calving intervals a cow's production performance is often measured by the weight of milk, standardized for butterfat content, produced in the first 305 days of a lactation. Smith (1971, 1973) has state variables for production in the previous two lactations, each one represented by 29 discrete values. A linear predicted-production equation was used in which the dependent variables were lactation number, calving interval and production in the previous two lactations.

However, because milk yields are affected by random factors such as health status and environmental conditions which change between lactations, even performance in the previous two lactations may not be an accurate indication of yield potential. It may be desirable to apply dynamic programming within an adaptive framework allowing the parameters of the probability distribution of yield potential to be re-evaluated at each decision stage along the lines discussed in Section 4.2.

7.2.3 Periodic revision of estimated yield potential

McArthur (1973) develops a dairy cow replacement model with previous production performance and lactation number as state variables. It is interesting as an example of adaptive dynamic programming. Previous production performance at stage i is measured by average production over all lactations prior to i and equals

$$\bar{q}_i = \sum_{j=1}^{l_i - 1} q_j/(l_i - 1)$$

where q_j is yield recorded for the j-th lactation. The yield potential of a cow is re-evaluated after each lactation drawing on the concept of repeatability used in the study of the genetics of livestock. Repeatability (ρ) takes a value between 0 and 1, and indicates the likelihood that a livestock unit will repeat an above or below individual average performance relative to the herd average performance (Lasley, 1978, Section 7.5). It is the fraction of the total variance amongst all individual performances in the herd which is due to permanent differences between individuals (Lush, 1945, p. 175). The repeatability factor increases with the number of lactation records, $l_i - 1$ at stage i, and equals

$$r\{l_i\} = (l_i - 1)\rho/(1 + (l_i - 2)\rho)$$

where ρ is the repeatability factor given one previous record. That is, $\rho = r\{1\}$. The repeatability factor for a range of $l_i - 1$ and ρ is shown in Table 7.4.

TABLE 7.4
Repeatability by Number of Lactation Records

Number of lactation records $l_i - 1$	*Repeatability factor* $r\{l_i\}$				
	$\rho = 1{\cdot}0$	$\rho = 0{\cdot}4$	$\rho = 0{\cdot}5$	$\rho = 0{\cdot}6$	$\rho = 0{\cdot}0$
1	1·00	0·40	0·50	0·60	0·00
2	1·00	0·57	0·67	0·75	0·00
3	1·00	0·67	0·75	0·82	0·00
4	1·00	0·73	0·80	0·86	0·00
5	1·00	0·77	0·83	0·88	0·00
6	1·00	0·80	0·86	0·90	0·00

Repeatability (ρ) is specific to animal type, character trait and production environment. McArthur set $\rho = 0{\cdot}6$ for milk yield in his application to Jersey cows in New Zealand. The repeatability factor is applied to the deviation of previous production performance from the herd average performance, $\bar{Q}$. The expected production of an individual cow at stage i starting its l_i-th lactation, and averaging $\bar{q}_i$ over previous lactations is

$$\hat{q}_i\{l_i, \bar{q}_i\} = r\{l_i\}(\bar{q}_i - \bar{Q}) + \bar{Q}$$

Note that $\hat{q}_i$ varies between $\bar{Q}$ for $\rho = 0$ and $\bar{q}_i$ for $\rho = 1$. Table 7.4 shows to what extent the accuracy of the estimate of yield potential made at stage i will be improved at stage $i + 1$ once a revised individual average performance, $\bar{q}_{i+1}$, has been calculated.

It is reasonable to assume that q_i is normally distributed with mean $\hat{q}_i$ and standard deviation dependent on the herd standard deviation of milk yields per lactation and $r\{l_i\}$. The assumption makes it possible to find the conditional probability that the k-th milk yield will be recorded in lactation i, $p\{q_i^k \mid l_i, \bar{q}_i\}$ (see Appendix Sections 7.A.3 and 7.A.4 for a rationale). An adaptive process can be set up whereby the conditional probability of all possible milk yields is revised from lactation to lactation in the light of historical milk yields. The recursive equation for the adaptive problem is obtained by modifying (7.1a to d) to include production performance as a state variable and conditional probabilities as follows:

$$V_i\{l_i, \bar{q}_i\} = \max\,[K_i'\{l_i, \bar{q}_i\}, R_i'\{l_i, \bar{q}_i\}] \qquad (i = n, \ldots, 1)$$

with

$$V_{n+1}\{l_{n+1}, \bar{q}_{n+1}\} = b_{n+1}\{l_{n+1}\}$$

where

$$K_i'\{l_i, \bar{q}_i\} = \sum_{k=1}^{m} p\{q_i^k \mid l_i, \bar{q}_i\}[a_i'\{l_i, q^k\} + \alpha p^s\{l_i\}V_{i+1}\{l_i + 1, \bar{q}_{i+1}^k\}]$$
$$+ \alpha[(p^f\{l_i\} + p^d\{l_i\})(V_{i+1}\{1,0\} - c_{i+1}) + p^f\{l_i\}b_{i+1}\{l_i + 1\}]$$

$$R_i'\{l_i, \bar{q}_i\} = \sum_{k=1}^{m} p\{q_i^k \mid 1,0\}[a_i'\{1,0\} + \alpha p^s\{1\}V_{i+1}\{2, \bar{q}_{i+1}^k\}]$$
$$+ \alpha[(p^f\{1\} + p^d\{1\})(V_{i+1}\{1,0\} - c_{i+1}) + p^f\{1\}b_{i+1}\{2\}]$$

with $\bar{q}_{i+1}^k = ((l_i - 1)\bar{q}_i + q_i^k)/l_i$ and m equal to the number of possible production levels.

Further background on the concept of repeatability and a Bayesian rationale for its use are given in Appendix 7.A.

7.3 Combined Feeding and Replacement Decisions

Some applications of dynamic programming to the optimal management of the weight gains of beef cattle, pigs and broilers are summarized in Tables 7.5 and 7.6. Bonnieux (1969), Nelson (1969), Meyer and Newett (1970) and Kennedy (1972) first formulated the problem of finding the optimal sequence of weight gains as a dynamic programming problem apparently independently. In this section an example problem is introduced and solved using dynamic programming. Extensions of the basic problem are discussed and related to reported applications in Section 7.4.

TABLE 7.5
Summary of Dynamic Programming Applications to Beef Production

Reference	*Interval between stages*	*Number of stages*	*State(s)*[a]	*Decision(s)*	*Objective function maximized*[b]	*Comments*
Bonnieux (1969, 1972)	14 days	Variable	Number of animals in each weight category (stoch.)	One ration from a range of alternatives	ENR	Management of a veal calf fattening operation; liveweight transition stochastic
Nelson (1969); Nelson and Eisgruber (1970)	1 month	36	By pen: Grade and average liveweight (180–540 kg)	By pen: (i) Sales (ii) Purchases—grade and average liveweight (iii) Weight gain	PV of NR	A trading constraint determines numbers in each pen; Bayesian updating of cattle prices; feed costs from response surface fitted to LP solutions
Meyer and Newett (1970)	Depends on decision	3	(i) Liveweight (230–360 kg) (ii) Time since start	(i) Weight gain (ii) Interval between stages	NR	Choice of 17 least-cost rations determined by LP; model adaptable to feeding operations for any type of animal
Kennedy (1972, 1973, 1983)	28 days	13–26	Liveweight (100–500 kg)	(i) Weight gain (ii) Replacement liveweight	PV of NR	LP least-cost feed rations; feedlot and/or grazing; periodic re-evaluation of gain potential
Clark and Kumar (1978)	3 months	8	Liveweight (234–639 kg)	Graze, sell and replace	NR	Different cattle breeds
Topham (1979)	Irregular	6 over 1 year	Liveweight of calves (37–333 kg) and cows	(i) Calf liveweight gain (ii) Beef cow ration to produce milk for calf	NR	Beef cow–calf feeding system; feeds are homegrown and purchased grains, pasture and hay, and milk; LP least-cost feed rations
Glen (1980*a*)	4 days	17	Liveweights (100–500 kg)	Liveweight gain	Minus total feeding costs	Linearization and parameterization used to obtain least-cost feed rations efficiently by LP (see also Glen, 1980*b*)
Yager *et al.* (1980)	30 days	12	(i) Liveweight (350–600 kg) (stoch.) (ii) Cow price (stoch.)	(i) Liveweight gain (ii) Sell	PV of ENR	Least-cost feed rations; feed requirements depend on time of year

[a] Stochastic variables are denoted by '(stoch.)'.
[b] 'PV of ENR' stands for 'Present value of expected net revenue'.

TABLE 7.6
Summary of Dynamic Programming Applications to Intensive Pig and Broiler Production

Reference	*Interval between stages*	*Number of stages*	*State(s)*	*Decision(s)*	*Objective function maximized*[a]	*Comments*
Glen (1983)	5 days	17	(i) Liveweight (20–100 kg) (ii) Carcase composition	(i) Weight gain (ii) Protein content of liveweight gain	Minus total feeding costs	Pig production; extension to maximizing return per animal per period considered; LP least-cost feed rations
Hochman and Lee (1972); Hochman (1973); Rausser and Hochman (1979)	1 week	52	Age of flock (7–12 weeks)	Price below which keep on *ad libitum* ration instead of replacing flock	ENR	Broiler production; seasonal variation in broiler prices allowed for—with prices (i) normally and (ii) uniformly distributed; returns from meat sales adjusted for meat quality
Kennedy *et al.* (1976)	1 week	∞	(i) Bird weight per 100 birds surviving (ages 0–12 weeks) (ii) Age	(i) Liveweight gain (ii) Replace	PV of NR	Broiler production; feed costs from linear interpolation of LP solutions; Lagrange multiplier used for also maximizing NR per unit of floor space

[a] ‘PV’ stands for ‘Present value’; ‘ENR’ stands for ‘Expected net revenue’.

7.3.1 The optimal ration sequence: an example

Consider a simple example problem of how best to fatten a steer over time which initially weighs 300 kg. Suppose that a ration which will result in one of six daily weight gains has to be chosen at each decision stage. The six possible daily gains range from 0·25 to 1·50 kg, at intervals of 0·25 kg. The time between decision stages is 28 days. Dependent on the ration decision made at the first stage, the steer will weigh 307, 314, 321, 328, 335 or 342 kg at the second stage. By stage 4, after three selected rations have been fed, the steer will have a minimum possible weight of $300 + (3 \times 7) = 321$ kg, and a maximum possible weight of $300 + (3 \times 42) = 426$ kg. Other possible weights lie between 321 and 426 kg at intervals of 7 kg. The number of possible liveweight paths across stages is $6 \times 6 \times 6 = 216$. Suppose steers at stage 4 fetch \$3 per kilogram liveweight, and ration costs by liveweight and liveweight gain are as

TABLE 7.7
Ration Costs (\$/day) by Liveweight and Liveweight Gain

Liveweight (x) (kg)	*Liveweight gain (u) (kg/day)*					
	0·25	*0·50*	*0·75*	*1·00*	*1·25*	*1·50*
300	0·45	0·85	1·38	2·03	2·81	3·71
307	0·46	0·86	1·40	2·05	2·83	3·74
314	0·47	0·88	1·41	2·08	2·86	3·78
321	0·48	0·89	1·43	2·10	2·89	3·81
328	0·48	0·91	1·45	2·12	2·92	3·84
335	0·49	0·92	1·47	2·15	2·95	3·88
342	0·50	0·93	1·49	2·17	2·98	3·91
349	0·51	0·95	1·51	2·20	3·01	3·95
356	0·52	0·96	1·53	2·22	3·04	3·98
363	0·53	0·98	1·55	2·24	3·07	4·01
370	0·54	0·99	1·57	2·27	3·10	4·05
377	0·55	1·00	1·59	2·29	3·12	4·08
384	0·56	1·02	1·60	2·32	3·15	4·12
391	0·57	1·03	1·62	2·34	3·18	4·15
398	0·57	1·05	1·64	2·36	3·21	4·18
405	0·58	1·06	1·66	2·39	3·24	4·22
412	0·59	1·07	1·68	2·41	3·27	4·25
419	0·60	1·09	1·70	2·44	3·30	4·29
426	0·61	1·10	1·72	2·46	3·33	4·32
433	0·62	1·12	1·74	2·48	3·36	4·35
440	0·63	1·13	1·76	2·51	3·39	4·39

$28(u^2 + x(0.2+u)/350)$

x = liveweight
u = liveweight gain.

shown in Table 7.7 (the same at all stages). The problem is to find which of the 216 possible paths yields the highest net revenue.

The problem can be solved by dynamic programming using the recursive functional equation:

$$V_i\{x_i\} = \max_{u_i} [-a_i\{x_i, u_i\} + V_{i+1}\{x_i + 28u_i\}] \qquad (i = 3, 2, 1) \; (7.2a)$$

subject to

$$u_i \in (0{\cdot}25, 0{\cdot}50, \ldots, 1{\cdot}50) \quad \text{and} \quad x_i + 28u_i \leq 440$$

$$x_1 = 300$$

with

$$V_4\{x_4\} = 3x_4 \qquad (7.2b)$$

where $a_i\{x_i, u_i\}$ is the ration cost for liveweight x_i and daily gain u_i shown in Table 7.7.

Using the GPDP programs the optimal sequence of daily weight gains over the three 28-day periods was found to be 1·0, 1·0, 1·0 kg, leading to an optimal liveweight sequence 300, 328, 356, 384 kg. The associated net return is

973·64.

$$\$((384 \times 3) - (2{\cdot}03 + 2{\cdot}12 + 2{\cdot}22) \times 28) = \$974$$

The problem could instead be solved by total enumeration. The net revenue associated with each path could be calculated, and the maximum-net-revenue path identified. However, the problem is solved with less computation by dynamic programming. The number of possible liveweight states at decision stages 1, 2 and 3 are 1, 6 and 11, respectively. Given that six decisions are evaluated at each combination of state and stage, $(1 + 6 + 11) \times 6 = 108$ paths are evaluated. Thus for the three-stage problem, computation by dynamic programming is roughly half that by total enumeration. The reduction becomes much greater as the number of decision stages is increased. Consider a planning horizon of one year, or 13 periods of 28 days. The number of paths which would have to be examined by total enumeration is just over 77 000. Note that large though this is, the number of paths is considerably less than 6^{13} because the maximum liveweight is 440 kg. The number of paths examined by dynamic programming is 734, a reduction factor greater than 100.

7.3.2 Maximizing net returns per unit of time

If the planning horizon is finite, and livestock feeding and marketing policies are to be evaluated in terms of net revenue, a question is often raised as to whether net revenue or net revenue per unit of time should be maximized (see, e.g., Heady *et al.*, 1976; Melton *et al.*, 1978; Glen, 1980*a*). Initially suppose for simplicity that the ration policy has already been determined, and that Fig. 7.1 represents net revenue as a function of the length of the feeding period. The decision problem is how long to feed the livestock unit. The answer depends on the earning opportunities available at the end of the feeding period. If there are none, the relevant objective is maximum net revenue, achieved with a feeding period equal to T_3 or the length of the planning horizon, whichever is shorter. If a

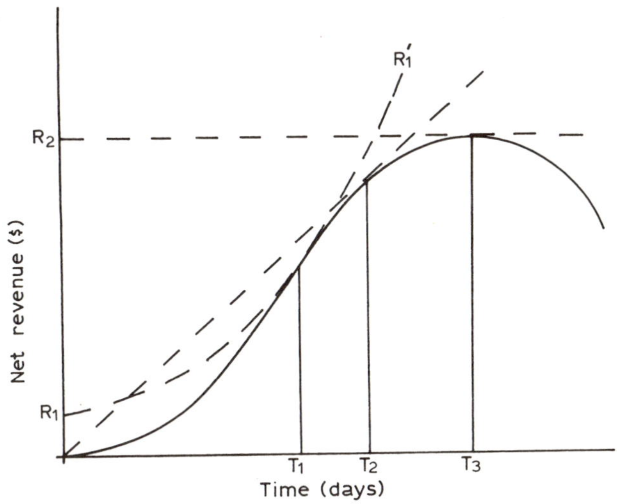

FIG. 7.1. Net revenue from feeding a livestock unit.

rate of return can always be obtained on invested funds then this opportunity cost of funds invested in the beef enterprise should be taken into account. Imagine in Fig. 7.1 a series of convex iso-present-value curves, the intercept on the vertical axis showing present value. If R_1–R'_1 is the highest iso-present-value curve attainable by feeding the beef animal for a period and then selling, drawn for some rate of return r, the optimal feeding period is T_1. The dynamic programming problem can be reformulated to allow for sale of the beef animal and investment of the proceeds by adding to the right-hand side of (7.2a) the optional value $3x_i(1 + r)^{4-i}$, and setting $V_i\{x_i\}$ to the higher value.

If there are no outside investment opportunities but the livestock unit can be replaced with another unit which can be fed under exactly the same conditions, the relevant objective is maximum net revenue per day. For an indefinite planning horizon the optimal feeding period is T_2. However, in the second case, the optimal solution is obtained by maximizing net revenue over the planning horizon instead of net revenue per day if replacement is treated as an option in the decision model. In Section 7.3.3 it is shown how a dynamic programming problem can be formulated to include the replacement decision. This method has the advantage that there need be no presumption that conditions for feeding periods after the first will be the same as for the first.

Dynamic programming can still be used to solve the combined problem of ration formulation and length of feeding period if replacement is not specified as an option but the objective is to maximize net return per day. Glen has discussed how this may be done in problems of beef production (1980*a*) and pig production (1983). He suggests using the optimal returns for different planning horizons obtained by dynamic programming to calculate returns per day, and then selecting the optimal plan by inspection. Heady *et al.* (1976) and Melton *et al.* (1978) consider similar types of problems in pig and beef production, respectively, but which were not solved by dynamic programming. The authors reported difficulty in obtaining solutions for the objective of maximum net revenue per day. In the pig production problem, resort was made to solution by total enumeration, though because the number of alternative paths were relatively small the task was not onerous. However, in the beef production problem, only approximate solutions were obtained. Because both of these problems could have been solved using dynamic programming, but with recursive equations different from those encountered so far, the problem type and solution procedure are briefly discussed with reference to the beef production study by Melton *et al.* (1978). The procedure is an alternative to the one suggested by Glen.

A steer is to be fattened from an initial weight of 196 kg to a final weight of 295 kg. The initial-to-final weight range is divided into five intervals. For each interval i isoquants have been estimated giving the amount of concentrate (c_i) required in the ration for any level of silage (s_i) so as to complete the weight interval. The concentrate consumption function is denoted by $c_i\{s_i\}$. The number of days $d_i\{s_i\}$ that the ration must be fed to complete the i-th weight interval is determined by the proportion of silage in the ration. Feed costs in each interval are feed quantities consumed times feed price, and are denoted by $a_i\{s_i\}$. The return at the end of the fifth interval is (295–196) kg times the price of weight gain, p^g. The problem is defined as

$$\max_{s_1,\ldots,s_5}\left[p^g(295\text{–}196)-\sum_{i=1}^{5}a_i\{s_i\}\right]\Big/\sum_{i=1}^{5}d_i\{s_i\}$$

This problem is not the usual dynamic programming problem because of the denominator, $\sum d_i\{s_i\}$, and cannot be solved using FDP in GPDP without modification. It can be solved using the recursive equations

$$V_i\{t_i\}=\max_{s_i}[-a_i\{s_i\}/D_{i+1}\{t_i+d_i\{s_i\}\}+V_{i+1}\{t_i+d_i\{s_i\}\}] \quad (7.3a)$$

$$D_i\{t_i\}=D_{i+1}\{t_i+d_i\{s_i^*\}\} \qquad (i=5,\ldots,1) \quad (7.3b)$$

with

$$V_6\{t_5+d_5\{s_5\}\}=p^g(295\text{–}196)/(t_5+d_5\{s_5\}) \quad (7.3c)$$

$$D_6\{t_5+d_5\{s_5\}\}=t_5+d_5\{s_5\} \quad (7.3d)$$

$$t_1=0 \quad (7.3e)$$

where t_i is the number of days of feeding before the start of weight interval i, and s_i^* in (7.3b) is the optimal s_i which solves (7.3a). $D_i\{t_i\}$ represents the optimal length of the total feeding period, from 196 to 295 kg, if at the start of the i-th weight interval the animal has been fed t_i days from when its weight was 196 kg. Consider the calculations for the last weight interval, with $i=5$. The optimal length of total feeding period is $t_5+d_5\{s_5^*\}$. From equations (7.3a, c and d) it follows that the optimal per day net return resulting from the last weight interval is

$$V_5\{t_5\}=(-a_5\{s_5^*\}+p^g(295\text{–}196))/(t_5+d_5\{s_5^*\})$$

Equation (7.3b) ensures that for $i=4,\ldots,1$, $D_{i+1}\{t_i+d_i\{s_i\}\}$ in (7.3a) is the same total feeding period as was used in calculating $V_{i+1}\{t_i+d_i\{s_i\}\}$. The maximum net return per day is given by $V_1\{0\}$, and the optimal total feeding period by $D_1\{0\}$.

7.3.3 Replacement a decision option

The simple ration problem can be readily extended to include replacement decisions. Suppose now that the steer can be sold at any decision stage. If it is sold, it is always replaced by a 300-kg steer at the subsequent stage. Each sale and replacement incurs a marketing charge of $25. The planning horizon is n 28-day periods. The discount rate is 10 per cent per

```
100     REM----------BEEF PROBLEM----------
110     REM
120     LET N$="BEEF.DAT"
130     OPEN N$ FOR OUTPUT AS FILE #1
140     REM
150     LET N=13
160     LET D2=.7358
170     DEF FNR(G,W)=-(W*.2/350+G^2+W*G/350)*28
180     LET C$=","
190     LET P=1
200     REM
210     REM-----WRITE BLOCK A DATA - PROBLEM PARAMETERS-----
220     PRINT #1, N
230     PRINT #1, D2
240     PRINT #1, "D"
250     PRINT #1, "Y"
260     PRINT #1, "Y"
270     REM
280     REM-----WRITE BLOCK B DATA - TERMINAL VALUES-----
290     PRINT #1, "0,0"
300     REM
310     REM-----WRITE BLOCK C DATA-----
320     PRINT #1, " T                I   D    R                    J    P"
330     REM
340     FOR I=1 TO 21
350     LET W=300+(I-1)*7
360     LET D=0
370        FOR D3=1 TO 7
380           IF D3>1 THEN 430
390           LET R=3.0*(W-300)-25
400           LET J=1
410           LET D=1
420           GO TO 480
430           LET G=(D3-1)*.25
440           IF (W+G*28)>440 THEN 500
450           LET R=FNR(G,W)
460           LET J=I+(D3-1)
470           LET D=D+1
480           PRINT #1, N;C$,I;C$;D;C$;R;C$,J;C$;P
490        NEXT D3
500     NEXT I
510     PRINT #1, N,",0,0,0,0,1"
520     REM
530     REM-----WRITE BLOCK D DATA - LABELS-----
540     PRINT #1, "WEIGHT"
550     FOR I=1 TO 21
560        PRINT #1, "W"+STR$(300+(I-1)*7)
570     NEXT I
580     PRINT #1, "LABEL";C$;7
590     FOR D=1 TO 7
600        IF D>1 THEN 630
610        PRINT #1, "REPLACE"
620        GO TO 640
630        PRINT #1, "K/G="+STR$((D-1)*.25)
640     NEXT D
650     REM
660     PRINT "DATA WRITTEN TO ";N$
670     CLOSE #1
680     END
```

FIG. 7.2. Program for writing data for the beef feeding and replacement problem to the file BEEF.DAT.

annum, which implies a 28-day discount factor α equal to 0·993. The recursive functional equation becomes

$$V_i\{x_i\} = \max\,[b_i\{x_i\}, c_i\{x_i\}] \qquad (i = n, \ldots, 1) \tag{7.4a}$$

with

$$V_{n+1}\{x_{n+1}\} = 3x_{n+1} \tag{7.4b}$$

where

$$b_i\{x_i\} = \max_{u_i}\,[-a_i\{x_i, u_i\} + \alpha V_{i+1}\{x_i + 28u_i\}] \tag{7.4c}$$

for the decision to continue feeding the steer until at least stage $i+1$; and

$$c_i\{x_i\} = [3(x_i - 300) - 25 + \alpha V_{i+1}\{300\}] \tag{7.4d}$$

for the decision to sell the current steer and replace it with a 300-kg steer at stage $i+1$.

This problem was also solved using the GPDP programs. The BASIC program BEEF.BAS for writing the problem data to the file BEEF.DAT is shown in Fig. 7.2. The ration cost data are generated using the equation

$$a\{x, u\} = 28(u^2 + x(0{\cdot}2 + u)/350)$$

which is the basis of the ration costs in Table 7.7.

The results of running FDP, reading from BEEF.DAT, for 13 decision stages are displayed in Fig. 7.3. The optimal liveweight sequence for all possible initial liveweights and details of the sequence for a steer initially weighing 328 kg are shown. The infinite-stage solution, obtained by running IDP, is presented in Fig. 7.4. The optimal stage cycle shows that whatever the starting liveweight, it is optimal to eventually pursue indefinitely the cycle with the liveweight sequence 300, 321, 342, 363, 384, 412, 440, 300 kg.

A further extension is to eliminate the delay of 28 days before replacement steers are installed, and to allow the weight of the replacement steer to be selected. The recursive functional equation becomes

$$V_i\{x_i\} = \max\,[b_i\{x_i\}, d\{x_i\}] \qquad (i = n, \ldots, 1)$$

with

$$V_{n+1}\{x_{n+1}\} = s_{n+1}\{x_{n+1}\}$$

$$d_i\{x_i\} = \max_{r_i}\,[s_i\{x_i\} - s_i\{r_i\} - 25 + b_i\{r_i\}]$$

where r_i is the selected weight of the replacement steer, $s_i\{x_i\}$ is the market price of a steer weighing x_i, and $b_i\{x_i\}$ is as defined in (7.4c). Because

DETERMINISTIC DYNAMIC PROGRAMMING SOLUTION
PROBLEM ---BEEF---

PROBLEM PARAMETERS:
NO. OF DECISION STAGES = 13
RATE OF DISCOUNT (PER CENT) = .7358

OPTIMAL STATE SEQUENCE FOR ALL INITIAL STATES

STAGE 1	2	3	4	5	6	7	8	9	10	11	12	13	14	VALUE
1	4	7	10	13	17	21	1	4	7	11	15	19	1	159.695
2	5	8	11	15	19	1	4	7	10	13	17	21	1	176.915
3	6	9	12	16	20	1	4	7	10	13	17	21	1	194.265
4	7	10	13	17	21	1	4	7	10	13	17	21	1	211.614
5	8	11	14	17	21	1	4	7	10	13	17	21	1	229.006
6	9	12	15	18	21	1	4	7	10	13	17	21	1	246.283
7	10	13	17	21	1	3	6	9	12	15	18	21	1	261.966
8	11	14	17	21	1	3	6	9	12	15	18	21	1	280.021
9	12	15	18	21	1	3	6	9	12	15	18	21	1	297.961
10	14	18	1	5	9	13	17	1	5	9	13	17	1	317.297
11	15	19	1	5	9	13	17	1	5	9	13	17	1	336.652
12	16	1	4	7	11	15	19	1	5	9	13	17	1	356.039
13	17	1	4	7	11	15	19	1	5	9	13	17	1	376.214
14	18	1	4	7	11	15	19	1	5	9	13	17	1	396.388
15	19	1	4	7	11	15	19	1	5	9	13	17	1	416.563
16	20	1	4	7	11	15	19	1	5	9	13	17	1	436.738
17	1	4	7	11	15	19	1	4	7	11	15	19	1	457.697
18	1	4	7	11	15	19	1	4	7	11	15	19	1	478.697
19	1	4	7	11	15	19	1	4	7	11	15	19	1	499.697
20	1	4	7	11	15	19	1	4	7	11	15	19	1	520.697
21	1	4	7	11	15	19	1	4	7	11	15	19	1	541.697

DETAILS OF OPTIMAL PATH
ENTER INITIAL STATE NO. (OR '0' TO FINISH) ? 5

STAGE NO.	STATE NO.	STATE WEIGHT	DECISION NO.	DECISION LABEL	DISCOUNTED STAGE RETURN	VALUE
1	5	W328	4	K/G=.75	-40.678	229.006
2	8	W349	4	K/G=.75	-41.9652	271.668
3	11	W370	4	K/G=.75	-43.2315	316.252
4	14	W391	4	K/G=.75	-44.477	362.772
5	17	W412	5	K/G=1	-65.5999	411.242
6	21	W440	1	REPLACE	380.783	482.317
7	1	W300	4	K/G=.75	-36.8911	87.9594
8	4	W321	4	K/G=.75	-38.1378	127.44
9	7	W342	4	K/G=.75	-39.3643	168.819
10	10	W363	4	K/G=.75	-40.5709	212.111
11	13	W384	5	K/G=1	-60.2789	257.328
12	17	W412	5	K/G=1	-62.3184	324.563
13	21	W440	1	REPLACE	361.736	395
14	1	W300	0		0	0

FIG. 7.3. The one-year horizon solution to the beef feeding and replacement problem obtained by running FDP with data from BEEF.DAT.

DETERMINISTIC DYNAMIC PROGRAMMING SOLUTION
PROBLEM ---BEEF---

PROBLEM PARAMETERS:
INFINITE NO. OF STAGES
RATE OF DISCOUNT (PER CENT) = .7358

NO. OF ITERATIONS TO FIND SOLUTION = 6

CURRENT STATE		DECISION		NEXT STATE		PRESENT VALUE
NO.	WEIGHT	NO.	LABEL	NO.	WEIGHT	
1	W300	4	K/G=.75	4	W321	1758.4
2	W307	4	K/G=.75	5	W328	1775.47
3	W314	4	K/G=.75	6	W335	1792.82
4	W321	4	K/G=.75	7	W342	1810.17
5	W328	4	K/G=.75	8	W349	1827.9
6	W335	4	K/G=.75	9	W356	1845.92
7	W342	4	K/G=.75	10	W363	1863.93
8	W349	4	K/G=.75	11	W370	1882.33
9	W356	4	K/G=.75	12	W377	1901.01
10	W363	4	K/G=.75	13	W384	1919.69
11	W370	5	K/G=1	15	W398	1938.76
12	W377	5	K/G=1	16	W405	1958.12
13	W384	5	K/G=1	17	W412	1977.48
14	W391	4	K/G=.75	17	W412	1996.87
15	W398	5	K/G=1	19	W426	2017.02
16	W405	5	K/G=1	20	W433	2037.19
17	W412	5	K/G=1	21	W440	2057.37
18	W419	1	REPLACE	1	W300	2077.55
19	W426	1	REPLACE	1	W300	2098.55
20	W433	1	REPLACE	1	W300	2119.55
21	W440	1	REPLACE	1	W300	2140.55

OPTIMAL STAGE CYCLE
ENTER INITIAL STATE NO. (OR '0' TO FINISH) ? 5

STAGE NO.	STATE		DECISION	
	NO.	WEIGHT	NO.	LABEL
1	5	W328	4	K/G=.75
2	8	W349	4	K/G=.75
3	11	W370	5	K/G=1
4	15	W398	5	K/G=1
5	19	W426	1	REPLACE
6	1	W300	4	K/G=.75
7	4	W321	4	K/G=.75
8	7	W342	4	K/G=.75
9	10	W363	4	K/G=.75
10	13	W384	5	K/G=1
11	17	W412	5	K/G=1
12	21	W440	1	REPLACE
13	1	W300	4	K/G=.75

FIG. 7.4. The infinite-stage solution to the beef feeding and replacement problem obtained by running IDP with data from BEEF.DAT.

ration costs for any weight gain increase with liveweight (see Table 7.7), if the market price of steers were directly proportional to liveweight as was assumed before, the optimal replacement weight would be the lowest possible—300 kg in the example. However, there must be other price functions, $s\{x\}$, for which other replacement weights are optimal.

The recursive functional equation is solved efficiently by implementing the following process at each stage i:

(i) use (7.4c) to solve for $b_i\{x_i\}$, $x_i \in (300, 307, \ldots, 440)$;
(ii) find $r_i = r_i^*$ for which $(b_i\{r_i\} - s_i\{r_i\})$ is a maximum, $r_i \in (300, 307, \ldots, 440)$; and
(iii) select the larger of the best return from continuing to feed the steer of weight $x_i, b_i\{x_i\}$, and the best return from replacing the steer, $s_i\{x_i\} - s_i\{r_i^*\} - 25 + b_i\{r_i^*\}$ and set this equal to $V_i\{x_i\}$.

Because maximization is performed across all liveweights three times during this process, the GPDP programs FDP and IDP, which allow for maximization across all states only once, would have to be modified to solve the problem in this way.

Of the studies referenced in Tables 7.5 and 7.6, those by Nelson (1970) and Kennedy (1972) allow for replacement as a decision option. Glen (1980*a* and 1983) allows for variable replacement weights by solving the dynamic programming problem several times with different replacement weights and determining the optimal replacement weight by inspection.

7.4 Extensions to the Combined Feeding and Replacement Problem

7.4.1 The number of livestock

If production from livestock is the main business activity, the decision on the number of livestock to run is conceptually straightforward. If the objective is to maximize expected net revenue, it will generally be optimal to follow the policy optimal for a single unit for as many livestock units as possible, subject to capacity constraints. Net returns to the most limiting constraint will be maximized, which may be building capacity, labour or financial resources.

Nelson (1969) and Nelson and Eisgruber (1970) show how problems can be solved heuristically using dynamic programming when a constraint on the number of cattle which may be bought or sold at any stage is introduced. If, for example, no more than a third of the total livestock units a feedlot can hold are able to be bought or sold, cattle are categorized into three pens, and decisions made for each pen at each stage. By exploiting the nature of the problem, they are able to obtain dynamic programming solutions without having to allow for the state variables to take all possible combinations of cattle numbers and pen.

In the case of broiler production, the amount of floor space available constrains not only the number of broilers produced but also their final weight. The space required per bird, sufficient to ensure no increase in crowding stress with growth, is a convex function of bird weight. The introduction of a space constraint may therefore make it optimal to feed birds for lower rates of gain, or to sell the flock at an earlier age. Kennedy *et al.* (1976) use the Lagrange multiplier procedure described in Section 4.6.5 for solving a broiler production problem in which space available is limiting and in which the cumulative mortality of birds is a function of their age and weight.

If intensively fed livestock units, all on the same ration, show a significant dispersion of individual weight gains, it may be desirable to formulate the dynamic programming problem with a state variable representing the number of livestock units at each possible liveweight. Bonnieux (1969, 1972) formulates a veal production problem in which the change in calf numbers by liveweight from stage to stage is represented by a first-order Markov process. A ration strategy applies to all calves, regardless of liveweight category. Kennedy (1972, 1973, 1983) for the case of a beef feedlot considers the possibility of using Bayesian methods to categorize cattle by gain potential, and to follow different ration strategies for groups of cattle differing by gain potential.

If the livestock activity competes with other farm activities for resources such as labour and land for grazing and growing feed, it may be more difficult to determine the optimal number of livestock units to carry, as well as the optimal feeding and marketing strategies. One possibility would be to run linear programming models of the farm activities excluding the livestock activity, but allowing the resources required for different livestock numbers to be drawn off. The net returns from the linear programming models would be used in determining stage returns for alternative livestock numbers in a dynamic programming model. The approach would be similar to that suggested for the tractor replacement decision in Section 5.2.

7.4.2 Variable livestock prices

Because of seasonal variation in the availability of feeds, there may be significant seasonal variations in both feed and livestock prices. This means that for any given liveweight, the optimal target weight gain may vary from stage to stage, as may the optimal ration composition for any target weight gain. If a seasonal cycle of prices can be identified, then infinite-stage solutions for the seasonal livestock problem can be obtained in the way explained in Chapter 3 for the seasonal crop-irrigation problem.

The prices received for finished livestock may be a function of other variables besides liveweight, such as age and the composition of the carcase in terms of fat and lean meat. If all states at stage one in a dynamic programming model represent animals of the same age, and replacement is not an option, stage number indicates age, and can be used with liveweight to determine expected carcase price. If replacement is allowed for, age may have to be included as a state variable. Kennedy *et al.* (1976), in a study of broiler production which allows for replacement of the flock, include age as a state variable. Age together with liveweight help to determine the weight gain, energy intake, cumulative mortality and killing-out percentage of broilers.

Carcase composition is perhaps more important in determining returns from pig rather than cattle carcases. Glen (1983) develops a dynamic programming model for pig production with the protein content as well as the weight of carcase as state variables.

7.4.3 Stochastic livestock prices

The prices which will be eventually received for the fattened livestock are uncertain in practice unless contracts to supply have been made or livestock prices are guaranteed by government. Consider again the first feeding and replacement problem in Section 7.3.3. Suppose that price per kilogram, instead of being fixed at \$3, takes one of m values, q_i, at stage i. The probability distribution is identical for all stages. The probability of the k-th price prevailing at stage i, q_i^k, is denoted by $p_i\{k\}$. Suppose for the moment that the probability distributions are independent across stages. If a decision is made at stage i to retain and fatten the livestock unit at least until stage $i+1$, the return is $b_i\{x_i\}$ as before in (7.4c). If the decision is to sell, the return becomes

$$c_i'\{x_i, q_i\} = [q_i\{x_i - 300\} - 25 + \alpha V_{i+1}\{300\}]$$

If q_i is sufficiently high for $c_i'\{x_i, q_i\}$ to exceed $b_i\{x_i\}$, so that selling is optimal, then $c_i'\{x_i, q_i\} > b_i\{x_i\}$ for higher values of q_i as well. This insight leads to the formulation of the problem in terms of finding the minimum value of q_i for which $c_i'\{x_i, q_i\} > b_i\{x_i\}$. This way of introducing price in the decision variable, rather than as an additional state variable, avoids increasing the dimensionality of the problem. Price does not have to be included as a state variable because q_i does not affect q_{i+1}. However, the recursive equation system has to be changed, so that problems formulated in this way cannot be solved using the GPDP programs without modification. The problem posed is

$$\min_{j;1,\ldots,m} q_i^j = q_i^{j^*} \qquad (i = n, \ldots, 1) \tag{7.5}$$

subject to

$$c_i'\{x_i, q_i^{j^*}\} > b_i\{x_i\}$$

with

$$V_i\{x_i\} = \sum_{k=1}^{j^*-1} p_i\{k\}b_i\{x_i\} + \sum_{k=j^*}^{m} p_i\{k\}c_i'\{x_i, q_i^k\}$$

$$V_{n+1}\{x_{n+1}\} = \sum_{k=1}^{m} p_{n+1}\{k\}q_{n+1}^k x_{n+1}$$

Hochman and Lee (1972) and Rausser and Hochman (1979) solve a broiler replacement problem using this method, although rate of weight gain is not a decision variable. Results are compared from assuming first that prices are uniformly distributed, and secondly that they are normally distributed.

If prices are not independent over stages, price state variables are necessary. Suppose prices follow a first-order Markov process. Revise the previous notation so that $p_i'\{k \mid q_{i-1}\}$ denotes the probability of the k-th price prevailing at stage i given the price prevailing at stage $i-1$ was q_{i-1}. The recursive equation takes the usual form

$$V_i\{x_i, q_i\} = \max\,[Eb_i'\{x_i, q_i\}, Ec_i'\{x_i, q_i\}] \qquad (i = n, \ldots, 1) \tag{7.6a}$$

with

$$V_{n+1}\{x_{n+1}, q_{n+1}\} = q_{n+1}x_{n+1} \tag{7.6b}$$

where

$$Eb_i'\{x_i, q_i\} = \max_{u_i}\left[-a_i\{x_i, u_i\} + \alpha \sum_{k=1}^{m} p_{i+1}'\{k \mid q_i\}V_{i+1}\{x_i + 28u_i, q_{i+1}^k\}\right] \tag{7.6c}$$

$$Ec_i'\{x_i, q_i\} = q_i(x_i - 300) - 25 + \alpha \sum_{k=1}^{m} p_{i+1}'\{k \mid q_i\}V_{i+1}\{300, q_{i+1}^k\} \tag{7.6d}$$

Yager *et al.* (1980) use this method to solve a problem of the optimal feeding and replacement of beef cows. Because this type of problem can be solved using the GPDP programs, they can also be used to solve the problem for independent price distributions. The equations are the same as (7.6a to d) except that '$p'_{i+1}\{k \mid q_i\}$' is replaced by '$p_{i+1}\{k\}$'.

7.4.4 Ration formulation systems

The ration costs used in the dynamic programming calculations should be minimum ration costs if the objective is to maximize net returns. It can be seen from Tables 7.5 and 7.6 that in most studies the rations and corresponding costs are determined by linear programming. The nutritional requirements by liveweight and liveweight gain are generally those recommended by the National Research Council (NRC) in the US or the Agricultural Research Council (ARC) in the UK. The two councils differ in their systems for determining energy requirements, the NRC using the net energy system and the ARC the metabolizable energy system. However, an important advantage of the dynamic programming approach to weight gain management, incorporating linear programming for ration formulation, is that standard estimates of nutritional requirements can be used. There is no need to carry out further feeding trials to obtain the data necessary to estimate functions such as isoquants.

If a linear programming problem has to be solved for every combination of liveweight and weight gain considered in the dynamic programming calculations, the amount of computation may be prohibitive. There are various ways of reducing the computational requirements. One way is to estimate a least-cost-ration response surface over a range of combinations of liveweight and weight gain. The combinations are selected according to the principles of experimental design. The response surface can then be used in the dynamic programming model. Nelson (1969) adopts this method for a feedlot problem. A disadvantage of the method is that information about the composition of the rations is not available in the dynamic programming results. A similar approach which obviates this problem is to determine least-cost rations for a limited number of liveweight and weight gain combinations over the required range, and to use linear interpolation to approximate ration costs and ration composition for other combinations considered in the dynamic programming routine (see, e.g., Kennedy, 1972; Yager *et al.*, 1980).

Another possibility is to use parametric linear programming whenever possible. Parameterization of the right-hand side of the linear programming ration problem is not straightforward because changes in liveweights or weight gain alter various nutritional requirements by different proportions. Glen (1980*b*) discusses in detail a method of parameterizing the energy concentration of rations for beef cattle based on the metabolizable energy system. In another study he shows how the nutritional requirements of beef cattle of various liveweights for each possible weight gain may be parameterized in the case of rations based on the net energy system (Glen, 1980*a*). He also uses parameterization in a pig production study (Glen, 1983).

In some production systems livestock obtain all of their feed by grazing pasture. Clark and Kumar (1978) use dynamic programming to model a grazing system in which the decision problem is whether to keep cattle grazing, or to sell them, with or without replacement, over a two-year period. Under other grazing systems feed from pasture is supplemented with other sources of feed, depending on the time of year. Topham (1979) models a beef cow–calf system. Cows produce calves, most of which are fattened and then sold, whilst some heifer calves are kept on to replace culled cows. Linear programming is used to determine least-cost rations for the calves and cows, taking account of requirements for alternative weight gains, maintenance and milk production for calf feeding. Allowance is made for the changing availability of pasture and hay, and homegrown and purchased feed grains throughout the year. Kennedy (1972, 1983) models a system in which beef cattle are yarded during the winter months and put out to graze with provision for supplementary feeding during the summer months. Least-cost rations are determined for each month. For the summer months assumptions are made about the pasture available for each grazing animal, dependent on the pasture management policy and stocking rate. Feed from pasture is available at zero cost in the linear programming sub-problem, but an annual charge is made against the livestock enterprise which takes account of pasture production costs and the opportunity cost of the land.

7.5 References

Bonnieux, F. (1969), Dynamic programming and Markov chains analysis applied to management of a veal calf fattening operation (in French), *Recherches d'Economie et de Sociologie Rurales*, **2**, 29–40.

Bonnieux, F. (1972), Modèle de conduite d'un élevage de veaux de boucherie, *Revue de Statistique Appliquée*, **20**(3), 71–84.

Brookhouse, J.K. and Low, E.M. (1967), Dynamic programming and the selection of replacement policies in commercial egg production, *Journal of Agricultural Economics*, **18**(3), 339–50.

Clark, J. and Kumar, S. (1978), Planning beef production: an application of dynamic programming, *Review of Marketing and Agricultural Economics*, **46**(3), 315–26.

Fisher, I.H. (1974), Resource optimization in arid grazing systems, Unpublished PhD thesis, University of New South Wales, Sydney.

Giaever, H. (1966), Optimal dairy cow replacement policies, Unpublished PhD thesis, University of California, Berkeley.

Glen, J.J. (1980*a*), A mathematical programming approach to beef feedlot optimization, *Management Science*, **26**(5), 524–35.

Glen, J.J. (1980*b*), A parametric programming method for beef cattle ration formulation, *Journal of the Operational Research Society*, **31**, 689–98.

Glen, J.J. (1983), A dynamic programming model for pig production, *Journal of the Operational Research Society*, **34**(6), 511–19.

Halter, A.N. and White, W.C. (1962), *A replacement decision process: an application to a caged layer enterprise*, Agricultural Experiment Station Bulletin 677, University of Kentucky.

Heady, E.O., Sonka, S.T. and Dahm, F. (1976), Estimation and application of gain isoquants in decision rules for swine producers, *Journal of Agricultural Economics*, **27**(2), 235–42.

Henderson, R.A. and Toft, H.I. (1979), Field testing of the Toft and O'Hanlon Drought Tactic Model, *Review of Marketing and Agricultural Economics*, **47**(3), 207–12.

Hochman, E. (1973), An optimal stopping problem of a growing inventory, *Management Science*, **19**(11), 1289–91.

Hochman, E. and Lee, I.M. (1972), *Optimal decision in the broiler producing firm: a problem of growing inventory*, Giannini Foundation Monograph 29, California Agricultural Experiment Station, Berkeley.

Jenkins, K.B. and Halter, A.N. (1963), *A multi-stage replacement decision model*, Agricultural Experiment Station, Technical Bulletin 67, Oregon State University, Corvallis.

Karp, L. and Pope, A. (III) (1984), Range management under uncertainty, *American Journal of Agricultural Economics*, **66**(4), 437–46.

Kennedy, J.O.S. (1972), A model for determining optimal marketing and feeding policies for beef cattle, *Journal of Agricultural Economics*, **23**(2), 147–59.

Kennedy, J.O.S. (1973), Control systems in farm planning, *European Review of Agricultural Economics*, **1**(4), 415–33.

Kennedy, J.O.S. (1983), A control system for intensive beef production, in *Planning and Decision in Agribusiness: Principles and Experiences*, C.H. Hanf and G. Schiefer (eds), Elsevier Science, Amsterdam.

Kennedy, J.O.S., Rofe, B.M., Greig, I.D. and Hardaker, J.B. (1976), Optimal feeding policies for broiler production: an application of dynamic programming, *Australian Journal of Agricultural Economics*, **20**(1), 19–32.

Lasley, J.F. (1978), *Genetics of Livestock Improvement*, 3rd edition, Prentice-Hall, Englewood Cliffs, New Jersey.

Lindley, D.V. (1965), *Introduction to Probability and Statistics from a Bayesian Viewpoint*, Cambridge University Press, Cambridge.

Lush, J.L. (1945), *Animal Breeding Plans*, 3rd edition, Iowa State University Press, Iowa.

McArthur, A.T.G. (1973), Application of dynamic programming to the culling decision in dairy cattle, *Proceedings of the New Zealand Society of Animal Production*, **33**, 141–7.

Melton, B.E., Heady, E.O., Willham, R.L., Hoffman, M.P. (1978), The impact of alternative objectives on feedlot rations for beef steers, *American Journal of Agricultural Economics*, **60**(4), 683–8.

Meyer, C.F. and Newett, R.J. (1970), Dynamic programming for feedlot optimization, *Management Science*, **16**(6), 410–26.

Nelson, A.G. (1969), The feasibility of an information system: the beef feedlot case, Unpublished PhD thesis, Purdue University, Lafayette.

Nelson, A.G. and Eisgruber, L.M. (1970), A dynamic information and decision system for beef feedlots, *Proceedings of the Western Agricultural Economics Association Meeting*, Tucson, Arizona, 19–22 July, 96–102.

Pouliquen, A. (1970), A dynamic programming model applied to the replacement of laying hens and to the economic study of two forced moulting processes (in French), *Recherches d'Economie et de Sociologie Rurales*, **3**, 93–127.

Rausser, G.C. and Hochman, E. (1979), *Dynamic Agricultural Systems: Economic Prediction and Control*, North-Holland, New York.

Redman, J.C. and Kuo, L.D. (1969), *Replacement of dairy cows: a multi-stage decision making problem*, University of Kentucky Agricultural Experiment Station Research Report No. 1, Kentucky.

Scobie, G.M. (1967), *An Economic Study of Replacement Policies in Merino Sheep Flocks*, Bureau of Agricultural Economics, Canberra, December.

Smith, B.J. (1971), *The dairy cow replacement problem: an application of dynamic programming*, Florida Agricultural Experiment Station Bulletin 745.

Smith, B.J. (1973), Dynamic programming of the dairy cow replacement problem, *American Journal of Agricultural Economics*, **55**(1), 100–4.

Stewart, H.M., Burnside, E.B., Wilton, J.W. and Pfeiffer, W.C. (1977), A dynamic programming approach to culling decisions in commercial dairy herds, *Journal of Dairy Science*, **60**(4), 602–17.

Throsby, C.D. (1964), Some dynamic programming models for farm management research, *Journal of Agricultural Economics*, **16**(1), 98–110.

Toft, H.I. and O'Hanlon, P.W. (1979), A dynamic programming model for on-farm decision-making in a drought, *Review of Marketing and Agricultural Economics*, **47**(4), 5–16.

Topham, M.R. (1979), A model of beef cow and calf feeding: a preliminary study, *Canadian Journal of Agricultural Economics*, **27**(3), 26–36.

White, W.C. (1959), The determination of an optimal replacement policy for a continually operating egg production enterprise, *Journal of Farm Economics*, **41**(5), 1535–42.

Yager, W.A., Greer, R.C. and Burt, O.R. (1980), Optimal policies for marketing cull beef cows, *American Journal of Agricultural Economics*, **62**(3), 456–67.

7.A Appendix: Yield Repeatability and Adaptive Dynamic Programming

McArthur (1973) was the first to apply adaptive dynamic programming to the dairy cow replacement problem. He used equations originally suggested by animal scientists for updating at stage i a cow's most probable yield on the basis of average yield calculated to stage i. In this appendix the concept of repeatability used by animal scientists for guidance on culling is briefly explained, and then shown to be consistent with a Bayesian approach to updating expected individual yields.

The adaptive dynamic programming approach to the livestock replacement decision has ready application to other intensive livestock enterprises besides milk production provided certain requirements are met. These are that it be economic to monitor individual livestock yields, to replace the livestock on an individual basis, and to keep all livestock animals in the group being monitored under similar environmental conditions as far as possible. The advent of devices linked to a computer for monitoring individual yields increases the scope for the approach. An adaptive dynamic programming routine can be run on the computer and used to make replacement or feeding decisions for individual animals. Other enterprises to which it would be worth considering applying the approach are beef feedlots, intensive pig production and egg production.

7.A.1 The concept of yield repeatability

Different animals in a group show different yields, partly because of variations in genetic yield potentials, partly because of variations in reaction to permanent environmental factors, and partly because of random environmental factors which impinge on the animals individually. If v_p denotes the variance in yields due to permanent factors, and v_t that due to temporary factors, and assuming the p and t factors are independent, total variance is $v_p + v_t$. Repeatability is defined as

$$\rho = v_p/(v_p + v_t) \tag{7A.1}$$

If there were no temporary factors at work, $\rho = 1$ and individual yields would show no change over time. If there were no variation in permanent factors, all variation would be due to temporary factors and $\rho = 0$.

Although repeatability is not a fundamental biological constant because it is determined partly by environmental factors, animal scientists quote repeatability values for specific animal and yield types. It may be possible to apply these values to a commercial enterprise with caution. Alternatively

v_t and v_p may be calculated from the yield records of all individuals in the group by analysis of variance. See, for example, Lasley (1978).

7.A.2 Repeatability of average yield

If the average yield over n periods instead of yield for one period is calculated, temporary variance is v_t/n. Variance v_p is not affected if there has been no change in the number of animals in the group. Repeatability using yields averaged over n periods is

$$r_n = v_p/(v_p + v_t/n) \tag{7A.2}$$

From (7A.1) it follows that

$$v_t = v_p(1 - \rho)/\rho$$

Substituting for v_t in (7A.2) leads to

$$r_n = \rho n/(1 + (n - 1)\rho) \tag{7A.3}$$

The repeatability of average yield is important for revising expected individual yield in the light of updated average individual yield, and average herd yield.

7.A.3 Expected yield given average individual and herd yields

Lush (1945) defined the 'most probable producing ability' of an individual $\hat{q}$ as a weighted average of the individual's average yield $\bar{q}$ and the herd's average yield $\bar{Q}$, so that

$$\hat{q} = r_n\bar{q} + (1 - r_n)\bar{Q} = r_n(\bar{q} - \bar{Q}) + \bar{Q} \tag{7A.4}$$

This can be justified on Bayesian grounds.

Suppose the individual's yield is normally distributed with unknown mean θ but known variance v_t. The individual's yield distribution $N(\theta, v_t)$ is referred to as the sampling distribution. The distribution of θ may be assumed to be given by $N(\bar{Q}, v_p)$ if θ is calculated from a large enough sample for it to be reasonable to ignore the effect of temporary factors. The distribution $N(\bar{Q}, v_p)$ is referred to as the prior distribution. Suppose a sample of size n of the individual's yields is taken, with mean $\bar{q} = \sum_{j=1}^{n} q_j/n$. Lindley (1965, Section 5.1), using Bayes' theorem, has shown by combining information from the sampling and prior distributions that the revised or posterior distribution of θ is $N(\bar{Q}_n, v_{pn})$, where

$$\bar{Q}_n = u\bar{q} + (1 - u)\bar{Q} \tag{7A.5}$$

$$v_{pn} = (1 - u)v_p \tag{7A.6}$$

with $u = n/(n + v_t/v_p) = v_p/(v_p + v_t/n) = r_n$. Expected individual yield, $\bar{Q}_n$, is therefore equal to $\hat{q}$ in (7A.4).

It is thus possible to update expected individual yield after each successive yield record by recalculating the individual's average yield, $\bar{q}$, and then weighting $\bar{q}$ and $\bar{Q}$ according to (7A.5) using the revised weighting factor $u = r_n$ given by (7A.3). The variance of the individual's expected yield is likewise readily updated using (7A.6) and (7A.3).

Both Q_n and v_{pn} can be used to calculate the probability that the $(n+1)$-th recorded yield will have some particular value.

7.A.4 *Yield probabilities conditional on recorded average yields*

Suppose that the range of yields which could occur with at least some small probability is represented by yields q^k equally spaced, $k = 1, \ldots, m$. The standard deviate for q^k is

$$z^k = (q^k - \bar{Q}_n)/v_{pn}^{1/2} \qquad (7A.7)$$

Given that the ordinate of the standard normal curve for z^k is

$$f\{z^k\} = (2\pi)^{-1/2} \exp(-(z^k)^2/2)$$

the probability of yield falling in the k-th interval is approximately

$$p\{z^k\} = \exp(-(z^k)^2/2) \Big/ \sum_{j=1}^{m} \exp(-(z^j)^2/2) \qquad (7A.8)$$

Equation (7A.8) is used to calculate the state transition probabilities in adaptive dynamic programming models. State variables are the number of individual yield records and the average yield calculated from them. If another record is obtained for the individual by the subsequent decision stage showing a yield of q_k, the number of yield records is increased by one, average yield is recalculated, and the probability of reaching the new state combination is given by $p\{z^k\}$ based on (7A.7) and (7A.8). The process of revision of state transition probabilities in adaptive dynamic programming is illustrated in Section 7.2.3 for a dairy cow replacement problem.

PART IV

Dynamic Programming Applications to Natural Resources

Chapter 8

LAND MANAGEMENT

In this chapter three applications of dynamic programming to the management of inanimate natural resources are discussed. In the first section the optimality conditions derived analytically using the dynamic programming approach in Chapter 1 are applied to the decision on how much of an exhaustible resource to mine through time. The optimality conditions for mining subject to uncertainty, and for combined mining and exploration, are also considered.

Whilst a dynamic programming approach can be used to give analytical insights into the nature of the solution to a problem, another advantage of the approach is that it can be used to obtain numerical solutions to problems which are analytically intractable. In Section 8.2 a production and pollution problem with discontinuous functions is introduced and is solved numerically.

The final section is devoted to a discussion of the relevance of stochastic dynamic programming to aid in making decisions which have irreversible effects. The context of the discussion is whether to take an irreversible decision to develop wilderness or agricultural land as a site for some project such as a dam. It is argued that it would probably be more fruitful to treat the decision problem as a stochastic dynamic programming problem than to work with the expected values associated with the different decisions, and then to have to attempt to adjust the expected values by including quasi-option values.

8.1 The Theory of Exhaustible Resources

The optimality conditions for the exploitation of an exhaustible resource were derived for the simple deterministic problem in Chapter 1 using the inductive logic of dynamic programming. In this section the optimality conditions are interpreted further, and conditions derived for extensions to the simple problem. The conditions can also be derived by solving a two-period maximization problem, by obtaining a solution to the n-period problem with dynamic Lagrange multipliers, or by using control theory. A two-period formulation, with one period representing the final period and the other the preceding period of indefinite length, has the merit of simplicity. However, the generalization to the solution of the n-period problem may not be obvious, and the shadow prices of the resource through time are not obtained. These limitations are avoided if approaches using dynamic Lagrange multipliers or control theory are taken, but at the expense of an immediate intuitive understanding of the optimality conditions. The dynamic programming approach also has the advantage that it is simple to derive the optimality conditions for stochastic problems.

8.1.1 The simple theory of the mine

In Chapter 2 a simple mining problem was formulated with discrete state and decision variables. It was used to illustrate Bellman's Principle of Optimality. The same problem is taken up here, but with the state and decision variables continuous. The state variable x_i represents the stock of the exhaustible resource and u_i the quantity mined, both at stage i. The net return from mining u_i over the period from stage i to stage $i+1$ is $a_i\{x_i, u_i\}$. If the mining firm is one out of many, and x_i is not the total resource base but only the stock of the resource which the firm owns, the firm may be taken to be perfectly competitive, and

$$a_i\{x_i, u_i\} = q_i u_i - c_i\{x_i, u_i\}$$

where q_i is the price of the resource at stage i, and $c_i\{x_i, u_i\}$ is the cost of mining and selling the resource. It is assumed that the cost of mining increases at an increasing rate with the amount mined ($\partial c_i/\partial u_i > 0$; $\partial^2 c_i/\partial u_i^2 > 0$) and increases at an increasing rate as the resource base is diminished ($\partial c_i/\partial x_i < 0$; $\partial^2 c_i/\partial x_i^2 > 0$).

If the mining firm has a monopoly over the resource, so that x_i represents the total resource base, the firm's sales depress price, and

$$a_i'\{x_i, u_i\} = q_i'\{u_i\}u_i - c_i\{x_i, u_i\}$$

with

$$\mathrm{d}q_i'/\mathrm{d}u_i < 0$$

Suppose the firm's problem is how to set $u_1, \ldots, u_n$ so as to maximize the present value of net returns. The problem can be expressed as

$$\max_{u_1,\ldots,u_n} \sum_{i=1}^{n} \alpha^{i-1} a_i\{x_i, u_i\}$$

assuming that any resource remaining at stage $n+1$ has zero value. The recursive equation is

$$V_i\{x_i\} = \max_{u_i} [a_i\{x_i, u_i\} + \alpha V_{i+1}\{x_i - u_i\}]$$

$$= a_i\{x_i, u_i^*\} + \alpha V_{i+1}\{x_i - u_i^*\} \qquad (i = n, \ldots, 1) \qquad (8.1a)$$

subject to

$$0 \le u_i \le x_i \qquad (8.1b)$$

$$x_1 \text{ given}$$

with

$$V_{n+1}\{x_{n+1}\} = 0$$

Assuming that the solution to (8.1) is an interior solution ($0 < u_i^* < x_i$, $i = 1, \ldots, n$), the following condition holds for u_i^*:

$$\partial a_i/\partial u_i - \alpha \, \mathrm{d}V_{i+1}/\mathrm{d}x_{i+1} = 0 \qquad (i = 1, \ldots, n) \qquad (8.2)$$

Differentiating (8.1a) gives

$$\mathrm{d}V_i/\mathrm{d}x_i = \partial a_i/\partial x_i + \alpha(\mathrm{d}V_{i+1}/\mathrm{d}x_{i+1}) \qquad (i = 1, \ldots, n) \qquad (8.3)$$

Substituting $\partial a_i/\partial u_i$ for $\alpha(\mathrm{d}V_{i+1}/\mathrm{d}x_{i+1})$ from (8.2) in (8.3) gives

$$\mathrm{d}V_i/\mathrm{d}x_i = \partial a_i/\partial x_i + \partial a_i/\partial u_i \qquad (i = 1, \ldots, n) \qquad (8.4)$$

and under perfect competition

$$= q_i - (\partial c_i/\partial u_i + \partial c_i/\partial x_i)$$

Equation (8.4) states that for u_i to be optimal, the value of an additional unit of the resource stock, or the shadow price of the resource unit *in situ*, equals the reduced cost of mining if the resource base were one unit larger plus the net revenue which could be obtained by extracting the additional unit. It is perhaps surprising that the optimal shadow price of the resource stock at stage *i* can be equated with partial derivatives at only stage *i*, without apparent reference to subsequent stages or the rate of interest. The reason why this can be done is that the shadow price is based on optimal values of $u_i, \ldots, u_n$ and hence of $x_{i+1}, \ldots, x_{n+1}$.

Another optimality condition which follows from (8.2) after substituting for $\mathrm{d}V_{i+1}/\mathrm{d}x_{i+1}$ using (8.4) is

$$\partial a_i/\partial u_i = (\partial a_{i+1}/\partial x_{i+1} + \partial a_{i+1}/\partial u_{i+1})/(1+r) \qquad (i = 1, \ldots, n) \qquad (8.5)$$

where r is the rate of interest. The partial derivative $\partial a_i/\partial u_i$ is sometimes termed the scarcity rent of the resource. It represents marginal rent which is positive if the resource is scarce, but otherwise zero. If there were no stock effect, ((i.e.), $\partial a_{i+1}/\partial x_{i+1} = 0$), (8.5) would stipulate Solow's (1974, pp. 2–3) 'fundamental principle of the economics of exhaustible resources', namely that the scarcity rent of a resource *in situ* should rise at a rate equal to the rate of interest. If the stage return function is the same at stage i as at stage $i+1$, this implies that more should be mined at stage i than at stage $i+1$ ((i.e.) that $u^*_{i+1} < u^*_i$). If there is a stock effect and $\partial c_i/\partial x_i < 0$, this counters the effect of r so that whether or not more should be mined at stage i depends on r and the net return functions.

8.1.2 Risky possession and risk aversion

There is usually some risk that an investor will lose the wealth embodied in a stock of a resource. For example, the resource may be accidently destroyed, it may be expropriated, or demand for it may fall drastically due to the introduction of a cheaper substitute. Let the probability that the resource base at stage i will still be intact at stage $i+1$ be p_i, and the probability of loss of value be $(1-p_i)$.

Suppose that the mine owner is risk averse and has utility at stage i

$$U_i = a_i\{x_i, u_i\} - b(a_i\{x_i, u_i\})^2$$

where b (>0) is a risk aversion coefficient. Assume that for all values of a_i encountered, $a_i < 1/(2b)$ so that the marginal utility of net return $\mathrm{d}U_i/\mathrm{d}a_i = 1 - 2ba_i\{\cdot\} > 0$. Suppose also that total utility over all i stages is separable in U_i and equals

$$W = \sum_{i=1}^{n} U_i \delta^{i-1}$$

where δ is the factor for discounting utility. The problem of finding $u_1, \ldots, u_n$ so that $E(W)$ is maximized can be defined by the recursive equation:

$$V_i\{x_i\} = \max_{u_i} [a_i\{x_i, u_i\} - b(a_i\{x_i, u_i\})^2 + \delta p_i V_{i+1}\{x_i - u_i\}]$$

$$(i = n, \ldots, 1)$$

with

$$V_{n+1}\{x_{n+1}\} = 0$$

subject to (8.1b) and x_1 given, where $V_i\{x_i\}$ is the expected utility at stage i from pursuing an optimal policy.

Following the same procedure as before results in the optimality condition

$$\partial a_i/\partial u_i = \delta p_i(\partial a_{i+1}/\partial x_{i+1} + \partial a_{i+1}/\partial u_{i+1}) \times (1 - 2ba_{i+1}\{\cdot\})/(1 - 2ba_i\{\cdot\}) \qquad (i = n, \ldots, 1) \qquad (8.6)$$

Condition (8.6) can be compared with (8.5). The introduction of p_i acts as a discount factor additional to δ used for discounting utility. A similar result is obtained when the forest rotation problem is considered in Chapter 9. The introduction of risk aversion leads to the right-hand side of (8.5) being multiplied by the ratio of marginal utilities,

$$(1 - 2ba_{i+1}\{x_{i+1}, u_{i+1}^*\})/(1 - 2ba_i\{x_i, u_i^*\})$$

This factor is greater than one if the optimal stage return is less at stage $i+1$ than at stage i. Whether or not it is depends on the functions involved.

8.1.3 Exploration

The resource base, x_i, can only be the known resource base, or the best estimate of the resource base, at stage i. Although the resource may be exhaustible, the known resource base may be increased as a result of exploration activity. Let the cost of effort required to find an additional z_i units of the resource at stage i be denoted by $e_i\{x_i, z_i\}$. It is assumed that the cost of exploration effort increases at an increasing rate with the amount discovered ($\partial e_i/\partial z_i > 0$; $\partial^2 e_i/\partial z_i^2 > 0$) and increases at an increasing rate as the resource base is diminished ($\partial e_i/\partial x_i < 0$; $\partial^2 e_i/\partial x_i^2 > 0$). Abstracting from risky possession and risk aversion, and assuming the objective is to maximize the present value of net returns, the recursive equation is

$$V_i\{x_i\} = \max_{u_i, z_i} [a_i\{x_i, u_i\} - e_i\{x_i, z_i\} + \alpha V_{i+1}\{x_i - u_i + z_i\}] \qquad (i = n, \ldots, 1) \quad (8.7)$$

with

$$V_{n+1}\{x_{n+1}\} = 0$$

and the same constraints as before.

Again assuming an interior solution, the conditions for the right-hand side of (8.7) to be a maximum are

$$\alpha(\mathrm{d}V_{i+1}/\mathrm{d}x_{i+1}) = \partial a_i/\partial u_i = \partial e_i/\partial z_i$$

The conditions state that for optimality the present value of the shadow price of the resource stock at stage $i+1$ should equal both the net revenues from mining an extra unit of the resource and the cost of finding an extra unit of the resource at stage i.

The model developed here closely follows one formulated by Fisher (1981, Section 2.7), although he uses a control theory framework. He has argued persuasively that scarcity rent is a good economic indicator of scarcity, superior to measures of known reserves, estimated reserves, market price of the resource, or cost of extraction. He has proposed that one way of estimating scarcity rent $\partial a_i/\partial u_i$ which is often not measurable is by estimating $\partial e_i/\partial z_i$.

Fisher justifies a deterministic exploration function in two ways (p. 57):

> First, formally modeling the exploration uncertainty would involve the more advanced mathematical models we have been trying to avoid. Second, and more important, in this particular case the sense of the results will not be affected by introducing uncertainty.

A simple way of introducing exploration uncertainty is to redefine $e_i\{x_i, z_i\}$ as the cost of effort required to find an amount z_i with probability p_i, the probability of finding nothing being $(1-p_i)$. The last term in (8.7) becomes

$$\alpha(p_i V_{i+1}\{x_i - u_i + z_i\} + (1-p_i)V_{i+1}\{x_i - u_i\})$$

For an interior solution, $\partial a_i/\partial u_i = (\partial e_i/\partial z_i)/p_i$. Thus a measure of resource scarcity is given by marginal exploration cost divided by the probability of success. For more complicated probability distributions it does become difficult to relate optimal $\partial a_i/\partial u_i$ to optimal $\partial e_i/\partial z_i$. For other considerations related to the optimal level of exploration under uncertainty, see Fisher (1981) and Pindyck (1978).

8.2 A pollution problem

8.2.1 Pollution as a stock variable

In the previous section optimality conditions were obtained for mining problems using dynamic programming analytically, assuming differentiable return and cost functions, and that solutions were interior solutions. For problems with discontinuous functions it may be difficult to gain an insight into the optimal time paths of state and decision variables by relying on analytical methods. It may be easier to formulate the problem as a numerical dynamic programming problem. A range of solutions can then be obtained for different values of problem parameters.

As an example, consider the type of pollution problem discussed by Pearce (1976) and Dasgupta (1982, Ch. 8). Waste generated in period i may not only lead to adverse external effects in period i, but in subsequent periods as well. The persistence of the effect depends on the type of pollutant and the assimilative capacity of the environment. For example, heavy metals such as cadmium and mercury do not degrade through time, so that heavy metal pollutants are cumulative in their effect. Waterways above some threshold level of purity are able to carry away bio-degradable pollutants without external costs. For levels below the threshold the waterway cannot absorb pollutants costlessly within one period, and perhaps not over many periods. Pearce points out that the usual Pigouvian prescription of setting taxes on pollutants equal to their marginal external cost within a period may be suboptimal when their effect is viewed over a longer time span. A dynamic optimization approach must be adopted.

8.2.2 A numerical example

Functions which characterize a simple problem of pollution stocks are presented in Fig. 8.1. The decision variable D_i is the annual production level set at stage i, and the state variable I_i is the stock of pollution at stage i. The capacity of the environment to disperse the pollutant stock is high when the stock is low. However, for high stock levels most of the stock cannot be assimilated naturally and is carried over to the next period. The pollutant stock carried over from stage i is denoted by $A\{I_i\}$. Production results in immediate benefits $B\{D_i\}$, but also in waste $W\{D_i\}$. The pollution stock at stage $i+1$ equals $A\{I_i\}$ plus waste generated from production at stage i. Society experiences external costs $C\{I_i\}$ dependent on the pollution stock. The problem is to find, for some initial stock of pollution, the optimal sequence of production decisions so that the present value of benefits net of costs is maximized. State and decision variables are restricted to integer values. In symbols, the problem is

$$\max_{D_1,\dots,D_n} \sum_{i=1}^{n} \alpha^{i-1}[B\{D_i\} - C\{I_i\}]$$

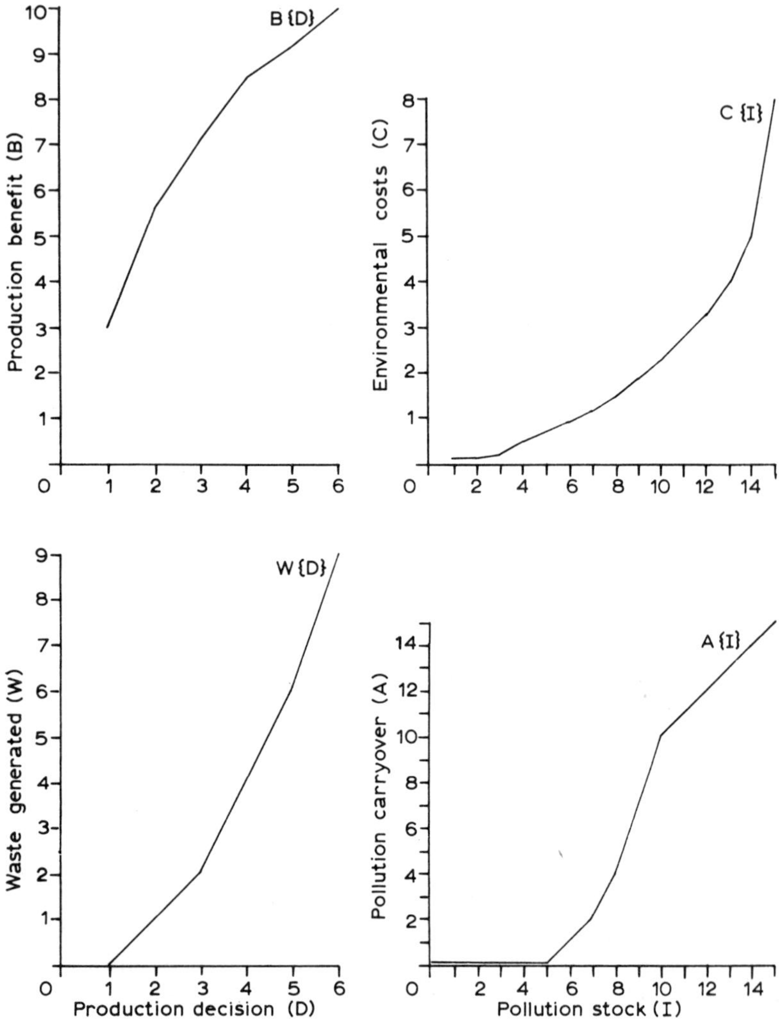

FIG. 8.1 The effect of production decisions and pollution stock on benefits and costs.

subject to

$$\left.\begin{aligned} &I_{i+1} = A\{I_i\} + W\{D_i\} \\ &D_i \in \text{integer}\,(1, 2, \ldots, 6) \\ &I_i \in \text{integer}\,(1, 2, \ldots, 15) \\ &I_1 \text{ given} \end{aligned}\right\} \quad (i = 1, \ldots, n)$$

The recursive equation for solving this problem is

$$V_i\{I_i\} = \max_{D_i} [B\{D_i\} - C\{I_i\} + \alpha V_{i+1}\{A\{I_i\} + W_i\{D_i\}\}] \qquad (i = n, \ldots, 1)$$

subject to the integer constraints with

$$V_{n+1}\{x_{n+1}\} = 0$$
$$I_1 \text{ given}$$

A BASIC program for writing the pollution problem data to the file POL.DAT is shown in Fig. 8.2. For the finite-stage problem, the number of decision stages is set equal to 7 (line 150). The data for the schedules B, C, A and W which are presented diagrammatically in Fig. 8.1 may be found in line numbers 180 to 250. The stage return is calculated at line 440, and the state transition at line 450. The output from running FDP (reading data from POL.DAT) is shown in Fig. 8.3, and that from running IDP for the infinite-stage solution is shown in Fig. 8.4. Details of the optimal time paths of pollution stocks and production decisions are shown in the lower parts of the two figures for an example initial pollution stock of 9.

The qualitative nature of the optimal time paths for an infinite planning horizon can be observed from Fig. 8.4. If the pollution stock reaches 10 or above, the stock cannot be reduced in subsequent periods (as is evident from the SE quadrant diagram in Fig. 8.1) but it is not optimal to increase the pollution stock by raising production above the minimum level. For all pollution stocks below 10, it is optimal to set production so that after a few stages the pollution stock is either 5 or 6. Pollution stock then oscillates between 5 and 6 indefinitely. Production is set at 5 when pollution stock is 5, which leads to a pollution stock of 6 at the next stage. Production is set at 4 when pollution stock is 6, which leads to a pollution stock of 5 at the next stage, and so on. The optimal infinite-stage policy is presented diagrammatically in Fig. 8.5, with the segment of oscillation ringed.

The optimal time paths of pollution stock and production for a planning horizon of 7 decision stages starting with a pollution stock of 9 are

```
100     REM----------POLLUTION PROBLEM----------
110     REM
120     LET N$="POL.DAT"
130     OPEN N$ FOR OUTPUT AS FILE #1
140     REM
150     LET N=7
160     LET D2=10
170     DIM B(6),C(15),A(15),W(6)
180     MAT READ B
190     DATA 3.,5.6,7.3,8.5,9.2,10.
200     MAT READ C
210     DATA .1,.2,.3,.5,.7,.9,1.2,1.5,1.9,2.3,2.8,3.3,4.,5.,8.
220     MAT READ A
230     DATA .0,.0,.0,.0,.0,1.,2.,4.,7.,10.,11.,12.,13.,14.,15.
240     MAT READ W
250     DATA .0,1.,2.,4.,6.,9.
260     LET C$=","
270     LET P=1
280     REM
290     REM-----WRITE BLOCK A DATA - PROBLEM PARAMETERS-----
300     PRINT #1, N
310     PRINT #1, D2
320     PRINT #1, "D"
330     PRINT #1, "Y"
340     PRINT #1, "Y"
350     REM
360     REM-----WRITE BLOCK B DATA - TERMINAL VALUES-----
370     PRINT #1, "0,0"
380     REM
390     REM-----WRITE BLOCK C DATA-----
400     PRINT #1, " T                 I   D    R                 J    P"
410     REM
420     FOR I=1 TO 15
430        FOR D=1 TO 6
440           LET R=B(D)-C(I)
450           LET J=A(I)+W(D)
460           IF J>15 THEN 480
470           PRINT #1, N;C$,I;C$;D;C$;R;C$,J;C$;P
480        NEXT D
490     NEXT I
500     PRINT #1, N,",0,0,0,0,1"
510     REM
520     REM-----WRITE BLOCK D DATA - LABELS-----
530     PRINT #1, "LABEL"
540     FOR I=1 TO 15
550        PRINT #1, "POLL"+STR$(I)
560     NEXT I
570     PRINT #1, "LABEL";C$;6
580     FOR D=1 TO 6
590        PRINT #1, "PROD"+STR$(D)
600     NEXT D
610     REM
620     PRINT "DATA WRITTEN TO ";N$
630     CLOSE #1
640     END
```

FIG. 8.2. Program for writing data for the pollution problem to the file POL.DAT.

DETERMINISTIC DYNAMIC PROGRAMMING SOLUTION
PROBLEM ---POL---

PROBLEM PARAMETERS:
NO. OF DECISION STAGES = 7
RATE OF DISCOUNT (PER CENT) = 10

OPTIMAL STATE SEQUENCE FOR ALL INITIAL STATES

STAGE 1	2	3	4	5	6	7	8	VALUE
1	6	5	6	5	6	7	11	44.6705
2	6	5	6	5	6	7	11	44.5705
3	6	5	6	5	6	7	11	44.4705
4	6	5	6	5	6	7	11	44.2705
5	6	5	6	5	6	7	11	44.0705
6	5	6	5	6	5	6	10	43.5999
7	6	5	6	5	6	7	11	42.8705
8	6	5	6	5	6	7	11	41.3705
9	8	6	5	6	5	6	10	37.2453
10	10	10	10	11	12	13	15	9.59753
11	11	11	11	11	12	13	15	5.90069
12	12	12	12	12	12	13	15	2.03992
13	13	13	13	13	13	14	15	-2.83771
14	14	14	14	14	14	14	15	-9.24289
15	15	15	15	15	15	15	15	-26.7763

DETAILS OF OPTIMAL PATH
ENTER INITIAL STATE NO. (OR '0' TO FINISH) ? 9

STAGE NO.	STATE NO.	STATE LABEL	DECISION NO.	DECISION LABEL	DISCOUNTED STAGE RETURN	VALUE
1	9	POLL9	2	PROD2	3.7	37.2453
2	8	POLL8	3	PROD3	5.27273	36.8999
3	6	POLL6	4	PROD4	6.28099	34.2099
4	5	POLL5	5	PROD5	6.38618	29.2708
5	6	POLL6	4	PROD4	5.1909	22.8479
6	5	POLL5	5	PROD5	5.27783	16.7727
7	6	POLL6	6	PROD6	5.13671	9.1
8	10	POLL10	0		0	0

FIG. 8.3. Finite-stage solution to the pollution problem.

```
------------------------------------------------------------
        DETERMINISTIC DYNAMIC PROGRAMMING SOLUTION
                   PROBLEM ---POL---

PROBLEM PARAMETERS:
INFINITE NO. OF STAGES
RATE OF DISCOUNT (PER CENT)  =  10
------------------------------------------------------------

NO. OF ITERATIONS TO FIND SOLUTION =  4
```

CURRENT STATE NO.	CURRENT STATE LABEL	DECISION NO.	DECISION LABEL	NEXT STATE NO.	NEXT STATE LABEL	PRESENT VALUE
1	POLL1	5	PROD5	6	POLL6	89.3857
2	POLL2	5	PROD5	6	POLL6	89.2857
3	POLL3	5	PROD5	6	POLL6	89.1857
4	POLL4	5	PROD5	6	POLL6	88.9857
5	POLL5	5	PROD5	6	POLL6	88.7857
6	POLL6	4	PROD4	5	POLL5	88.3143
7	POLL7	4	PROD4	6	POLL6	87.5857
8	POLL8	3	PROD3	6	POLL6	86.0857
9	POLL9	2	PROD2	8	POLL8	81.9597
10	POLL10	1	PROD1	10	POLL10	7.7
11	POLL11	1	PROD1	11	POLL11	2.2
12	POLL12	1	PROD1	12	POLL12	-3.3
13	POLL13	1	PROD1	13	POLL13	-11
14	POLL14	1	PROD1	14	POLL14	-22
15	POLL15	1	PROD1	15	POLL15	-55

```
OPTIMAL STAGE CYCLE
ENTER INITIAL STATE NO. (OR '0' TO FINISH) ? 9
```

STAGE NO.	STATE NO.	STATE LABEL	DECISION NO.	DECISION LABEL
1	9	POLL9	2	PROD2
2	8	POLL8	3	PROD3
3	6	POLL6	4	PROD4
4	5	POLL5	5	PROD5
5	6	POLL6	4	PROD4

FIG. 8.4. Infinite-stage solution to the pollution problem.

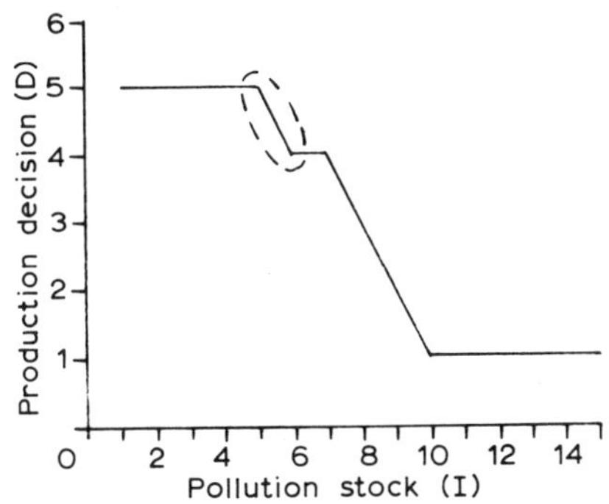

FIG. 8.5. Optimal infinite-stage production policy.

presented diagrammatically in Fig. 8.6. From stage 3 pollution stock oscillates between 6 and 5 until the last stage, when pollution jumps to 10. It is optimal to increase production to 6 at stage 7 because there are no costs attached to the resulting increase in the pollution stock at stage 8. The optimal infinite-stage path is joined at stage 3 and left at stage 6.

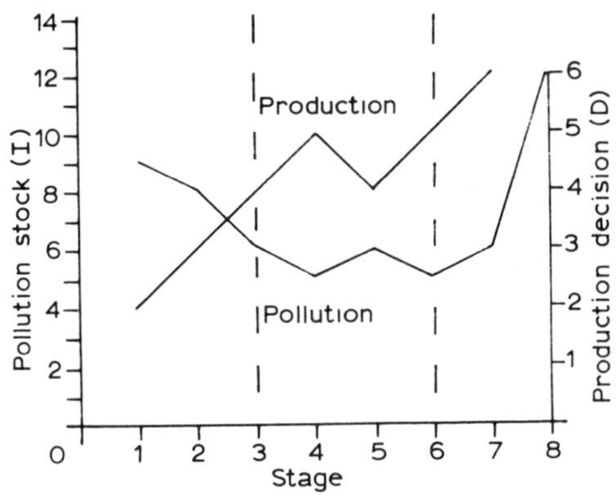

FIG. 8.6. Optimal time profiles of pollution stock and production decisions.

8.3 Rules for Making Irreversible Decisions Under Uncertainty

8.3.1 Irreversible decisions and quasi-option value

In the pollution problem examined in the previous section, if when pollution stock was low a production decision led to a pollution stock of

10 or more, it became impossible subsequently to make any production decision which would reduce the pollution stock to its former level. Such a production decision is an example of an irreversible decision with consequent net returns known with certainty.

Other examples of irreversible decisions in land management are development decisions such as to build dams and roads, and to strip mine, which also lead to the destruction of wilderness areas, natural river systems, unique species of flora and fauna or achaeological sites. The benefits from the natural state are typically uncertain, and are likely to change over time. If a decision not to develop is taken now, it can always be reversed later if the net benefits of development over preservation become more attractive. However, if a decision to develop is taken now, the decision cannot be reversed later, no matter how desirable the attractions of preservation over development might become in future. If it is known that more information on the future benefits of preservation and development will come to light by later stages there is likely to be some advantage from postponing any decision to develop until more information is available. Fisher and Krutilla (1975) refer to the expected value of continuing to keep options open as quasi-option value. They stress that quasi-option value applies even if the decision maker is risk neutral.

However, it is important not to overemphasize the usefulness of the concept of quasi-option value. It is needed only if the decision-making approach is not consistent with the objective of the decision making. It would be more useful to emphasize the correct decision-making approach, which is based on decision-tree analysis or stochastic dynamic programming.

It has been shown by Arrow and Fisher (1974) and Henry (1974) that if the irreversible decision were to be made now on a once-and-for-all basis of selecting the option showing maximum expected present value, ignoring the possibilities of postponing an irreversible development decision and of waiting for more information on the benefits of development and preservation, then a quasi-option value is associated with the reversible preservation decision. Henry (1974) proves this proposition using stochastic dynamic programming. However, if these possibilities were recognized in the decision-making framework in the first place, there would be no need to invent the concept of quasi-option value. The implications of some suggestions that have been made are that cost–benefit analysis should be conducted using an incorrect decision-making framework, and then to in some way correct for possible error by applying a quasi-option value. For example, Arrow and Fisher (1974, p. 319) state: 'Essentially, the point is that the expected benefits of an irreversible decision should be adjusted to reflect the loss of options it entails'. However, there are no general guides as to the size of the adjustment factor, despite attempts by Henry (1974) to provide some. Porter (1982), describing a new approach to wilderness preservation through cost–benefit analysis, appears to believe that a decision-making approach consistent with the nature of an irreversible decision under uncertainty is beyond the scope of cost–benefit analysis. He states (p. 60) '... decision making under uncertainty intrudes critically when an irreversible decision is considered where the foregone future wilderness values are currently unclear; benefit–cost analysis, in this area as elsewhere, is not yet equipped to handle uncertainty explicitly and comfortably'. The way in which stochastic dynamic programming may be used to solve such problems is described below. There is no reason why the technique should not be viewed as one of the tools of the cost–benefit analyst.

8.3.2 A numerical example

Consider as an example problem a modification of one posed by Hodge (1984). At the beginning of each year a decision has to be made whether to preserve or develop a given area of land, unless the land has already been developed, in which case it must remain developed. The annual return D from a development decision is the same for all years. The annual return from preservation P in year i is subject to a stochastic growth rate g_i, which may be positive, zero or negative. The growth rate equals the percentage rate of growth achieved in year $i-1$, plus or minus one per cent. More specifically, g_i is equally likely to be one of $(g_{i-1}-0{\cdot}01)$, g_{i-1} and $(g_{i-1}+0{\cdot}01)$. The objective is to maximize the present value of expected returns. The time horizon is infinite.

There are two state variables for this problem. One is x_i, representing the status of the land by 1 if the land has been developed prior to stage i, and 0 if the land is still undeveloped. The other state variable is g_{i-1}. The rate of growth which will be recorded for the period starting at stage i but which is uncertain at stage i is $g_i^k = g_{i-1} + 0{\cdot}01(k-2)$ where k is a random integer variable taking the values 1, 2 or 3. The expected stage returns $a\{x_i, g_{i-1}, x_{i+1}\}$ can be read off the payoff matrix in Table 8.1. The entry '$-\infty$' for the transition from state '1' to state '0' effectively makes the transition infeasible if the aim is maximum expected present value. Thus the irreversibility of the problem is evident because, within the confines of the binary 0/1 states, it is always possible to move to the

TABLE 8.1
Expected Stage Returns for Preservation and Development Decisions

State (x_i)	*Decision* (x_{i+1})	
	0	*1*
0	$P(1+g_{i-1})$	D
1	$-\infty$	D

same state or a higher state, but never to a lower state. To make results easier to interpret, the annual return from the development decision is the same whether the land is undeveloped or developed. However, it would be more realistic to allow for the capital costs of development, making the NE quadrant of Table 8.1 negative for capital costs, and leaving the SE quadrant positive for subsequent returns.

The recursive equation for solving the problem is

$$V\{x_i, g_{i-1}\} = \max_{x_{i+1}=0 \text{ or } 1} \left[a\{x_i, g_{i-1}, x_{i+1}\} + \alpha \sum_{k=1}^{3} p\{k\} V\{x_{i+1}, g_i^k\} \right] \quad (8.8)$$

Let P and D in Table 8.1 both be \$1. If a once-and-for-all decision were made at stage 1 based on maximum present value of expected return, a preservation decision would be made for any g_0 greater than zero. The aim of solving the stochastic dynamic programming problem is to discover how low the threshold g_0 should be when the actual decision-making structure is taken into account.

The growth rate g_{i-1} is restricted to one of 25 values from -3 to $+3$ per cent at intervals of 0·25 per cent. If g_i^k in (8.8) falls outside these bounds it is reset to the closer bound. The state of the system can therefore assume one of 26 values, numbered 1 to 25 for $x_i = 0$ (land not developed) and associated g_{i-1}, and the last, 26, for $x_i = 1$. The rate of discount is initially set at 5 per cent. A BASIC program for writing the problem data to the file LAND.DAT is presented in Fig. 8.7.

The infinite-stage solution obtained by running IDP is shown in Fig. 8.8. It is optimal to preserve the land for any $g_{i-1} \geq -1{\cdot}25$ per cent, an indication of the bias required in favour of preservation if decisions were made on a once-and-for-all basis. As expected, the bias is reduced if the rate of discount is increased to 10 per cent. The benefits of keeping options open whilst waiting for further information are reduced. The

```
100 REM----------LAND----------
110 REM
120 LET N$="LAND.DAT"
130 OPEN N$ FOR OUTPUT AS FILE #1
140 REM
150 LET N=20
160 LET D1=5
170 LET F1=1/(1+D1/100)
180 LET P5=1.
190 LET D5=1.
200 LET T=1
210 DIM G(50)
220 LET G9=25
230 LET S=G9+1
240 LET S2=S/2
250 FOR I=1 TO G9
260     LET G(I)=(S2-I)*.0025
270 NEXT I
280 LET C$=","
290 REM
300 REM-----WRITE BLOCK A DATA - PROBLEM PARAMETERS-----
310 PRINT #1, N
320 PRINT #1, D1
330 PRINT #1, "S"
340 PRINT #1, "Y"
350 PRINT #1, "Y"
360 REM
370 REM-----WRITE BLOCK B DATA - TERMINAL STATE VALUES-----
380 PRINT #1, "0,0"
390 REM-----WRITE BLOCKS CS AND ES DATA-----
400 PRINT #1, " T                 I    D     R                      J     P"
410 FOR I=1 TO S
420     FOR D=1 TO 2
430         IF I<S THEN 490
440         IF D>1 THEN 470
450         LET R=-1000000
460         GO TO 530
470         LET R=D5*F1
480         GO TO 530
490         IF D=2 THEN 520
500         LET R=P5*(1+G(I))*F1
510         GO TO 530
520         LET R=D5*F1
530         FOR P9=1 TO 3
540             IF D=2 THEN 610
550             IF I=S THEN 610
560             LET J=I-(P9-2)*4
570             IF ABS(J-S2)<S2 THEN 590
580             LET J=I
590             LET P2=1/3
600             GO TO 630
610             LET J=S
620             LET P2=1
630             PRINT #1, T;C$,I;C$;D;C$;R;C$,J;C$;P2
640             IF J=S THEN 660
650         NEXT P9
660     NEXT D
670 NEXT I
680 PRINT #1, T,",0,0,0,0,1"
690 REM-----WRITE BLOCK D DATA - LABELS-----
700 PRINT #1, "LABEL"
710 FOR I=1 TO G9
```

FIG. 8.7. Program for writing data for the land development problem to the file LAND.DAT.

STOCHASTIC DYNAMIC PROGRAMMING SOLUTION
PROBLEM ---LAND---

PROBLEM PARAMETERS:
INFINITE NO. OF STAGES
RATE OF DISCOUNT (PER CENT) = 5

NO. OF ITERATIONS TO FIND SOLUTION = 5

CURRENT STATE		DECISION		NEXT STATE		PRESENT VALUE
NO.	LABEL	NO.	LABEL	NO.	LABEL	
1	L.03	1	PRESERVE	?		20.2854
2	L.0275	1	PRESERVE	?		20.2475
3	L.025	1	PRESERVE	?		20.2105
4	L.0225	1	PRESERVE	?		20.178
5	L.02	1	PRESERVE	?		20.2383
6	L.0175	1	PRESERVE	?		20.2022
7	L.015	1	PRESERVE	?		20.1672
8	L.0125	1	PRESERVE	?		20.1373
9	L.01	1	PRESERVE	?		20.167
10	L.0075	1	PRESERVE	?		20.1348
11	L.005	1	PRESERVE	?		20.1039
12	L.0025	1	PRESERVE	?		20.0797
13	LO	1	PRESERVE	?		20.0908
14	L-.0025	1	PRESERVE	?		20.0652
15	L-.005	1	PRESERVE	?		20.0413
16	L-.0075	1	PRESERVE	?		20.0266
17	L-.01	1	PRESERVE	?		20.0283
18	L-.0125	1	PRESERVE	?		20.0128
19	L-.015	2	DEVELOP	?		20
20	L-.0175	2	DEVELOP	?		20
21	L-.02	2	DEVELOP	?		20
22	L-.0225	2	DEVELOP	?		20
23	L-.025	2	DEVELOP	?		20
24	L-.0275	2	DEVELOP	?		20
25	L-.03	2	DEVELOP	?		20
26	D.000	2	DEVELOP	?		20

FIG. 8.8 Infinite-stage solution to the land development problem.

threshold g_{i-1} is 1·00 per cent. Because the selection of discount rates is controversial, it is worth considering the discounting procedure further.

8.3.3 The discounting procedure

Factors such as income and company tax drive a wedge between the opportunity cost of funds (ρ) and the rate of time preference (r). The question arises as to whether α in (8.8) should be set at $1/(1+\rho)$, $1/(1+r)$ or some other value. Whilst some (e.g. Baumol, 1968) have argued that on the grounds of promoting economic efficiency, the rate of discount should be ρ, others (e.g. Bradford, 1975) have argued for a synthetic approach which takes account of both ρ and r. Under the synthetic approach, all future consumption and investment flows generated by the costs and returns of the project being evaluated are considered. All consumption flows are assumed to earn a return of r, and all investment flows a return of ρ. It is also assumed that money flows to and from the project are split between consumption and investment in the proportions $(1-\theta)$ and θ, respectively. The value of θ may be made dependent on the source of the money flow. Further, the reinvestment of each monetary return may entail the evaluation of an infinite stream of concomitant annual consumption and investment flows.

In evaluating a development project it may be necessary to estimate some flows which are non-monetary consumption flows, such as the recreation benefits from the preservation of a wilderness area. Porter (1982, p. 72) has argued:

> Whichever the rationale, the wilderness preservationist should urge that two different discount rates (ρ and r) be used in assessing a potential development project, depending upon whether investment or consumption flows are affected. This dichotomization of the discount factor can make a critical difference in the evaluation of wilderness development projects.

A natural question is how the solution equation (8.8) has to be modified to take account of two relevant discount rates. The answer is to work in consumption flows only and to use $\alpha = 1/(1+r)$ as the discount factor in (8.8). All monetary flows such as those originating from the development decision are converted to equivalent consumption flows. The conversion factor depends on the assumptions made about reinvestment behaviour. For every \$1 money flow generated by the project, suppose $\$\theta$ are invested at the rate ρ. The following period the funds available from the original money flow are $\$\theta(1+\rho)$, of which $\$\theta$ can be thought of as the capital component, and $\$\theta\rho$ as the interest component. One alternative is to assume that $(1-\theta)$ of the *interest* is consumed in the following period, equal to $\$(1-\theta)\theta\rho$, leaving $\$\theta$ capital and $\$\theta\rho^2$ unconsumed interest equal to $\$\theta(1+\theta\rho)$ for reinvestment. If this process continues indefinitely, and $\theta\rho < r$, eventually the funds available for reinvestment tend toward zero. The present value of the consumption stream, which is also the required conversion factor, is

$$c = 1 + \theta(\rho - r)/(r - \rho\theta)$$

For example, if $\theta = 0{\cdot}2$, $r = 0{\cdot}05$ and $\rho = 0{\cdot}10$, then $c = 1{\cdot}333$.

An alternative assumption is that $(1-\theta)$ of the interest *and capital* is consumed in the following period, equal to $\$(1-\theta)\theta(1+\rho)$, leaving only $\$\theta^2(1+\rho)$ for reinvestment. If this process continues indefinitely, and $\theta(1+\rho)<(1+r)$, the conversion factor tends to

$$c' = (1-\theta)(1+r)/(1+r-\theta(1+\rho))$$

For the same parameter values, $c' = 1{\cdot}012$.

For further discussion of the synthetic approach to discounting see Bradford (1975), Lind (1982), Mendelsohn (1981) and Mishan (1975). The effects of alternative discounting procedures on the irreversibility decision problem are examined in more detail in Kennedy (1985).

8.4 References

Arrow, K.J. and Fisher, A.C. (1974), Environmental preservation, uncertainty and irreversibility, *Quarterly Journal of Economics*, **88**(2), 312–19.

Baumol, W.J. (1968), On the social rate of discount, *American Economic Review*, **58**(4), 788–802.

Bradford, D.F. (1975), Constraints on government investment opportunities and the choice of discount rate, *American Economic Review*, **65**(6), 887–99.

Dasgupta, P. (1982), *The Control of Resources*, Basil Blackwell, Oxford.

Fisher, A.C. (1981), *Resource and Environmental Economics*, Cambridge University Press, Cambridge.

Fisher, A.C. and Krutilla, J.V. (1975), Resource conservation, environmental preservation and the rate of discount, *Quarterly Journal of Economics*, **89**(3), 358–70.

Henry, C. (1974), Investment decisions under uncertainty: the 'irreversibility effect', *American Economic Review*, **64**(6), 1006–12.

Hodge, I. (1984), Uncertainty, irreversibility and the loss of agricultural land, *Journal of Agricultural Economics*, **35**(2), 191–202.

Kennedy, J.O.S. (1985), Time preference, the opportunity cost of capital and irreversible decisions, Unpublished paper, School of Economics, La Trobe University, Melbourne.

Lind, R.C. (1982), A primer on the major issues relating to the discount rate for evaluating national energy options in Lind, R. *et al.* (1982), *Discounting for Time and Risk in Energy Policy*, Resources for the Future, Washington DC.

Mendelsohn, R. (1981), The choice of discount rates for public projects, *American Economic Review*, **71**(1), 239–41.

Mishan, E.J. (1975), 2nd edition, *Cost–Benefit Analysis*, George Allen and Unwin, London.

Pearce, D.W. (1976), Limits of cost–benefit analysis as a guide to environmental policy, *Kyklos*, **29**(1), 97–112.

Pindyck, R.S. (1978), Optimal exploration and production of a renewable resource, *Journal of Political Economy*, **86**(5), 841–62.

Porter, R.C. (1982), The new approach to wilderness preservation through benefit–cost analysis, *Journal of Environmental Economics and Management*, **9**(1), 59–80.

Solow, R.M. (1974), The economics of resources or the resources of economics, *American Economic Review*, **64**(2), 1–14.

Chapter 9

FORESTRY MANAGEMENT

9.1 Problems in Forestry Management

If land has been assigned indefinitely to timber production, the age at which trees should be clearcut and the land replanted must be determined. This is an optimal replacement problem, similar to that encountered in livestock management. However, just as in the production of livestock for meat, there may be other management decisions taken between replacement decisions which affect final biomass at replacement. In the case of livestock production the decisions are usually feed input decisions, whereas in the case of timber production the decisions are output decisions on partial harvesting before clearcutting.

The problem of the optimal timing of the replacement of a forest is referred to as the *rotation problem* and the problem of optimal partial harvesting as the *thinning problem*. The rotation and thinning problems are usually interdependent, and need to be solved together. The problems can be formulated as multistage decision processes, and solved using dynamic programming.

An important difference between the livestock problem and the forestry problem is the length of the productive life of the biological unit. Forest rotations are of the order of 30 to 60 years, which has various implications for the rotation decision. The optimal rotation period is much more sensitive to the discount rate than is the optimal period between livestock replacements. The correct principles to use for determining the relevant discount rate are controversial. Even when the principles are agreed on, using them to estimate rates 20 years or so ahead is also difficult. Of course, there is no reason why the same discount rate should be used for all years, or why the rate should not be treated as a random variable.

Another consequence of the length of the rotation period is that decisions are subject to uncertainty. The longer the rotation period the greater the chance of losing the forest stand through disease, fire, drought or expropriation. Long-term stand growth and quality, timber prices, harvesting costs and planting costs are all highly uncertain. The uncertain planning environment makes it prudent to employ an adaptive approach to decision making. Information about the optimal rotation period is better seen as a by-product of determining whether to thin or clearcut now rather than an accurate forecast of the harvest age of the next stand of timber.

In the next section the condition for the optimal rotation period is derived using a dynamic programming approach, first for the deterministic case, and then for a stochastic case. The conditions are simple to derive by this approach, and give insight into the determinants of the optimal rotation period. In Section 9.3 the combined rotation and thinning problem is examined, and numerical dynamic programming applications are reviewed. Extensions to the basic rotation and thinning problem are considered in the final section.

9.2 The Optimal Rotation Period

In the deterministic case, all relevant information is known with certainty. The rules for this case are derived first. In the stochastic problem examined thereafter, the successful transition of the stand from one year to the next is probabilistic. The transition may be unsuccessful because the stand is destroyed through fire or some other calamity. Though this is an important case, it is only one stochastic problem out of many which could be considered.

9.2.1 Deterministic problems

Suppose a given area of land has been committed indefinitely to timber production. If there are trees already standing, the decision to be made at every stage is whether to allow the trees to grow at least until the next decision stage, or to clearcut the stand and replant. In this simplest possible case, the only state determinant of the decision is the age of the trees. All trees in the stand are the same age. If the land is bare the only possible action is to plant trees.

Planting costs are denoted by k, and are incurred at the beginning of the period in which the land is bare. The return from clearcutting and selling the timber, net of all harvesting, transport and selling costs, is a function of the age of the stand at harvesting, and is referred to as net stumpage value, $s\{t\}$. Below a certain age t^0 the trees have no commercial value so that $s\{t\}=0$ for $t<t^0$. For $t \geq t^0$, $s\{t\}$ is positive, initially increasing with t but eventually decreasing with t for $t>T$. It is assumed that a decision to clearcut is implemented instantaneously, and that clearcutting is followed immediately by planting another stand. The stand must be clearcut if it reaches T years of age.

The objective of the decision making is to maximize the present value of net income streams to infinity. The planting costs k and rate of discount r are constant through time, and the stumpage value function $s\{t\}$ is stationary. Because the prospects for a stand of trees aged t years in all rotations are identical, the optimal decision and present value of net income, $V\{t\}$, are the same in all rotations.

A dynamic programming specification of $V\{t\}$ and the decision alternatives for each age of stand, t, are:

$$V\{t\} = \begin{cases} -k + \alpha V\{1\} & \text{(plant)} \quad \text{for } t=0 \\ \max\,[\alpha V\{t+1\}, s\{t\} + V\{0\}] & \text{(keep or clearcut)} \quad \text{for } 0<t<T \\ s\{t\} + V\{0\} & \text{(clearcut)} \quad \text{for } t=T \end{cases} \tag{9.1}$$

where α is the discount factor.

The typical optimal policy for this problem can be defined in terms of the optimal rotation period $t^* < T$ as follows

plant	if $t=0$
keep	if $t<t^*$
clearcut	if $t \geq t^*$

The present value function for this policy is

$$V\{t\} = \begin{cases} -k + \alpha V\{1\} & \text{(plant)} \quad \text{for } t=0 \\ \alpha V\{t+1\} & \text{(keep)} \quad \text{for } 0<t<t^* \\ s\{t\} + V\{0\} & \text{(clearcut)} \quad \text{for } t \geq t^* \end{cases} \tag{9.2}$$

Now t^* is the rotation period for which $V\{t\}$ is maximized. Let $\hat{V}\{t\}$ denote the present value of net income to infinity derived from a stand of age t if the rotation period is instead $\hat{t}$. It follows that for a sufficiently small interval between stages, $\hat{V}\{t\} - V\{t\}$ would be inconsequentially small for all t if $\hat{t} = t^* + 1$. It is of particular interest that the forest manager would be indifferent between $\hat{V}\{0\}$ and $V\{0\}$, and between $\hat{V}\{t^*\}$ and $V\{t^*\}$. Another way of putting this is that when the age of the forest stand is t^* years he is indifferent between clearcutting to receive $s\{t^*\} + V\{0\}$, and postponing clearcutting one year, the present value of which is $\alpha V\{t^* + 1\} = \alpha(s\{t^* + 1\} + V\{0\})$. That is,

$$V\{t^*\} = s\{t^*\} + V\{0\} = \alpha(s\{t^* + 1\} + V\{0\})$$

Substituting $1/(1 + r)$ for α and rearranging, this becomes

$$s\{t^* + 1\} - s\{t^*\} = r(s\{t^*\} + V\{0\}) = rV\{t^*\} \tag{9.3}$$

This is a discrete-time version of the Faustmann formula[1] which specifies the condition for the optimal rotation period t^*. The term $V\{0\}$ is the capitalized value of the land when it is bare immediately prior to planting. It is referred to in the forestry literature as *site value* or *soil expectation*, and can be thought of as the opportunity cost of the land on which the trees are grown. The Faustmann formula is another way of expressing indifference between a rotation of t^* years and one of $t^* + 1$ years when t^* is the optimal rotation period. The increase in net stumpage value from postponing clearcutting must be exactly offset by the interest foregone by not having $(s\{t^*\} + V\{0\})$ to invest for one year.

The site value $V\{0\}$ can be expressed in terms of k, $s\{t^*\}$ and α. From (9.2) it follows that

$$\begin{aligned} V\{0\} &= -k + \alpha V\{1\} = -k + \alpha^2 V\{2\} = \cdots \\ &= -k + \alpha^{t^*}(s\{t^*\} + V\{0\}) \\ &= (-k + \alpha^{t^*} s\{t^*\})/(1 - \alpha^{t^*}) \end{aligned} \tag{9.4}$$

Substituting for $V\{0\}$ in (9.3) gives the following alternative version of the Faustmann formula

$$s\{t^* + 1\} - s\{t^*\} = r(s\{t^*\} - k)/(1 - \alpha^{t^*}) \tag{9.5}$$

The condition for the optimal rotation is given a diagrammatic interpretation in Fig. 9.1. The net stumpage value is added to site value to give the curve marked $s\{t\} + V\{0\}$. The vertical intercept of the curve marked $\alpha^{t^*-t}(s\{t^*\} + V\{0\})$ shows the present value at $t = 0$ of net income

[1] To be rigorous the stage interval must be approaching zero. For historical background and a full exposition of the analytics of the optimal rotation, see Samuelson (1976).

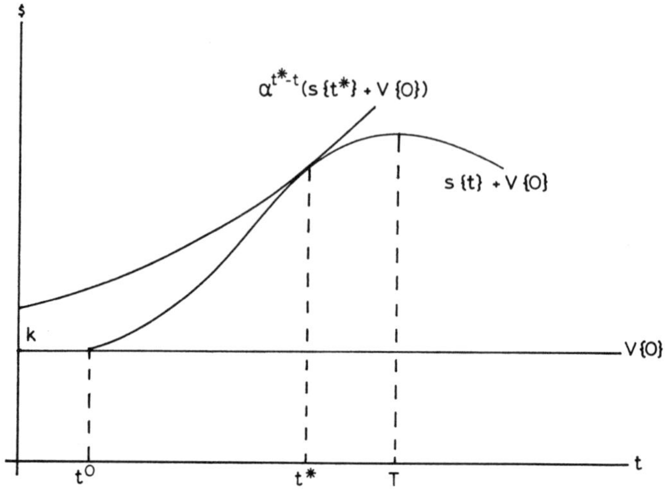

FIG. 9.1. Condition for the optimal rotation period, t^*.

from clearcutting. From (9.2), $V\{t\} = \alpha^{t^*-t}(s\{t^*\} + V\{0\})$ for $t \le t^*$, and $s\{t\} + V\{0\}$ for $t \ge t^*$.

At time $t = 0$, before planting, the forest manager is indifferent between selling the land for its maximum capitalized opportunity cost, $V\{0\}$, and alternatively planting trees at a cost of k, obtaining net income from clearcutting at the end of the optimal rotation period, and then having bare land on hand again worth $V\{0\}$.

The two curves are tangential to one another at $t = t^*$. The slope of the curve marked $s\{t\} + V\{0\}$ is approximately $s\{t^* + 1\} - s\{t^*\}$. The slope of the other curve is approximately $rV\{t\}$ for $t < t^*$, which follows from $V\{t\} = \alpha V\{t+1\}$ in (9.2) for the 'keep' decision. Thus at t^*, $s\{t^* + 1\} - s\{t\} = rV\{t^*\} = r(s\{t^*\} + V\{0\})$ which is the Faustmann formula (9.3).

Given data for a practical problem, how should t^* be determined? Equation (9.5) can be solved for t^* numerically, or graphically as shown in Clark (1976, p. 260). It could be found by solving the dynamic programming problem (9.1). However, probably the simplest method is to use a search procedure. The present value of net income from bare land dedicated to forestry production with a rotation period of t is $\Pi\{t\} = (-k + \alpha^t s\{t\})/(1 - \alpha^t)$ by (9.4). The problem can be solved by finding t for which $\Pi\{t\}$ is a maximum. This is the same problem as finding $V\{0\}$ and t^* in (9.1) and (9.2).

9.2.2 Stochastic problems

Suppose that the rotation problem is the same as before except that there is no longer certainty that trees aged t left for a year survive intact over the year. Let the probability that they do survive be p_t. Assume that if the trees are involuntarily destroyed, the return from salvaging destroyed trees and clearing the land nets out to zero, and that at the next stage the land is ready for planting. The optimal value of a stand of timber aged t is now an expected value, and using the stochastic dynamic programming approach is

$$V\{t\} = \begin{cases} -k + \alpha(p_t V\{1\} + (1-p_t)V\{0\}) & \text{for } t = 0 \\ \max\,[\alpha(p_t V\{t+1\} + (1-p_t)V\{0\}), s\{t\} + V\{0\}] & \text{for } 0 < t < T \\ s\{t\} + V\{0\} & \text{for } t = T \end{cases} \tag{9.6}$$

Arguing as before, when the age of the stand is t^* years, the forest manager is indifferent between clearcutting immediately or one year later, which means that

$$\begin{aligned} s\{t^*\} + V\{0\} &= \alpha(p_t(s\{t^*+1\} + V\{0\}) + (1-p_t)V\{0\}) \\ &= \alpha(p_t s\{t^*+1\} + V\{0\}) \end{aligned}$$

which after rearrangement becomes

$$s\{t^*+1\} - s\{t^*\} = ((1 + r - p_t)/p_t)s\{t^*\} + (r/p_t)V\{0\} \tag{9.7}$$

This is a stochastic version of the Faustmann formula (9.3), and is the same as (9.3) for $p_t = 1$.

Equation (9.6) can be used to find $V\{0\}$. Let the expected value at $t = 0$ from successfully harvesting the stand at age t^* years be

$$R^{s} = -k + \alpha^{t^*} \sum_{t=1}^{t^*} p_{t-1}(s\{t^*\} + V\{0\})$$

and the expected value of failure to reach planned cutting age be

$$R^{f} = V\{0\} \sum_{t=1}^{t^*} \alpha^{t}(1 - p_{t-1}) \prod_{i=1}^{t} p_{i-2}$$

defining $p_{-1} \equiv 1$. Then the expected value of a rotation of planned length t^* is $V\{0\} = R^s + R^f$, from which it follows that

$$V\{0\} = \frac{-k + \alpha^{t^*} s\{t^*\} \prod_{t=1}^{t^*} p_{t-1}}{1 - \alpha^{t^*} \prod_{t=1}^{t^*} p_{t-1} - \sum_{t=1}^{t^*} \alpha^t (1 - p_{t-1}) \prod_{i=1}^{t} p_{i-2}} \tag{9.8}$$

If the probability of failure is the same over all years so that $p_1 = p_2 = \cdots = p_T = p$, then (9.8) simplifies to

$$V\{0\} = ((1 + r - p)/r)(-k + (p/(1 + r))^{t^*} s\{t^*\})/(1 - (p/(1 + r))^{t^*})$$

Substituting for $V\{0\}$ in (9.7) gives the stochastic Faustmann formula for time-invariant survival probabilities

$$s\{t^* + 1\} - s\{t^*\} = r'(s\{t^*\} - k)/(1 - (1 + r')^{-t^*}) \tag{9.9}$$

where $r' = (r + 1 - p)/p$. Equation (9.9) is the same as (9.5) with a change of discount rate. In fact, (9.9) can also be viewed as (9.5) after substituting αp for α. The discount factor αp equals the factor $(1 + r')^{-1}$ which appears in (9.9). This result has been obtained by Burt (1965) for both the discrete-time problem examined here and the continuous-time problem, but using a different line of reasoning. Reed (1984) also obtains the result for the continuous-time problem.

The stochastic Faustmann formula (9.7) for time-dependent probabilities with $V\{0\}$ given by (9.8) may be used to find the optimal rotation period t^* in an empirical problem. Alternatively, a search for the t which maximizes $V\{0\}$ in (9.8) may be conducted. This method has been suggested by Burt (1965) using a formula equivalent to (9.8). A dynamic programming routine such as GPDP could also be used to obtain a solution. An example follows in the next subsection. If the survival probabilities are constant, it is easiest to solve the problem as if it were a deterministic problem using as the discount factor the market discount factor weighted by the survival probability.

9.2.3 A numerical example of a combined rotation and protection problem

It has so far been assumed that the forest manager is unable to influence the probability of the stand surviving intact from one year to the next.

However, it is generally possible to increase the survival probability at some cost. For example, it may be possible to reduce the probability of a fire destroying a stand by improving lookout services, or by spending more on a fire-fighting capability. In the problem posed here the forest manager is faced every 5 years with two decisions. One is whether to clearcut a stand. The other is how much to spend on increasing the probability of survival of the stand if it is kept, or the probability of survival of a newly planted stand if the existing stand is cut. There are other methods of increasing the probability of survival of a capital nature which could also be considered. For example, stand density could be reduced at the time of planting, or the size of fire breaks could be increased.

Suppose that the net stumpage value as a function of time is

$$s\{t\} = 1200 - 140t + 3{\cdot}834t^2 - 0{\cdot}01667t^3$$

and that planting costs are $k = \$10$. These values have been used by Howe (1979, p. 230) in an example of a deterministic rotation problem. The net stumpage value function is based on data for the growth of a typical stand of Douglas fir trees in British Columbia (Pearse, 1967). Values at 5-year age intervals are shown in Table 9.1.

Howe uses the Faustmann formula to show that the optimal rotations are 44 and 52 years for discount rates of 10 and 6 per cent, respectively.

Consider an extended problem in which the probability of the stand

TABLE 9.1
Net Stumpage Values by Age of Tree

Age t (*years*)	*Net stumpage value* $s\{t\}$ (\$)
30	1
35	282
40	668
45	1 145
50	1 701
55	2 324
60	3 002
65	3 721
70	4 469
75	5 234
80	6 003

TABLE 9.2
Probability of Survival for at Least 5 Years

Protection expenditure (u) $	0	3	6	9	12
Probability of survival ($p\{u\}$)	0·89	0·93	0·96	0·98	0·99

surviving at least until the next decision stage is $p\{u\}$, where u is a decision variable, expenditure on protection. Suppose decision stages are spaced at intervals of 5 years, and that $p\{u\}$ is the function shown in Table 9.2. In a more realistic problem the probability of survival would be a function of stand age as well as protection expenditure.

It might be expected that as the stand approaches maturity, it would be worth spending more on protection because there is more to lose if the stand is destroyed by chance. To illustrate the optimal rate of increase of protection expenditure with age, the problem of maximizing the present value of expected net returns to infinity is solved using GPDP.

The recursive equation is

$$V\{t\} = \begin{cases} -k + \max_u [-u + \alpha(p\{u\}V\{1\} + (1 - p\{u\})V\{0\})] & \text{for } t = 0 \\ \max\left[\max_u (-u + \alpha(p\{u\}V\{t+1\} + (1 - p\{u\})V\{0\}),\ s\{t\} + V\{0\}\right] & \\ & \text{for } 0 < t < T \\ s\{t\} + V\{0\} & \text{for } t = T \end{cases}$$

A BASIC program for writing the problem data to file is shown in Fig. 9.2. Solutions were obtained by running IDP with this data file. The solutions are shown in Figs 9.3 and 9.4 for discount rates of 10 and 5 per cent per annum, respectively. The optimal rotation periods for the two discount rates are 45 and 55 years, respectively.

For both discount rates it is optimal to increase expenditure on protection of the stand as the age of planned cutting approaches. It is optimal to take protective action at an earlier stand age in the case of the lower discount rate.

In the BASIC program shown in Fig. 9.2 planting costs have not been deducted in calculating net returns for the rotation starting with zero-aged trees. They have been deducted for all subsequent rotations (see line 300). Thus the site value $V\{0\}$ is $\$(8{\cdot}38 - 10{\cdot}00 = -1{\cdot}62)$ for a 10 per

```
100 REM----------FOREST----------
110 REM
120 LET N$="FOREST.DAT"
130 OPEN N$ FOR OUTPUT AS FILE #1
140 REM
150 LET N=3
160 LET D9=10
170 LET D1=((1+D9/100)^5-1)*100
180 DIM S(20),U(5),P(5)
190 MAT READ U
200 DATA 0,3,6,9,12
210 MAT READ P
220 DATA .89,.93,.96,.98,.99
230 REM
240 FOR I=1 TO 7
250    LET S(I)=-1E6
260 NEXT I
270 FOR I=8 TO 17
280    LET T=(I-1)*5
290    LET S(I)=1200-140*T+3.834*T^2-.01667*T^3
300    LET S(I)=S(I)-10
310 NEXT I
320 LET T=1
330 LET C$=","
340 REM
350 REM-----WRITE BLOCK A DATA - PROBLEM PARAMETERS-----
360 PRINT #1, N
370 PRINT #1, D1
380 PRINT #1, "S"
390 PRINT #1, "Y"
400 PRINT #1, "Y"
410 REM
420 REM-----WRITE BLOCK B DATA - TERMINAL STATE VALUES-----
430 PRINT #1, "0,0"
440 REM-----WRITE BLOCKS CS AND ES DATA-----
450 PRINT #1, " T                 I   D    R                 J    P"
460 FOR I=1 TO 17
470    FOR D=1 TO 10
480       FOR P9=1 TO 2
490          IF D>5 THEN 660
500 REM----------KEEP DECISION----------
510          LET R=-U(D)
520          IF P9=2 THEN 580
530 REM-----NO FAILURE-----
540          LET J=I+1
550          LET P2=P(D)
560          GO TO 600
570 REM-----FAILURE-----
580          LET J=1
590          LET P2=1-P(D)
600          IF I<17 THEN 750
610 REM-----TOO OLD TO LEAVE-----
620          LET R=-1E6
630          LET J=1
640          GO TO 750
650 REM----------FELL DECISION----------
660          LET R=S(I)-U(D-5)
```

(continued)

FIG. 9.2. Program for writing data for the forest rotation and protection problem to the file FOREST.BAS.

```
670             IF P9=2 THEN 730
680 REM-----NO FAILURE-----
690             LET J=2
700             LET P2=P(D-5)
710             GO TO 750
720 REM-----FAILURE-----
730             LET J=1
740             LET P2=1-P(D-5)
750             PRINT #1,  T;C$,I;C$;D;C$;R;C$,J;C$;P2
760         NEXT P9
770     NEXT D
780 NEXT I
790 PRINT #1,  T,",0,0,0,0,1"
800 REM-----WRITE BLOCK D DATA - LABELS-----
810 PRINT #1,  "AGE"
820 FOR I=1 TO 17
830     PRINT #1,  (I-1)*5
840 NEXT I
850 PRINT #1,  "-KF-/$U,10"
860 FOR D=1 TO 5
870     PRINT #1,  "KEEP/$"+STR$(U(D-1))
880 NEXT D
890 FOR D=6 TO 10
900     PRINT #1,  "FELL/$"+STR$(U(D-6))
910 NEXT D
920 PRINT "DATA WRITTEN TO ";N$
930 CLOSE #1
940 END
```

FIG. 9.2.—*contd.*

cent per annum discount rate. In other words, continuous timber production is not viable for this discount rate. For a 5 per cent per annum discount rate the site value is much greater, equal to \$(117·31 – 10·00 = 107·31).

Other runs with the model with a 10 per cent discount rate show that the optimal rotation period is still 45 years for the case of no threat to survival at any age ($p = 1$ for $u = \$0$). If no protective action could be taken and the survival probability was always 0·89, the optimal rotation period would be reduced to 40 years.

9.3 The Optimal Rotation and Thinning Problem

Forest stands are not always managed in the way considered so far. Instead of cutting all trees at the end of the rotation period, trees may be cut progressively over the rotation period. In other words, the stand may be progressively thinned before final clearcutting.

If those trees which are cut in the thinning process are of average size and quality, referred to as 'thinning from above', why is it desirable to

```
-----------------------------------------------------------------
          STOCHASTIC DYNAMIC PROGRAMMING SOLUTION
                    PROBLEM ---FOREST---

PROBLEM PARAMETERS:
INFINITE NO. OF STAGES
RATE OF DISCOUNT (PER CENT)  =  61.051
-----------------------------------------------------------------

NO. OF ITERATIONS TO FIND SOLUTION =  4
```

CURRENT STATE NO.	CURRENT STATE AGE	DECISION NO.	DECISION -KF-/$U	NEXT STATE NO.	NEXT STATE AGE	PRESENT VALUE
1	0	1	KEEP/$0	?		8.37783
2	5	1	KEEP/$0	?		14.1247
3	10	1	KEEP/$0	?		24.5241
4	15	1	KEEP/$0	?		43.3425
5	20	2	KEEP/$0	?		77.3954
6	25	3	KEEP/$3	?		138.593
7	30	4	KEEP/$6	?		242.222
8	35	5	KEEP/$9	?		412.681
9	40	5	KEEP/$9	?		690.777
10	45	6	FELL/$0	?		1143.18
11	50	6	FELL/$0	?		1699.63
12	55	6	FELL/$0	?		2322.76
13	60	6	FELL/$0	?		3000.06
14	65	6	FELL/$0	?		3719.03
15	70	6	FELL/$0	?		4467.17
16	75	6	FELL/$0	?		5231.97
17	80	6	FELL/$0	?		6000.94

FIG. 9.3. Solution to the forest rotation and protection problem for a 10 per cent per annum discount rate.

remove them but not the whole stand? It might be economic to progressively thin if the stage return were concave in the number of trees removed. For example, the marginal cost of logging might increase with logging, or the price received for timber might fall with the quantity of timber for sale. The other possibility is that the rate of growth in the total volume of timber on a stand may be a decreasing function of total volume. It is the latter aspect of thinning which appears to be the main motivation for thinning. To model this aspect of thinning, alternative or additional state variables are required in dynamic programming models of forest management.

Some numerical applications of dynamic programming to the combined rotation and thinning problem are summarized in Table 9.3. The different approaches which have been taken are reviewed in the following subsections.

```
STOCHASTIC DYNAMIC PROGRAMMING SOLUTION
PROBLEM ---FOREST---

PROBLEM PARAMETERS:
INFINITE NO. OF STAGES
RATE OF DISCOUNT (PER CENT)  =  27.6281

NO. OF ITERATIONS TO FIND SOLUTION =  4
```

CURRENT STATE		DECISION		NEXT STATE		PRESENT VALUE
NO.	AGE	NO.	-KF-/$U	NO.	AGE	
1	0	1	KEEP/$0	?		117.305
2	5	1	KEEP/$0	?		153.72
3	10	3	KEEP/$3	?		205.94
4	15	4	KEEP/$6	?		276.878
5	20	4	KEEP/$6	?		369.913
6	25	5	KEEP/$9	?		491.074
7	30	5	KEEP/$9	?		647.365
8	35	5	KEEP/$9	?		848.85
9	40	5	KEEP/$9	?		1108.6
10	45	5	KEEP/$9	?		1443.46
11	50	5	KEEP/$9	?		1875.16
12	55	6	FELL/$0	?		2431.69
13	60	6	FELL/$0	?		3108.99
14	65	6	FELL/$0	?		3827.96
15	70	6	FELL/$0	?		4576.1
16	75	6	FELL/$0	?		5340.9
17	80	6	FELL/$0	?		6109.87

FIG. 9.4. Solution to the forest rotation and protection problem for a 5 per cent per annum discount rate.

9.3.1 Stage intervals

It is generally the case that the shorter the interval between decision stages, the larger must be the number of discrete values of the state variables. In only two applications are the stage intervals as short as one year. Kao and Brodie (1979) discuss a technique of using 'neighbourhood storage locations' which helps to overcome the requirement for a fine grid of state values when the stage interval is short. The technique is described in the next subsection.

In the study conducted by Murphy *et al.* (1977) the interval between decision stages is not necessarily regular. The formulation is quite different from the others reported in Table 9.3. The problem considered is purely a rotation or replacement problem. Consequently, it is possible to make each stage represent the start of a rotation. In Chapter 7 a similar formulation by Pouliquen (1970) was noted in a problem of the optimal replacement of laying hens. The state variable is the number of years to the end of the planning horizon.

9.3.2 *State variables*

Because gross returns from the forest stand are a function of volume of timber harvested, and thinning affects the rate of growth of volume, it is essential that the state variables describe timber volume. Any other significant characteristics of the stand which affect volume growth should also be included as state variables. The study with most state variables is that reported by Riitters *et al.* (1982) with four state variables which describe tree age, basal area of the stand, number of trees and the time elapsed since the previous thinning. The authors explain how they obtained solutions despite the resulting computer storage problem by using random access files and by special packing of data.

Some state variables such as age and number of trees are discrete, and are naturally updated in discrete units from one stage to the next as a result of the stage decision. Other variables such as volume measures are continuous variables which are approximated by a range of feasible discrete values. It may be difficult to arrange for thinning decisions taken at one stage to lead to one of the discrete volume levels at the next stage. The problem arises because the growth functions are typically non-linear. A simple partial remedy suggested by Schreuder (1971) is to ensure that stage transitions are always between state nodes by making the level of thinning the requisite residual value. However, this may mean that the number of alternative thinning levels which can be considered is unrealistically low unless the number of discrete values of the state variables is very large.

Brodie and Kao (1979) and Kao and Brodie (1979) report that their use of a neighbourhood storage location technique successfully overcomes the problem of continuous variables. For the technique to be applicable forward recursion must be used. Usually under forward recursion the question asked at each state at stage i is: what is the best state at stage $i-1$ to *have come from*? Brodie and Kao suggest that from each state at stage $i-1$ the state reached at stage i over all decisions be calculated, without any restrictions on the decisions. Inevitably there is a scatter of approach paths around each state node at stage i. The idea is to specify neighbourhoods around each node which encompass the full range of the state variable. The path into each neighbourhood giving the highest return is identified. The exact state reached at stage i by this path is substituted

TABLE 9.3

Summary of Dynamic Programming Applications to Forestry Management

Reference	*Interval between stages*	*Number of stages*	*State(s)*[a]	*Decision(s)*	*Objective function maximized*[b]	*Comments*
Hool (1966)	2 years	8	(i) Stand volume (stoch.) (ii) Tree count	Thinning, harvest cutting or selection cutting	Expected merchantable yield	Policies of thinning from above compared with thinning from below
Amidon and Akin (1968)	5 years	13 and others	Stand volume	Intensity of thinning	PV of NR	Application to Loblolly Pine trees
Kilkki and Vaisanen (1969)	5 years	10	(i) Age (ii) Stand volume	Intensity of thinning	PV of NR	Application to Scotch Pine trees
Risvand (1969)	4 years	11	(i) Stand volume (ii) Mean diameter	Intensity of thinning	PV of NR	Forward recursive method used, giving plans for range of time horizons
Kingma (1974)	1 year	∞	(i) Merchantable volume per tree (ii) Tree count	Intensity of thinning	PV of NR	Solution obtained through value iteration and successively finer grids
Won (1974)	5 years	10	(i) Stand volume (ii) Mean diameter	Intensity of thinning	PV of NR	Application to Loblolly Pine trees
Lembersky and Johnson (1975)	10 years	∞	(i) Mean diameter (stoch.) (ii) Tree count (stoch.) (iii) Sawlog value (stoch.)	(i) No action (ii) Partial harvest (iii) Clear and plant—alternative tree numbers	PV of ENR	Uses a Markov decision process approach; many silvicultural practices besides thinning considered
Lembersky (1976)	10 years	∞	(i) Mean diameter (ii) Tree count (stoch.)	(i) No action (ii) Partial harvest (iii) Clear and plant—alternative tree numbers	Expected sustainable yield	A stochastic infinite-stage problem without discounting

Murphy *et al.* (1977)	Rotation length	*n*	Years to end of the planning horizon	Rotation length	FV of NR	*n* set so that total planning horizon is less than 100 years; forward recursive method; results for perfect and imperfect capital markets
Brodie *et al.* (1978)	10 years	∞	(i) Age (ii) Stand volume	Intensity of thinning	PV of NR	Contrasts advantages of forward and backward recursion
Brodie and Kao (1979)	10 years	∞	(i) Age (ii) Stand basal area (iii) Tree count	Intensity of thinning	PV of NR	Allows for accelerated diameter growth as a result of thinning; 'neighbourhood storage locations' used to overcome rounding errors due to discrete state levels
Kao and Brodie (1979)	1 year	∞	(i) Age (ii) Stand volume	Intensity of thinning	PV of NR	Solves the Brodie *et al.* (1978) problem using 'neighbourhood storage locations' to permit shorter intervals between stages; see Kao and Brodie (1980) for an alternative solution technique
Chen *et al.* (1979)	10 years	11	Stand basal area	Intensity of thinning	Physical harvest over one rotation	Calculus used for maximization across the continuous state variable
Riitters *et al.* (1982)	5 years	∞	(i) Age (ii) Stand basal area (iii) Tree count (iv) Time since previous thinning	Value of each state variable at subsequent stage	PV of NR from production of timber and beef	Investigates the advantages of joint timber and beef production; forage feed a function of basal area

[a] Stochastic variables are denoted by '(stoch.)'
[b] PV of (E)NR stands for 'Present value of (expected) net revenue'; FV stands for 'Future value'.

for the neighbourhood node. It is from these substituted states that the states which may be reached at stage $i+1$ are calculated.

This technique appears to be a major step forward in reducing the dimensionality problem. One drawback of the technique is that because it relies on forward recursion stochastic problems cannot be solved. A decision made at stage $i-1$ leads to more than one possible state at stage i, so it is not feasible at stage i to find the best state to have come from. Nevertheless, Kao (1982) does attempt a stochastic application using the neighbourhood technique and forward recursion.

Another technique which also tackles the problem of representing continuous variables by discrete variables has been suggested by Lewis (1977). The idea is to use linear interpolations between the values of the optimal value function $V\{x\}$ at discrete values of x as an estimate of intermediate values of $V\{x\}$. The method can still be used with backward recursion, and hence can be applied to stochastic problems. Lewis suggested the method in the context of a problem in fisheries management, and is therefore considered in more detail in Chapter 10.

9.3.3 *Decision variables*

In most studies the decision variable is intensity of thinning. The decision to clearcut is covered by the option of 100 per cent thinning. Lembersky and Johnson (1975) and Lembersky (1976) permit a wider range of options, including precommercial thinning and alternative planning densities of a new stand after clearcutting.

9.3.4 *Objective function*

The most common objective function is the present value of the infinite stream of net revenue from continuous timber production. Livestock graze on some forest land, in which case the forest has to be managed to take account of net revenue flows from joint production. In the study by Riitters *et al.* (1982) forest management policies to maximize the present value of net revenue from both timber and beef production are investigated.

In three of the 14 reported studies the objective function is in terms of physical yield. Lembersky (1977) argues that maximum sustainable yield might be the appropriate objective for managers of state-owned forests and maximum net present value for a private landowner. However, maximizing sustainable yield from forests is unlikely to be a socially desirable objective because the demand for and supply of timber, and the opportunity costs of planting, logging and land, are ignored. Technically, however, the problem of maximizing sustainable yield to infinity is of interest because as noted in Chapter 3 a different solution process is required to solve infinite-stage problems with no discounting.

With the notable exceptions of the early study by Hool (1966) and by Lembersky and Johnson (1975) and Lembersky (1976), the dynamic programming formulations have been deterministic. This is surprising in view of the inherent uncertainties in forestry production. However, Lembersky and Johnson (1975) allow quite fully for uncertainty, and adopt an expected net present value objective function. Timber prices as well as the physical state variables follow a first-order Markov process.

9.4 Extensions

To end this chapter some extensions to the basic thinning and rotation problem are considered. What problems arise for forest management if the forest consists of trees of different sizes and ages? There are also questions about the appropriate objective function to use in forestry models, and whether alternative uses of land besides timber production should be evaluated.

9.4.1 Allowance for distributions of tree sizes and ages

An important question in forestry policy is whether a forest should be managed to eventually consist of trees all of the same age, or whether there should be a distribution of ages so that there are always trees ready for cutting. The former case of an even-aged forest has been considered so far. Timber is clearcut every t^* years. In the latter case of a fully-regulated or synchronized forest ages are distributed so that the same number of trees are ready for cutting every year. The two cases are idealized states. In practice a forest is likely to consist of stands of different-aged trees, but not so distributed as to result in a perfectly even flow of timber output each year. This type of forest has been described as partially synchronized by Dasgupta (1982, p. 188) who considers the implications for all of these cases in more detail.

In order to find the optimal thinning and rotation policy for each stand in a partially synchronized forest, is it necessary to take account of the distribution of tree ages across all stands in the forest? If it were necessary, the number of state variables required would be very large. For example, the total number of discrete state values might be the product of the number of stands, the number of possible tree ages on each stand, and the number of possible stand volumes. Fortunately, the forest problem is often separable into stand problems. Stand problems are separable if harvesting and replanting on one stand does not affect returns from timber production on other stands, either physically or through any price effects. A similar problem arises in harvesting fish from a multicohort fishery. Must an optimal policy take account of numbers in all age categories, or can an optimal policy be determined separately for each cohort? This question is considered in Chapter 10, but the conditions for separability are much less likely to be met in the case of a fishery than in the case of a forest. Even in the case of a forest they may not be met if there are economic advantages in reducing the annual variability of labour requirements for logging and planting.

Another complicating factor is that there is usually a distribution of sizes of trees in a stand. If thinning is carried out, should the distribution of sizes of trees cut be the same as the distribution before thinning, or should only the below-average-size trees be cut (thinning from below)? Hool (1966) has used dynamic programming to evaluate these alternative policies.

9.4.2 Alternative objectives

The objective functions in stochastic dynamic programming applications to forest management have been either the present value of expected net revenue or expected sustainable yield. This implies that forest managers are risk neutral. In cases where this is not valid, there is scope for formulating utility a function of say expected net revenue and the variance of net revenue as the stage return.

It is widely recognized that there are many externalities, mainly advantageous, which flow from the existence of a forest, such as the benefits of low soil erosion, the prevention of water tables from rising and the psychic returns from recreation use. There is no reason why these wider social criteria should not be included in the objective functions.

Forest land can be put to alternative uses. One possibility is to allow livestock to graze forest land. The study by Riitters *et al.* (1982) in which dynamic programming is used to evaluate joint livestock and timber production has already been mentioned. Kingma (1974) developed a dynamic programming model within a simulation model to evaluate multiple-land-use policies.

There is also scope for using dynamic programming to determine whether at any stage agricultural land should be switched to timber production, or forest land should be switched to agricultural production.

McConnell *et al.* (1983) have analyzed this problem using a control theory approach. Empirical problems of this type could be solved by integrating dynamic programming models of forest thinning and rotation with linear programming models of farm activity mixes.

9.5 References

Amidon, E.L. and Akin, G.S. (1968), Dynamic programming to determine optimum levels of growing stock, *Forest Science*, **14**(3), 287–91.

Brodie, J.D. and Kao, C. (1979), Optimizing thinning in Douglas-fir with three-descriptor dynamic programming to account for accelerated diameter growth, *Forest Science*, **25**(4), 665–71.

Brodie, J.D., Adams, D.M. and Kao, C. (1978), Analysis of economic impacts on thinning and rotation for Douglas-fir (*Pseudotsugamen ziesii*) using dynamic programming, *Forest Science*, **24**(4), 513–22.

Burt, O.R. (1965), Optimal replacement under risk, *Journal of Farm Economics*, **47**(2), 324–46.

Chen, C.M., Rose, D.W. and Leary, R.A. (1980), Derivation of optimal stand density over time—a discrete stage, continuous state dynamic programming solution, *Forest Science*, **26**(2), 217–27.

Clark, C.W. (1976), *Mathematical Bioeconomics*, John Wiley, New York.

Dasgupta, P. (1982), *The Control of Resources*, Basil Blackwell, Oxford.

Hool, J.N. (1966), *A Dynamic Programming—Markov Chain Approach to Forest Production Control*, Forest Science Monograph 12.

Howe, C.W. (1979), *Natural Resource Economics*, John Wiley, New York.

Kao, C. (1982), Optimal stocking levels and rotation under risk, *Forest Science*, **28**(4), 711–19.

Kao, C. and Brodie, J.D. (1979), Determination of optimal thinning entry interval using dynamic programming, *Forest Science*, **25**(4), 672–4.

Kao, C. and Brodie, J.D. (1980), Simultaneous optimization of thinnings and rotation with continuous stocking and entry intervals, *Forest Science*, **26**(3), 338–46.

Kilkki, P. and Vaisanen, U. (1969), Determination of the optimal policy for forest stands by means of dynamic programming, *Acta Forestalia Fennica*, **102**, 100–12.

Kingma, O.T. (1974), Recursive economic systems, land tenure and multiple land use, Unpublished PhD thesis, University of New England, Armidale.

Lembersky, M.R. (1976), Maximum average annual volume for managed stands, *Forest Science*, **22**(1), 69–81.

Lembersky, M.R. (1977), The application of Markov decision processes to forest management, in *Dynamic Programming and its Applications*, M. Puterman (ed.), Academic Press, New York.

Lembersky, M.R. and Johnson, K.N. (1975), Optimal policies for managed stands—an infinite horizon Markov decision process approach, *Forest Science*, **21**(2), 109–22.

Lewis, T. R. (1977), Optimal fishery management under conditions of uncertainty, in *Economic Impacts of Extended Fisheries Jurisdiction*, L.G. Anderson (ed.), Ann Arbor Science, Ann Arbor, Michigan.

McConnell, K.E., Daberkow, J.N., and Hardie, I.W. (1983), Planning timber production with evolving prices and costs, *Land Economics*, **59**(3), 292–9.

Murphy, P.A., Fortson, J.C. and Bethune, J.E. (1977), Timber management decision making under imperfect capital markets, *American Journal of Agricultural Economics*, **59**(2), 302–10.

Pearse, P. (1967), The optimal forest rotation, *Forestry Chronicle*, **43**, June, 178–95.

Pouliquen, A. (1970), A dynamic programming model applied to the replacement of laying hens and to the economic study of two forced moulting processes (in French), *Recherches d'Economie et de Sociologie Rurales,* **3**, 93–127.

Reed, W.J. (1984), The effects of risk of fire on the optimal rotation of a forest, *Journal of Environmental Economics and Management*, **11**, 180–90.

Riitters, K., Brodie, J.D. and Hann, D.W. (1982), Dynamic programming for optimization of timber production and grazing in Ponderosa Pine, *Forest Science*, **28**(3), 517–26.

Risvand, J. (1969), Economic analysis of cutting programs applying dynamic programming, in *Readings in Forest Economics*, A. Svendsrud (ed.), Universitrsforlaget, Oslo.

Samuelson, P.A. (1976), Economics of forestry in an evolving society, *Economic Inquiry*, **17**(4), 466–92.

Schreuder, G.F. (1971), The simultaneous determination of optimal thinning schedule and rotation for an even-aged forest, *Forest Science*, **17**, 333–9.

Won, D.J. (1974), A dynamic programming approach to optimal management strategies for loblolly pine plantations in the Delmarva region, Unpublished PhD thesis, University of Maryland.

Chapter 10

FISHERIES MANAGEMENT

10.1 The Management Problem

The question of how best to exploit a fishery poses some of the most interesting and challenging problems in the management of living systems. There are challenging modelling problems at both the biological and economic levels. The biological system is complex to model, partly because a stock of fish interacts richly with other fish stocks as predators or prey, and with the changing currents, temperatures and other aspects of the marine environment. The biological system is also difficult to model because so little about it is known. Fish are mobile in three dimensions, making assessment of stocks hazardous. Whilst migration and spawning patterns, and rates of growth by age, are usually readily established, information on natural rates of mortality by age, and the relationship between recruitment of juveniles to the fishery and parental biomass is more difficult to obtain. Thus limited information about stock levels and some of the biological relationships make uncertainty a particularly important factor to contend with in fisheries management. Another feature is the lack of opportunities for control, which in open-seas fishing is often restricted to levels of harvesting inputs. There are no controls over feed inputs, reproduction or ambient temperature. The range of controls is greatly increased in aquaculture.

At the economic level, prices and costs are uncertain, as they are in most industries. The fishing industry can often be typified as competitive, though some harvesting is controlled by processor groups which exert some degree of monopoly power. One of the most important characteristics of fisheries, however, is the structure of property rights. Unlike in agriculture where a competitive structure can be assumed to promote economic efficiency, the common-property nature of the fishery under competitive harvesting leads to over-exploitation. Gordon (1954) pointed out in an early influential article that open access to the fishery leads to the dissipation of economic rents. It is true that in agriculture under perfect competition each farm's long-run profit is driven to zero. The sum of the actual or imputed costs of all inputs equals the return from outputs. However, since all inputs are owned, there is a return to all inputs. In the fishery with no property rights existing in the fish stock there is no return to the fish stock.

The failure of property rights in the fishery has led to the examination of alternative institutional arrangements for regulating the fishery, such as the imposition of total and individual quotas, taxes, and restrictions on the areas fished, the fishing season and vessel type. The benefits of the rents obtained under these arrangements have to be balanced against the costs of regulation. Thus optimal utilization of the fishery is made more difficult because it is usually not possible to rely on market forces for economic efficiency.

Many nations have responded to the common-property problem by declaring exclusive fishing zones. Although the potential for control over harvesting is thereby increased, each nation has still to determine the best method of managing the fish stocks within its zone. Another problem arises when fish spend only a part of their lives within an exclusive fishing zone. One party fishing within a zone affects the interests of another party fishing outside the zone through price, cost, stock and recruitment interactions, and *vice versa*. There is clearly scope for the application of game theory to the management of fisheries which are transboundary resources (see e.g. Munro, 1979).

In the next section some frequently encountered bioeconomic models of the fishery are introduced, which provide the background for the following two sections dealing with dynamic programming applications. Section 10.3 deals with the analytical use of dynamic programming to derive optimality conditions for deterministic and stochastic cases. Numerical approaches are explored in Section 10.4. An example problem is solved, and other numerical applications are reviewed.

10.2 Modelling Approaches

The aim in this section is to outline by way of background for subsequent sections some of the models which have been developed to study particular fisheries problems. In keeping with a dynamic programming approach, discussion is restricted to discrete-time models. Stages mark the beginning of annual harvesting seasons. For simplicity all functions are taken to be independent of stage. State transition functions are considered first, followed by stage return functions.

10.2.1 Stock dynamics

The simplest way of describing the state of a fishery at stage i is in terms of the biomass of the total stock of fish (x_i). This is simple because only one state variable is involved. The transition from one level of biomass to another depends on the initial biomass level, autonomous growth $g\{x_i\}$ and harvesting $h\{x_i, u_i\}$ as follows:

$$x_{i+1} = x_i + g\{x_i\} - h\{x_i, u_i\} \tag{10.1}$$

where u_i is harvesting effort exerted over the period from stage i to stage $i+1$.

The harvesting function is often assumed to be a Cobb–Douglas production function

$$h\{x, u\} = qx^{\beta}u^{\gamma} \tag{10.2}$$

where q is referred to as a catchability coefficient, dependent on the species of fish and fishing technology. The exponents β and γ are often set to unity, so that the function exhibits increasing returns to scale.

An example of a growth function is a logistic model used by Schaefer (1957) and others. It has the form

$$g\{x\} = bx(1 - x/c) \tag{10.3}$$

where b and c are parameters denoting natural growth rate and carrying capacity, respectively. Growth is positive for $0 < x < c$ and concave in x, reaching a maximum at $x = c/2$.

Such biological models are referred to as surplus-yield models because between stages any harvesting $h\{x_i, u_i\}$ less than autonomous growth $g\{x_i\}$ results in an increase in biomass. The model is unrealistic when applied to fish which survive more than one harvesting season in that there is no recognition that the biomass of the total stock is the sum of the biomasses of stocks of different ages, and that the productivity of the total stock depends on the age distribution of the stock. Growth $g\{x_i\}$ represents the net addition to biomass through recruitment (the entry of juvenile fish), natural mortality and the increase in the weight of surviving fish. Each of these processes is age-dependent, so that $g\{x_i\}$ can be expected to be quite different if x_i is the biomass of stock consisting wholly of adults compared with what it would be if the stock consisted wholly of juveniles. This type of model is also referred to as a lumped-parameter or general-production model.

An alternative is to describe the state of a fishery at stage i in terms of stock numbers in each age category or cohort. Such models are referred to as multicohort or dynamic-pool models, and stem from the early work of Beverton and Holt (1957). State transitions can then be described in terms of equations which simulate the process of recruitment, natural mortality and weight increase. For example, assuming that l is the maximum number of cohorts and that recruitment occurs at age two, the following equations may be used:

$$x_{j+1,i+1} = x_{j,i} - m_j\{x_{j,i}\} - h_j\{x_{j,i}, u_i\} \qquad (j = 2, \ldots, l-1) \qquad (10.4)$$

$$x_{1,i+1} = R\left\{\sum_{j=k}^{l} x_{j,i} w_j\right\} \qquad (10.5)$$

with catch equation

$$H_i = \sum_{j=2}^{l} h_j\{x_{j,i}, u_i\} w_j$$

where $x_{j,i}$ denotes the number of fish aged j years at stage i, $m_j\{\cdot\}$ is the number of fish which die of natural causes, $h_j\{\cdot\}$ is the number harvested given harvesting effort applied to all age cohorts indiscriminantly, k is the age of sexual maturity, w_j is the mid-year average weight of fish and R specifies recruitment as a function of parental biomass. In some fisheries, effort can be targeted at particular age groups, either through harvesting technique (e.g. mesh size of nets) or because the fish swim in different areas or at different depths as they age. In such cases, subscript j should be added to u_i in (10.4) and the catch equation.

If $m_j\{x_{j,i}\}$ is directly proportional to $x_{j,i}$ (which means that the rate of natural mortality is independent of $x_{j,i}$), and if recruitment is proportional to parental biomass, (10.4) and (10.5) can be replaced by the matrix equation

$$\mathbf{x}_{i+1} = L\mathbf{x}_i - \mathbf{h}_i \qquad (10.6)$$

where $\mathbf{x}_i$ is a vector of stock numbers by cohort, $\mathbf{h}_i$ is a vector of harvest levels by cohort, and L is a matrix of coefficients accounting for rates of natural mortality, recruitment and weight. Conrad (1982) points out that a matrix such as L was originally proposed by Lewis (1942) and Leslie (1945) independently. Mendelssohn (1976) uses stochastic dynamic programming to explore optimal yield policies for models with stochastic Leslie models. He emphasises the shortcomings of such models because of the lack of density-dependent effects.

For the special class of single-cohort fisheries, one state variable can be used to represent a fishery. Ricker (1954) has suggested the stock-recruitment function

$$x_{i+1} = (x_i - h_i)\exp(\beta - \gamma(x_i - h_i)) \tag{10.7}$$

where x_i denotes population numbers or biomass and β and γ are parameters. The level $(x_i - h_i)$ is sometimes referred to as escapement. It is applicable to fish species such as Pacific salmon which die after reproducing. The interval between stage i and $i + 1$ is the lifespan of one generation (see Walters, 1978).

10.2.2 Stage return

One reason for modelling the fishery is to predict the harvesting behaviour of fishermen under open access. If fishermen have no property rights in the fish stock, each fisherman can be assumed to harvest so as to maximize his rent over one harvesting season. Changes in the number of fishermen in the industry can be taken to be some positive function of industry rent generated.

Another reason for modelling the fishery is to find management policies which maximize net returns to society. These optimal policies will be implemented by some kind of central agency. Given that an overriding aim of centralized fishery management is to ensure that harvesting does not result in extinction of the fish, management of the fishery can be expected to continue indefinitely. It is therefore often appropriate to make the planning horizon unbounded. Stage returns may be harvest level, industry rent, the present value of industry rent, or the present value of social rent. By industry rent is meant the difference between total gross returns from harvesting and total harvesting costs. Social rent is a measure of the difference between society's valuation of the total catch less total harvesting costs. Society's valuation is the total willingness to pay for the catch, or the area under the demand curve for fish, to the quantity sold.

The price of fish is frequently assumed to be independent of the size of the catch. Cost of harvesting is taken to be a function of harvesting effort, very often a linear function. In some models allowance is made for set-up costs or the fixed costs of harvesting.

An example of a stage return in the case of a central management aiming to maximize the present value of social rents from the fishery is

$$a\{x,u\} = \int_0^{qxu} p\{H\}\,\mathrm{d}H - cu \tag{10.8}$$

where qxu is the harvest level from (10.2) with $\beta = \gamma = 1$, $p\{H\}$ is the price of fish when demand is H, and c is the unit cost of fishing effort.

The two types of modelling, behavioural and normative, can be combined in hierachical modelling. At the top level, policy makers aim to optimize some social objective function, subject to the bottom level of fishermen's decision making which takes account of the top-level policies. An example is the multilevel linear programming model of a fishery developed by Meuriot and Gates (1983). They specify non-linear stock dynamics, but do not allow for a positive discount rate in the social objective function.

10.2.3 Developments in analytical modelling

Theoretical developments in fisheries management are described and expounded in Anderson (1986), Munro and Scott (1984) and Waugh (1984). Since the 1950's the type of analysis has shifted from static to dynamic and from deterministic to stochastic. Parallel developments have occurred in farm management. In contrast to the development of management models for agriculture, there has perhaps been more emphasis on theory and less on applications.

The so-called static model originally suggested by Gordon (1954) and developed by Schaefer (1957) is useful for showing that maximum sustainable yield is likely to exceed the optimal economic sustainable yield when harvesting costs are taken into account. It can also be used to demonstrate the overexploitation of the open-access fishery. The model is static in the sense that the foregone investment opportunities arising from leaving fish in the sea are ignored, or assumed to be zero.

Resort must be made to techniques of dynamic optimization to obtain the optimal level of harvesting through time, taking account of stock dynamics and the financial opportunity cost of fish left in the sea. Dynamic optimization is also required to investigate the circumstances under which stocks would be driven to extinction, either for the open-access case or for the case of central management. One approach which has proved very fruitful and has been extensively used for obtaining solutions to a wide variety of deterministic problems is optimal control theory based on the maximum principle. A milestone in the application of control theory to fisheries management was the publication of Clark's *Mathematical Bioeconomics* in 1976 (Clark, 1976a). The nature of the optimal harvesting policy through time for a surplus-yield problem known as the Gordon–Schaefer model is investigated in detail. It is shown that if $a\{x, u\}$ is linear in u, and an optimal equilibrium stock x^* exists, then x^* should be approached from any $x \neq x^*$ at the fastest rate that any limits on u permit. In other words, the control policy is of the bang-bang type. Optimal policies for many other problems are also discussed. Important extensions are made by increasing the number of state variables considered. The complexities of multicohort and multispecies models are investigated. Another significant state variable in determining optimal harvesting is capital in the form of vessels and fishing gear. This means that optimal increases in capital through investment have to be considered. Clark (1976*a*) and Clark *et al.* (1979) have examined the optimal investment problem with control theory. One of the cases they consider is that where capital in the fishery is quasi-malleable. They mean by this that whilst capital can be increased by any amount, it can only be reduced with zero investment and reliance on depreciation through time. Capital at any point in time places an upper limit on fishing effort.

There have been fewer dynamic programming expositions of the theory of deterministic surplus-yield fisheries. Sancho and Mitchell (1975) show how dynamic programming can be used to obtain analytical solutions to the fishery problem when the stage return function is quadratic and the stock-dynamics equation is linear. This approach has been discussed in Section 4.3.1. As noted in Section 4.4, Burt and Cummings (1977) use dynamic programming to determine approximately-optimal harvesting policies analytically. Clark (1976*b*) uses dynamic programming to examine under what circumstances it is optimal to approach the optimal equilibrium stock of fish as fast as possible when recruitment occurs several periods after birth, instead of immediately after birth. Levhari *et al.* (1981) explain comprehensively and lucidly harvesting policies under open access and central management. They use the Lagrange multiplier approach to obtain optimality conditions, but use dynamic programming in the appendix to obtain some important general results.

Stochastic dynamic programming has been widely used to analyze optimal harvesting policies when growth and harvesting functions are stochastic. Mann (1970), Jaquette (1972), Reed (1974, 1978, 1979) and

Spulber (1982) study stochastic problems with $a\{x, u\}$ linear in u and with set-up costs. They deduce circumstances under which (S, s) inventory-type policies are optimal. Under an (S, s) policy, with $S < s$, stocks are fished down to the level S if stocks exceed s. If stocks are below s, there is no harvesting. The size of $(s - S)$ depends on the size of the set-up costs. If there are zero set-up costs, $s = S$, and the optimal policy is bang-bang control. With $a\{x, u\}$ linear in u, the deterministic solution is a special case of the stochastic solution.

Smith (1985) uses stochastic dynamic programming to find optimal policies for stochastic surplus-yield models with the stage return in terms of utility which is a concave function of catch. Pindyck (1984) illustrates how analytical solutions may be obtained for similar problems but with the price of fish endogenous, harvesting costs dependent on catch and stocks, and a particular random process specified in the stock updating equation. Mendelssohn (1978) uses stochastic dynamic programming to investigate under what conditions it is optimal to only harvest fish above a certain age in multicohort models with stochastic dynamics. Further generalisations are made by Lovejoy (1983). Other extensions of deterministic results based on stochastic dynamic programming are made by Charles and Munro (1985) for problems with irreversible investment in fishing fleets, and by Spulber (1985) for the multicohort fishery. Andersen and Sutinen (1984) comprehensively review studies of various types of stochastic fishery problem.

10.3 Analytical Dynamic Programming Approaches

10.3.1 Deterministic results

The aim of this section is to show how dynamic programming can be used to determine the condition which holds for the optimal steady-state stock of the surplus-yield model, assuming one exists. The condition has been referred to by Munro (1981, p. 134) as the Golden Rule of Resource Conservation. Optimality conditions for the approach path are also given. The analysis is a more formal presentation of analysis already presented in Sections 1.4 and 4.4.

Suppose the management problem is

$$\max_{u_i \geq 0} \sum_{i=1}^{\infty} \alpha^{i-1} a\{x_i, u_i\}$$

subject to

$$x_{i+1} = x_i + g\{x_i\} - h\{x_i, u_i\}$$
$$x_1 \text{ given}$$

Typical examples of $a\{x, u\}$, $g\{x\}$ and $h\{x, u\}$ have been introduced in equations (10.8), (10.3) and (10.2) respectively. Because the planning horizon is infinite, functions are independent of stage and the optimal value function, $V\{x\}$, is the same for all stages. The basic recursive equation is therefore

$$V\{x_i\} = \max_{u_i} [a\{x_i, u_i\} + \alpha V\{x_i + g\{x_i\} - h\{x_i, u_i\}\}] \tag{10.9}$$

If the right-hand side of (10.9) is concave in u_i and the solution is an interior solution, the optimal value of u_i, denoted u_i^*, satisfies

$$\left.\frac{\mathrm{d}V}{\mathrm{d}x}\right|_{x_{i+1}} = (1 + r)\frac{\partial a/\partial u}{\partial h/\partial u} \tag{10.10}$$

For notational convenience derivatives in (10.10) for which the point of evaluation has not been made explicit are evaluated at x_i and u_i^* as appropriate, a practice which is continued below. Equation (10.10) states that if the optimal harvesting policy is pursued for all time, the value of the marginal fish (the shadow price of fish) at stage $i+1$ equals the net returns from extracting an additional fish at stage i, compounded forwarded to stage $i+1$. Equation (10.9) can be rewritten

$$V\{x_i\} = a\{x_i, u_i^*\} + \alpha V\{x_i + g\{x_i\} - h\{x_i, u_i^*\}\}$$

which after differentiation gives

$$\left.\frac{\mathrm{d}V}{\mathrm{d}x}\right|_{x_i} = \frac{\partial a}{\partial x} + \alpha\left.\frac{\mathrm{d}V}{\mathrm{d}x}\right|_{x_{i+1}}\left(1 + \frac{\mathrm{d}g}{\mathrm{d}x} - \frac{\partial h}{\partial x}\right)$$

$$(1 + r)\left.\frac{\mathrm{d}V}{\mathrm{d}x}\right|_{x_i} = (1 + r)\frac{\partial a}{\partial x} + \left.\frac{\mathrm{d}V}{\mathrm{d}x}\right|_{x_{i+1}}\left(1 + \frac{\mathrm{d}g}{\mathrm{d}x} - \frac{\partial h}{\partial x}\right)$$

$$\left.\frac{\mathrm{d}V}{\mathrm{d}x}\right|_{x_{i+1}} - \left.\frac{\mathrm{d}V}{\mathrm{d}x}\right|_{x_i} = r\left.\frac{\mathrm{d}V}{\mathrm{d}x}\right|_{x_i} - (1 + r)\frac{\partial a}{\partial x} - \left.\frac{\mathrm{d}V}{\mathrm{d}x}\right|_{x_{i+1}}\left(\frac{\mathrm{d}g}{\mathrm{d}x} - \frac{\partial h}{\partial x}\right)$$

$$\frac{\left.\frac{\mathrm{d}V}{\mathrm{d}x}\right|_{x_{i+1}} - \left.\frac{\mathrm{d}V}{\mathrm{d}x}\right|_{x_i}}{\left.\frac{\mathrm{d}V}{\mathrm{d}x}\right|_{x_i}} + \frac{(1 + r)\frac{\partial a}{\partial x} + \left.\frac{\mathrm{d}V}{\mathrm{d}x}\right|_{x_{i+1}}\left(\frac{\mathrm{d}g}{\mathrm{d}x} - \frac{\partial h}{\partial x}\right)}{\left.\frac{\mathrm{d}V}{\mathrm{d}x}\right|_{x_i}} = r \tag{10.11}$$

The first term of (10.11) is the rate of change in the value of a unit addition to the fish stock. The second term is the rate of return on an investment in the fishery of one fish added to the fish stock at stage i: the numerator equals the reduced effort costs at stage i compounded forwarded to stage $i+1$, plus the value of the net increase in stock by stage $i+1$; the denominator is the value of the investment.

If the rate of return from the fishery is less than r, the level of investment is uneconomic. Suppose conditions are such that $V\{x\}$ is strictly increasing and concave (see Appendix 4.A). For (10.11) to hold, the rate of change of $\mathrm{d}V/\mathrm{d}x$ over time must be positive, which by concavity means that $x_{i+1} < x_i$. The level of investment falls, which must be because $h\{x_i, u_i^*\} > g\{x_i\}$. Conversely, if the rate of return from the fishery is greater than r, investment in fish stocks rises.

If the rate of return from the fishery equals r, for (10.11) to hold the rate of change in $\mathrm{d}V/\mathrm{d}x$ must be zero, which also implies that $x_{i+1} = x_i$ and $h\{x_i, u_i^*\} = g\{x_i\}$. There is no tendency for the level of fish stock to change, and a fishing rent of $a\{x_i, u_i^*\}$ is sustained indefinitely. The Golden Rule of Resource Conservation for the steady-state case follows immediately from (10.11). If for equilibrium $h\{x_i, u_i^*\} = g\{x_i\}$, $\mathrm{d}g/\mathrm{d}x = \partial h/\partial x$, and (10.11) becomes

$$\frac{(1+r)\,\partial a/\partial x}{\left.\dfrac{\mathrm{d}V}{\mathrm{d}x}\right|_{x_i}} = r \tag{10.12}$$

where $\partial a/\partial x$ is the change in *sustainable rent* occasioned by a one-unit increase in fish stock. Marginal sustainable rent, compounded one period and expressed as a return to investment in the fishery, must equal the market rate of interest.

The Golden Rule can be derived even more directly. Denote the optimal steady-state stock by x^*, and the corresponding steady-state effort level by u^*, for which $h\{x^*, u^*\} = g\{x^*\}$. By the recursive equation (10.9)

$$V\{x^*\} = a\{x^*, u^*\} + V\{x^*\}/(1+r)$$

$$rV\{x^*\} = (1+r)a\{x^*, u^*\} \tag{10.13}$$

Equation (10.12) follows immediately after differentiating (10.13) with respect to x. Equation (10.13) also gives the relationship which is to be expected between the optimal annuity $a\{x^*, u^*\}$ and the optimal value of the steady-state stock $V\{x^*\}$.

10.3.2 Stochastic results

It is difficult to find illuminating optimality conditions for stochastic problems. Results for two simple cases are given here. In the first case it is assumed that growth at stage $i+1$ is subject to a random variable e_i such as temperature which is observable at stage i. The probability that the random variable equals e^k, the k-th value of e, is denoted by p^k. The structure of the problem is similar to one posed by Smith and Silvert (1977). The growth function is changed to $g\{x_i, e_{i-1}\}$. Presentation is simplified by making harvest level, h, the control variable, instead of effort, u. The control problem is

$$\max_{h_i \geq 0} E\left(\sum_{i=1}^{\infty} \alpha^{i-1} a\{x_i, e_{i-1}, h_i\}\right)$$

subject to

$$\left.\begin{aligned} x_{i+1} &= x_i + g\{x_i, e_{i-1}\} - h_i \\ \sum_{k=1}^{m} p^k\{e^k\} &= 1 \end{aligned}\right\} (i = 1, \ldots) \qquad (10.14)$$

$$x_1 \text{ given}$$

The recursive equation is

$$V\{x_i, e_{i-1}\} = \max_{h_i \geq 0} [a\{x_i, e_{i-1}, h_i\} + \alpha \sum p^k V\{x_i + g\{x_i, e_{i-1}\} - h_i, e_i^k\}] \qquad (i = 1, \ldots) \qquad (10.15)$$

where $V\{x_i, e_{i-1}\}$ is the expected present value of the resource at stage i.

Assuming the right-hand side is concave in h_i, for optimal $h_i = h_i^*$

$$\alpha \sum p^k \left.\frac{\partial V}{\partial x}\right|_{x_{i+1}}^{e_i^k} = \left.\frac{\partial a}{\partial h}\right|_{x_i}^{h_i^*, e_{i-1}} \qquad (10.16)$$

Differentiating (10.15) with respect to x_i and using (10.16) gives

$$\begin{aligned} \left.\frac{\partial V}{\partial x}\right|_{x_i}^{e_{i-1}} &= \left.\frac{\partial a}{\partial x}\right|_{x_i}^{h_i^*, e_{i-1}} + \alpha \sum p^k \left.\frac{\partial V}{\partial x}\right|_{x_{i+1}}^{e_i^k} \left(1 + \left.\frac{\partial g}{\partial x}\right|_{x_i}^{e_{i-1}}\right) \\ &= \left.\frac{\partial a}{\partial x}\right|_{x_i}^{h_i^*, e_{i-1}} + \left.\frac{\partial a}{\partial h}\right|_{x_i}^{h_i^*, e_{i-1}} \left(1 + \left.\frac{\partial g}{\partial x}\right|_{x_i}^{e_{i-1}}\right) \end{aligned} \qquad (10.17)$$

Equation (10.17) gives a general value for $\partial V/\partial x$ which may be substituted back in (10.16) to yield

$$\left.\frac{\partial a}{\partial h}\right|_{x_i}^{h_i^*}\!\!{}_{e_{i-1}} = \alpha \sum p^k \left(\left.\frac{\partial a}{\partial x}\right|_{x_{i+1}}^{h_{i+1}^*}\!\!{}_{e_i^k} + \left.\frac{\partial a}{\partial h}\right|_{x_{i+1}}^{h_{i+1}^*}\!\!{}_{e_i^k} \left(1 + \left.\frac{\partial g}{\partial x}\right|_{x_{i+1}}^{e_i^k} \right)\right) \tag{10.18}$$

Note that though the problem is stochastic, because e_{i-1} is known at stage i it is always possible to adjust h_i to ensure any desired x_{i+1}. Assume that $a\{\cdot\}$ is linear in h and that a steady state stock x^* is optimal. With all derivatives evaluated at $x_i = x_{i+1} = x^*$, (10.18) simplifies to

$$\left.\frac{\partial a}{\partial h}\right|_{e_{i-1}}^{h_i^*} = \alpha \sum p^k \left(\left.\frac{\partial a}{\partial x}\right|_{e_i^k}^{h_{i+1}^*} + \left.\frac{\partial a}{\partial h}\right|_{e_i^k}^{h_{i+1}^*} \left(1 + \left.\frac{\partial g}{\partial x}\right|_{e_i^k} \right)\right) \tag{10.19}$$

Depending on the functions involved it may be possible to solve (10.18) for x^* given

$$h_i^* = g\{x^*, e_{i-1}\}$$

Many problems are more complex because the random disturbance factor cannot be observed. Indeed it is usually the case that the stock level itself is uncertain. Lewis (1977) has suggested how this problem may be dealt with in a stochastic dynamic programming framework.

In the second simple example problem the random variable is an additive term in the stock dynamics equation, but is not observed. The problem is the same as that formulated in (10.14) but with the stock dynamics equation replaced by

$$x_{i+1} = x_i + g\{x_i\} - h_i + e_i$$

Following the same procedures as before it can be shown that

$$\left.\frac{\partial a}{\partial h}\right|_{x_i}^{h_i^*} = \alpha \sum p^k \left(\left.\frac{\partial a}{\partial x}\right|_{x_{i+1}}^{h_{i+1}^*} + \left.\frac{\partial a}{\partial h}\right|_{x_{i+1}}^{h_{i+1}^*} \left(1 + \left.\frac{\mathrm{d}g}{\mathrm{d}x}\right|_{x_{i+1}} \right)\right)$$

10.4 Numerical Dynamic Programming Applications

Whilst analytical results are useful for studying the nature of equilibria and the optimal approaches to them, in order to obtain quantitative policy recommendations it is usually necessary to make use of some numerical procedure such as dynamic programming. However, the number of state variables in many fishery problems is potentially large. They include, for example, fish stocks in each cohort and capital in the fishing fleet.

One suggestion for reducing the curse of dimensionality by introducing linear approximation has been made by Lewis (1975, 1977). Suppose a harvesting decision at stage i would result in a stock x' at stage $i+1$, but that to include x' in the set of discrete values of x would be computationally burdensome. Denote the two discrete values of x which lie below and above x' by x^1 and x^2. If $x' = \beta x^1 + (1-\beta)x^2$ for $0 < \beta < 1$, it may be reasonable to approximate $V\{x'\}$ by $\beta V\{x^1\} + (1-\beta)V\{x^2\}$. If linear interpolation can be used without much error, the number of discrete values of x can be reduced, and made independent of the number of permissible values of the decision variable. The weights β and $(1-\beta)$ in a deterministic formulation play the same type of role as probabilities would play in a stochastic formulation if β and $(1-\beta)$ were the probabilities of reaching x^1 and x^2 at stage $i+1$ after making the harvesting decision at stage i. This means that it is possible to solve deterministic problems using linear interpolation with the stochastic dynamic programming routines of GPDP.

Lewis has used the linear interpolation method for solving both deterministic and stochastic problems relating to the Eastern Pacific yellowfin tuna fishery. In the next subsection the method is illustrated in a simple application to the southern bluefin tuna fishery.

10.4.1 An application to the southern bluefin tuna fishery

Southern bluefin tuna are exploited primarily by Australian and Japanese fleets in two distinctly different ways. As a broad generalization, the Australian fleet catches juveniles (fish up to the age of 7) using pole boats and purse seiners. The catch is either canned in Australia or exported to Europe. The price received is not sensitive to the size of the catch. On the other hand the Japanese fleet catches adults (aged between 8 and 20) using longliners. The Japanese catch is sold on the Japanese sashimi market as raw fish after special processing, where prices are up to 20 times the Australian price. The Japanese price is sensitive to sales. At present relatively little of the Australian catch reaches the Japanese market.

The biomass of the juvenile stock, net of natural mortality, increases rapidly in the absence of fishing mortality at a rate of about 37 per cent per annum. This contrasts with a rate of about 10 per cent per annum for the adult stock. The high rate of growth of the juvenile stock suggests that if the fishery were managed in the combined interests of Australia and Japan, only adult stock should be fished. Throughout the 1960's it was largely the adult stock that was harvested by Japan, peaking at 70 000 tonnes. By the beginning of the 1980's Japan's harvest had plummeted to under 20 000 tonnes despite increased fishing effort, and Australia's had risen to about 20 000 tonnes. Biologists were warning of the risk of recruitment failure if restrictions on harvesting were not introduced.

The following dynamic programming model of the southern bluefin tuna fishery has been developed for illustrative purposes. More detail on the background of the fishery, and on data and modelling approaches may be found in the relevant report by the Australian Industries Assistance Commission (1984). The base harvesting season is 1982/83. Given the structure of the fishery it is reasonable to segment the fishery for modelling purposes into Australian and Japanese sectors. There are two state variables x^{A} and x^{J} representing the biomasses (in thousands of tonnes) of the juvenile and adult stocks fished by Australia and Japan, respectively. There are two decision variables, u^{A} and u^{J}, representing levels of Australian and Japanese fishing effort as a proportion of effort in 1982/83.

Harvesting functions for each fleet (A and J) are of the Cobb–Douglas form

$$h\{x, u\} = qxu^{0\cdot 8} \tag{10.20}$$

with the catchability coefficients q^{A} and q^{J} calculated so that h^{A} and h^{J} equal the 1982/83 harvest levels (in thousands of tonnes) for the 1982/83 biomass and effort levels. (Here the effort exponent has been chosen arbitrarily to reflect diminishing returns to effort; under sensitivity analysis it is later increased to 1·2.) Thus for $h^{A} = 21$, $x^{A} = 198$, and $u^{A} = 1$, $q^{A} = 0{\cdot}106$; for $h^{J} = 17$, $x^{J} = 224$, and $u^{J} = 1$, $q^{J} = 0{\cdot}0759$.

Recruitment, the number of juvenile fish recruited to the fishery at the start of year $i + 1$ for any level of the parental biomass, x_i^{J} is

$$R\{x^{J}\} = \begin{cases} 34\,880x^{J}/(1 + (x^{J}/346)^{1\cdot 5}) & \text{if } x^{J} \leq 205 \\ 5\,285\,000 & \text{if } x^{J} > 205 \end{cases}$$

The function is one that has been used in other simulation models (Australian Industries Assistance Commission, 1984, p. 123). Recruitment is constant for $x^{J} > 205$ thousand tonnes. Recruitment numbers are converted to recruitment biomass by multiplying by 0·002 tonne—the mid-year average weight of fish of the recruitment age class.

The stock-dynamics equations are

$$x^{A}_{i+1} = 1{\cdot}37(x^{A}_{i} + 0{\cdot}002R\{x^{J}_{i}\} - h^{A}\{x^{A}_{i}, u^{A}_{i}\})(1 - 0{\cdot}225)$$

$$x^{J}_{i+1} = 1{\cdot}37(x^{A}_{i} + 0{\cdot}002R\{x^{J}_{i}\} - h^{A}\{x^{A}_{i}, u^{A}_{i}\})(0{\cdot}225) + 1{\cdot}10(x^{J}_{i} - h^{J}\{x^{J}_{i}, u^{J}_{i}\})$$

It is assumed that harvesting and recruitment occur instantaneously at the beginning of each year, and that the biomass of Australian and Japanese stocks increase at the rates of 37 and 10 per cent, respectively. The percentage of x^{A} which transfers to x^{J} each year is 22·5.

Linear inverse demand equations for Australian and Japanese harvests were estimated with the form

$$p = d - eh$$

where p is the price of tuna and d and e are parameters. Harvesting costs are assumed to be linear in effort. The cost coefficients were calculated for 1982/83 assuming that open access conditions prevailed. This is reasonable because significant harvesting restrictions were not introduced until the 1983/84 season. The social rent functions are thus of the form

$$a\{x, u\} = dh - 0{\cdot}5eh^2 - cu$$

where a is in millions of Australian dollars; d^{A}, e^{A} and $c^{A} = 0{\cdot}7$, $0{\cdot}0$ and $14{\cdot}7$, respectively; and d^{J}, e^{J} and $c^{J} = 28$, $0{\cdot}412$ and 238, respectively.

The problem is to find the effort levels, u^{A} and u^{J}, for each possible combination of stock levels, x^{A} and x^{J}, which maximize the present value of the sum of Australian and Japanese social rents over infinite years. The recursive equation is

$$V\{x^{A}_{i}, x^{J}_{i}\} = \max_{u^{A}_{i}, u^{J}_{i}} [a^{A}\{x^{A}_{i}, u^{A}_{i}\} + a^{J}\{x^{J}_{i}, u^{J}_{i}\} + \alpha V\{x^{A}_{i+1}, x^{J}_{i+1}\}] \qquad (i = 1, \ldots)$$

subject to

$$0 \leq u^{A}_{i}, u^{J}_{i} \leq 1{\cdot}2$$

$$x^{A}_{1} = 198, \qquad x^{J}_{1} = 224$$

and to the stock-dynamics equations. The discount factor is α, set equal to 1/1·1.

A program for writing the problem data to file for solution using the IDP routine is shown in Fig. 10.1. In lines 190 to 230 the number of discrete values of both x^{A} and x^{J} is limited to 5, and of both u^{A} and u^{J} to 7. There are therefore 25 possible combinations of x^{A} and x^{J} which

```
100 REM----------FISHING PROBLEM---TUNA----------
110 REM
120 LET N$="TUNA.DAT"
130 OPEN N$ FOR OUTPUT AS FILE #1
140 REM
150 DIM P3(2,2)
160 LET N=3
170 LET D9=10
180 REM---NO. OF X VALUES---
190 LET N1(1)=5
200 LET N1(2)=5
210 REM---NO. OF U VALUES---
220 LET N2(1)=7
230 LET N2(2)=7
240 REM---LIMITS ON X VALUES---
250 LET L1(1)=150000
260 LET H1(1)=400000
270 LET L1(2)=200000
280 LET H1(2)=400000
290 REM---LIMITS ON U VALUES---
300 LET L2(1)=0
310 LET H2(1)=1.2
320 LET L2(2)=0
330 LET H2(2)=1.2
340 REM---CATCHABILITY COEFFICIENTS---
350 LET Q(1)=.106
360 LET Q(2)=.0759
370 REM---EFFORT EXPONENTS---
380 LET G(1)=.8
390 LET G(2)=.8
400 REM---RECRUITMENT COEFFICIENTS---
410 LET B9=205000
420 LET A9=34.88
430 LET K9=345935
440 LET E9=1.5
450 LET R9=5285061
460 LET W9=.002
470 REM---GROWTH RATES---
480 LET G9(1)=1.3706
490 LET G9(2)=1.1035
500 REM---TRANSFER BETWEEN STOCK COEFFICIENT---
510 LET P9=.225
520 REM---DEMAND PARAMETERS---
530 LET D4(1)=700
540 LET E4(1)=0
550 LET D4(2)=28000
560 LET E4(2)=-.824
570 REM---COST COEFFICIENTS---
580 LET C4(1)=14.7E6
590 LET C4(2)=238E6
600 REM---RETURN WEIGHTINGS---
610 LET W(1)=1
620 LET W(2)=1
630 REM
640 REM---RETURN FUNCTION---
650 DEF FNR(H,U,D,E,C)=(D*H+.5*E*H^2-C*U)*1E-6
660 REM---STATE VALUE FUNCTION---
670 DEF FNS(I,L,H,N)
680 IF N>1 THEN 710
```

FIG. 10.1 Program for writing data for the tuna problem to the file TUNA.BAS.

```
690 LET FNS=L
700 GO TO 720
710 LET FNS=L+(H-L)*(I-1)/(N-1)
720 FNEND
730 REM---STATE NO. FUNCTION---
740 DEF FNT(I1,I2,N)=I2+(I1-1)*N
750 REM
760 C$=","
770 REM
780 REM-----WRITE BLOCK A DATA - PROBLEM PARAMETERS-----
790 PRINT #1, N
800 PRINT #1, D9
810 PRINT #1, "S"
820 PRINT #1, "Y"
830 PRINT #1, "Y"
840 REM
850 REM-----WRITE BLOCK B DATA - TERMINAL STATE VALUES-----
860 PRINT #1, "0,0"
870 REM-----WRITE BLOCK CS DATA-----
880 PRINT #1, " T                    I   D    R                  J    P"
890 FOR I1=1 TO N1(1)
900    LET I2(1)=I1
910    FOR I5=1 TO N1(2)
920       LET I2(2)=I5
930       LET I=FNT(I1,I5,N1(2))
940       FOR D1=1 TO N2(1)
950          LET D2(1)=D1
960          FOR D5=1 TO N2(2)
970             LET D2(2)=D5
980             LET D=FNT(D1,D5,N2(2))
990             LET R=0
1000             FOR K=1 TO 2
1010                LET X(K)=FNS(I2(K),L1(K),H1(K),N1(K))
1020                LET U(K)=FNS(D2(K),L2(K),H2(K),N2(K))
1030                LET H(K)=Q(K)*X(K)*U(K)^G(K)
1040                LET R=R+FNR(H(K),U(K),D4(K),E4(K),C4(K))*W(K)
1050             NEXT K
1060 REM---RECRUITMENT---
1070             IF X(2))B9 THEN 1100
1080             LET R1=(A9*X(2))/(1+(X(2)/K9)^E9)
1090             GO TO 1120
1100             LET R1=R9
1110 REM---STOCK UPDATING---
1120             LET X2(1)=(X(1)+R1*W9-H(1))*G9(1)*(1-P9)
1130             LET X2(2)=(X(2)-H(2))*G9(2)+P9*X2(1)/(1-P9)
1140 REM---J VALUES AND J WEIGHTINGS---
1150             FOR K=1 TO 2
1160                IF X2(K))L1(K) THEN 1180
1170                LET X2(K)=L1(K)
1180                IF X2(K)(H1(K) THEN 1200
1190                LET X2(K)=H1(K)
1200                FOR I9=2 TO N1(K)
1210                   LET X9=FNS(I9,L1(K),H1(K),N1(K))
1220                   IF X2(K)(X9 THEN 1240
1230                NEXT I9
1240                LET I3(K)=I9-1
1250                LET P3(K,1)=(X9-X2(K))*(N1(K)-1)/(H1(K)-L1(K))
1260                LET P3(K,2)=1-P3(K,1)
```

FIG. 10.1—*contd.*

```
1270                NEXT K
1280                FOR V9=1 TO 2
1290                    FOR U9=1 TO 2
1300                        LET J=FNT(I3(1)+U9-1,I3(2)+V9-1,N1(2))
1310                        LET P=P3(1,U9)*P3(2,V9)
1320                        PRINT #1, N;C$,I;C$;D;C$;R;C$,J;C$;P
1330                    NEXT U9
1340                NEXT V9
1350            NEXT D5
1360        NEXT D1
1370    NEXT I5
1380 NEXT I1
1390 PRINT #1, N,",0,0,0,0,1"
1400 REM-----WRITE BLOCK D DATA - LABELS-----
1410 PRINT #1, "X1 /X2"
1420 FOR I1=1 TO N1(1)
1430    FOR I5=1 TO N1(2)
1440        LET X1=INT(FNS(I1,L1(1),H1(1),N1(1))/1000)
1450        LET X5=INT(FNS(I5,L1(2),H1(2),N1(2))/1000)
1460        PRINT #1, STR$(X1)+"/"+STR$(X5)
1470    NEXT I5
1480 NEXT I1
1490 N9=N2(1)*N2(2)
1500 PRINT #1, "U1/U2,";N9
1510 FOR D1=1 TO N2(1)
1520    FOR D5=1 TO N2(2)
1530        LET U1=INT(FNS(D1,L2(1),H2(1),N2(1))*100)
1540        LET U5=INT(FNS(D5,L2(2),H2(2),N2(2))*100)
1550        PRINT #1, STR$(U1)+"/"+STR$(U5)
1560    NEXT D5
1570 NEXT D1
1580 PRINT "DATA WRITTEN TO ";N$
1590 END
```

FIG. 10.1—*contd.*

represent the state of the fishery. The method of linear interpolation suggested by Lewis (1975, 1977) is used, but in two dimensions instead of one. Thus if

$$x^{A'} = \beta x^{A1} + (1-\beta)x^{A2}$$

$$x^{J'} = \gamma x^{J1} + (1-\gamma)x^{J2}$$

where (x^{A1}, x^{J1}), (x^{A2}, x^{J1}), (x^{A1}, x^{J2}) and (x^{A2}, x^{J2}) are four of the 25 combinations of discrete values of x^A and x^J, and $(x^{A'}, x^{J'})$ is a combination of non-represented intermediate values, then

$$V\{x^{A'}, x^{J'}\} = \beta\gamma V\{x^{A1}, x^{J1}\} + (1-\beta)\gamma V\{x^{A2}, x^{J1}\}$$
$$+ \beta(1-\gamma)V\{x^{A1}, x^{J2}\} + (1-\beta)(1-\gamma)V\{x^{A2}, x^{J2}\}$$

The optimal effort policy obtained by running IDP is shown in Fig. 10.2. Although the results were obtained by the solution procedure normally used to solve stochastic problems, they are in fact results for a

STOCHASTIC DYNAMIC PROGRAMMING SOLUTION
PROBLEM ---TUNA---

PROBLEM PARAMETERS:
INFINITE NO. OF STAGES
RATE OF DISCOUNT (PER CENT) = 10

NO. OF ITERATIONS TO FIND SOLUTION = 4

CURRENT STATE		DECISION		NEXT STATE		PRESENT VALUE
NO.	X1 /X2	NO.	U1/U2	NO.	X1 /X2	
1	150/200	4	0/60	?		2572.62
2	150/250	5	0/80	?		2695.15
3	150/300	5	0/80	?		2782.36
4	150/350	5	0/80	?		2836.84
5	150/400	5	0/80	?		2863.45
6	212/200	4	0/60	?		2632.94
7	212/250	5	0/80	?		2735.14
8	212/300	5	0/80	?		2809.56
9	212/350	5	0/80	?		2854.71
10	212/400	5	0/80	?		2881.32
11	275/200	4	0/60	?		2683.76
12	275/250	5	0/80	?		2769.62
13	275/300	5	0/80	?		2838.4
14	275/350	12	20/80	?		2874.31
15	275/400	12	20/80	?		2900.92
16	337/200	4	0/60	?		2718.33
17	337/250	5	0/80	?		2799.9
18	337/300	12	20/80	?		2860.45
19	337/350	12	20/80	?		2895.36
20	337/400	12	20/80	?		2921.97
21	400/200	25	60/60	?		2745.56
22	400/250	33	80/80	?		2825.84
23	400/300	33	80/80	?		2881.62
24	400/350	33	80/80	?		2916.53
25	400/400	33	80/80	?		2943.13

FIG. 10.2 Infinite-stage solution to the tuna problem.

deterministic problem. The *U*1/*U*2 column shows the optimal effort combination u^A/u^J expressed as a percentage of the 1982/83 effort level. Permissible effort levels ranged from 0 to 120 per cent. For all biomass combinations *X*1/*X*2 (representing x^A and x^J in thousands of tonnes) the optimal effort level was less than 100 per cent, with *U*1 percentages more severely curtailed than *U*2 percentages. The 'value' column shows the present value of net social rents in millions of Australian dollars over an infinite planning horizon obtained by following the optimal effort policy.

The optimal effort policy is translated into the optimal harvesting policy shown under 'run 1' in Table 10.1, using the harvesting function (10.20).

TABLE 10.1
Southern Bluefin Tuna Harvest Policies for Alternative Model Parameters

Stocks (thousand tonnes)		*Harvest (thousand tonnes)*					
x^A	x^J	*Run 1*		*Run 2*		*Run 3*	
		h^A	h^J	h^A	h^J	h^A	h^J
150	200	0·0	10·1	0·0	15·2	0·0	15·2
150	250	0·0	15·9	0·0	19·0	0·0	19·0
150	300	0·0	19·0	0·0	22·8	0·0	22·8
150	350	0·0	22·2	0·0	26·6	0·0	26·6
150	400	0·0	25·4	0·0	30·4	0·0	30·4
212	200	0·0	10·1	0·0	15·2	0·0	15·2
212	250	0·0	15·9	0·0	19·0	0·0	19·0
212	300	0·0	19·0	0·0	22·8	0·0	22·8
212	350	0·0	22·2	0·0	26·6	0·0	26·6
212	400	0·0	25·4	0·0	30·4	0·0	30·4
275	200	0·0	10·1	8·0	15·2	0·0	15·2
275	250	0·0	15·9	8·0	19·0	0·0	19·0
275	300	0·0	19·0	8·0	22·8	0·0	22·8
275	350	8·0	22·2	8·0	26·6	0·0	26·6
275	400	8·0	25·4	8·0	30·4	0·0	30·4
337	200	0·0	10·1	8·0	15·2	0·0	15·2
337	250	0·0	15·9	8·0	19·0	0·0	19·0
337	300	9·9	19·0	8·0	22·8	0·0	22·8
337	350	9·9	22·2	8·0	26·6	0·0	26·6
337	400	9·9	25·4	8·0	30·4	0·0	30·4
400	200	28·2	10·1	35·5	15·2	52·8	15·2
400	250	35·5	15·9	35·5	19·0	52·8	19·0
400	300	35·5	19·0	35·5	22·8	52·8	22·8
400	350	35·5	22·2	35·5	26·6	52·8	26·6
400	400	35·5	25·4	35·5	30·4	52·8	30·4

For reference, the 1982/83 x^A/x^J estimated biomass levels were 198/224 thousand tonnes, and the 1982/83 h^A/h^J harvesting levels were 21/17 thousand tonnes. Results for two other runs are included in Table 10.1. In runs 2 and 3 the objective function is changed so that zero weighting is placed on Japan's social rent, $a^J\{x^J, u^J\}$. Japan's effort level is held arbitrarily fixed at the 1982/83 level equal to one. The aim is to find the optimal Australian harvesting levels assuming Japan does not react by changing her effort level. Clearly there is scope for designing game theoretic dynamic programming models along the lines suggested in Section 4.5.3.

In run 2 the exponent on the effort variable in both harvesting functions (10.20) is kept at 0·8, whereas in run 3 the exponent is changed to 1·2 to allow for increasing instantaneous harvesting returns to fishing effort. Compared with the results for run 1, the harvesting policy for run 2 permits positive levels of Australian harvesting at lower levels of x^A. Results for run 3 show that a policy of pulse-fishing is optimal. Australia would only fish once x^A reaches very high levels, and then would fish using maximum effort. Effort levels u^A equal the maximum value of 1·2 for positive run-3 harvests in Table 10.1.

The major limitation of the model is the representation of the state of the fishery in terms of only two categories of biomass. For example, the relatively high growth rates of the biomasses of recruits added at any stage i to the juvenile and adult biomasses are not reflected at subsequent stages. The rate of growth of biomass is constant, regardless of the age distribution of fish making up the biomass. However, more state variables must be used in a multicohort model. A multicohort model of the southern bluefin tuna fishery is reviewed in the final subsection, along with other numerical applications of dynamic programming.

10.4.2 A review of applications

A summary of some reported applications of numerical dynamic programming to commercial fisheries is presented in Table 10.2. Eight of the eleven applications are stochastic. Although stochastic models are more realistic than deterministic models to the extent that they reflect uncertainty in the real world, it should be noted that optimal harvesting policies need not be very different from those for the equivalent deterministic model. Lewis (1975) investigates the effect of price, growth and catch uncertainty on optimal policies. If harvesting costs are a particular quadratic function of effort, there is little difference between the stochastic and deterministic results. However, there are significant differences for other types of stage return function. Walters (1975) finds that optimal policies are insensitive to the probability distribution of the recruitment parameter in a variant of the Ricker model (10.7). Hutchinson and Fischer (1979) find for a model with different stochastic formulations of the harvesting function (10.2) and the logistic growth function (10.3) that optimal policies result in significantly higher expected profits compared with policies optimal for deterministic versions of the model.

TABLE 10.2
Summary of Dynamic Programming Applications to Fisheries Management

Reference	*Interval between stages*	*Number of stages*	*State(s)*[a]	*Decision(s)*	*Objective function maximized*[b]	*Comments*
Lewis (1975)	1 year	∞	Population biomass (stoch.)	Harvest	PV of ESUR	Application to East Pacific Yellowfin Tuna; emphasis on uncertainty about current biomass, yield and price; state value interpolation used; increasing returns to harvesting effort results in pulse fishing; see also Lewis (1977)
Walters (1975)	1 year	∞	Stock numbers (stoch.)	Harvest	Expected long-term harvest or minus the sum of variances around desired harvest	Application to Skeena River Sockeye Salmon; optimal policy insensitive to variability in the Ricker stock production parameter; see also Walters (1978)
Hilborn (1976)	1 year	∞	Numbers of each of two stocks (stoch.)	Harvest	Expected long-term harvest	Application to salmon harvesting; investigates conditions under which a constant escapement policy is suboptimal
United States Department of Commerce (1978)	1 year	∞	Population biomass	Harvest	PV or SR	Application to the Northern Anchovy fishery; results obtained for alternative US shares of the total harvest; see also Huppert (1981)
Anderson *et al.* (1979)	1 year	∞	(i) Species (ii) Biomass	Harvest effort	PV of SR	Optimal allocation of harvest effort to numbers of different vessel types determined by LP or QP subroutine; biological

						simulator used for determining effort required for state transitions; see Custis (1979) for a user's guide
Hutchinson and Fischer (1979)	1 year	5	Population biomass (stoch.)	Harvest effort	PV of ENR	Application to the US Atlantic Sea-Scallop fishery; model run for different formulations of stochastic growth and harvest functions; performance of open-loop deterministic, and stochastic control evaluated by simulation
Dudley and Waugh (1980)	1 month	12	Stock numbers (stoch.)	Harvest effort	ENB	Application to the Australian Exmouth Gulf Prawn fishery; simulation used to determine state transition probabilities, evaluate policies, and find optimal investment in vessels; extensive experimentation with alternative stochastic formulations; see also Waugh (1984)
Mendelssohn (1980)	1 year	∞	(i) Population size (stoch.) (ii) Previous harvest	Harvest	Weighted sum of expected harvest and MAD of current from previous harvest	Application to salmon runs off Bristol Bay, Alaska; stock next year a random function of escapement
Mendelssohn (1982)	1 year	∞	Recruitment population (stoch.)	Harvest	PV of EUH	Application to salmon runs off Bristol Bay, Alaska; investigates the effect on policies of the rate of discount and the degree of risk aversion

(continued)

TABLE 10.2—*contd.*

Reference	*Interval between stages*	*Number of stages*	*State(s)*[a]	*Decision(s)*	*Objective function maximized*[b]	*Comments*
Lovejoy (1984)	1 year	∞	(i) Biomass of Surf Clams (stoch.) (ii) Biomass of Ocean Quahogs (stoch.) (iii) Number of vessels *N* out of a total of *K* geared to catch Surf Clams	Number of vessels to switch from catching one species to the other	PV of ENB	Application to the US Surf Clam Ocean Quahog fishery; solutions to problems of low dimensionality provide tight upper and lower bounds for the solution to the original problem of high dimensionality
Kennedy and Watkins (1986)	1 year	∞	Stock numbers in each of 4 age categories	Harvest levels from each age category by each of 2 fleets	PV of SR	Application to the Australian/Japanese Southern Bluefin Tuna fishery; value function approximated by a quadratic response surface to combat high dimensionality

[a] Stochastic variables are denoted by '(stoch.)'

[b] Key: PV—present value; ESUR—expected social utility of rent; SR—social rent; ENR—expected net revenue; ENB—expected net benefit; MAD—mean absolute deviation; and EUH—expected utility of harvest.

Several studies allow for the social decision maker to be risk averse. Social utility functions are posited by Lewis (1975, 1977) and Mendelssohn (1982). Others have incorporated some measure of harvest dispersion in the objective function (see Walters, 1975 and Mendelssohn, 1980).

Because uncertainty is so pervasive in fishery management, there is important scope for the application of adaptive dynamic programming. Walters and Hilborn (1976) consider some of the computational problems of adaptive models with active learning.

The power of dynamic programming as a solution technique is increased if, for each possible state transition, alternative ways of making the transition are evaluated, and the way which optimizes the stage return is selected. Anderson *et al.* (1979) extend the scope of dynamic programming to fishery management by incorporating a linear or quadratic programming subroutine to determine the optimal size and composition of fishing fleets in transferring from one stock size to another.

None of the studies has capital in the fishing fleet as a state variable. However, experiments with alternative levels of harvesting capital available are conducted in the studies reported by the United States Department of Commerce (1978) and by Dudley and Waugh (1980).

In most applications fish stocks are described by a single state variable. In a model of the southern bluefin tuna fishery, Kennedy and Watkins (1986) use four state variables to represent stock numbers in four age categories, although ideally at least twelve state variables would be required to represent numbers in each year class. Because they model harvesting annually, the stock dynamics equations hold only approximately. In order to save on computer processing and memory requirements the value function $V\{x\}$ in the four state variables was approximated at each stage iteration by a quadratic response function.

Under certain circumstances the optimal harvesting policy for a multicohort fishery can be found by solving the problem of how best to harvest a single cohort over the time it takes to pass from the youngest to the oldest age category. The optimal steady-state annual harvest from each cohort in the multicohort fishery is the same as the optimal harvest for each corresponding year in the single-cohort model. With one state variable, the single-cohort problem is easily solved using numerical dynamic programming. Unfortunately the circumstances under which the two solutions are the same are highly restrictive. Recruitment must be independent of parental biomass, it must be possible to selectively harvest from each year class, and the price of fish from a cohort can only be a function of harvest from that cohort.

However, Lovejoy (1983) has argued that where the conditions for separability of cohorts do not hold, solving a simpler problem which is separable may give a useful bound for the optimal harvesting policy of the original problem. He demonstrates rigorously for an example ten-cohort fishery with stochastic survival rates and stock-dependent recruitment that a lower bound for the optimal age at first capture can be found very simply by solving the deterministic, constant recruitment problem. He has developed the theory for applying this approach quite generally. In doing so he shows how the curse of dimensionality can be skilfully avoided. Walters (1978, p. 244), in a similar context, also argues that experience with dynamic programming applications to fisheries has shown that 'for many of these problems much of the complexity is basically irrelevant (at least for decision making) and could have been ignored in the first place.'

In this chapter discussion has been limited to dynamic programming used for determining optimal harvesting policies for free-roaming commercial fisheries. Intraseasonal applications have been considered by Lord (1973), and applications to aquaculture by Allen *et al.* (1984).

10.5 References

Allen, P.G., Botsford, L.W., Schuur, A.M. and Johnston, W.E. (1984), *Bioeconomics of Aquaculture*, Elsevier, Amsterdam.

Andersen, P. and Sutinen, J.G. (1984), Stochastic bioeconomics: A review of basic methods and results, *Marine Resource Economics*, **1**(2), 117–36.

Anderson, L.G. (1986), *The Economics of Fisheries Management*, 2nd edition, Johns Hopkins University Press, Baltimore.

Anderson L.G., Ben-Israel, A., Custis, G. and Sarabur, C.C. (1979), Modeling and simulation of interdependent fisheries, and optimal effort allocation using mathematical programming, in *Applied Operations Research in Fishing*, K.B. Haley (ed.), Plenum Press, New York.

Australian Industries Assistance Commission (1984), *Southern bluefin tuna*, Report No. 349, Australian Government Printing Service, Canberra.

Beverton, R. and Holt, S. (1957), *On the dynamics of exploited fish populations*, Ministry of Agriculture, Fisheries and Food, Fisheries Investigation Series **2**(19), London.

Burt, O. R. and Cummings, R. G. (1977), Natural resource management, the steady state, and approximately optimal decision rules, *Land Economics*, **53**(1), 1–22.

Charles, A.T. and Munro, G.R. (1985), 'Irreversible investment and optimal fisheries management: a stochastic analysis', *Marine Resource Economics*, **1**(3), 247–64.

Clark, C.W. (1976*a*), *Mathematical Bioeconomics: The Optimal Management of Renewable Resources*, Wiley-Interscience, New York.

Clark, C.W. (1976*b*), A delayed-recruitment model of population dynamics, with an application to Baleen whale population, *Journal of Mathematical Biology*, **3**, 381–91.

Clark, C.W., Clarke, F.H. and Munro, G.R. (1979), The optimal exploitation of renewable resource stocks: problems of irreversible investment, *Econometrica*, **47**(1), 25–47.

Conrad, J.M. (1982), Management of a multiple cohort fishery: the hard clam in Great South Bay, *American Journal of Agricultural Economics*, **64**(3), 463–74.

Custis, G. (1979), *DYFISH: Program and Users Guide*, University of Delaware.

Dudley, N. and Waugh, G. (1980), Exploitation of a single cohort fishery under risk: a simulation–optimization approach, *Journal of Environmental Economics and Management*, **7**, 234–55.

Gordon, H.S. (1954), The economic theory of a common property resource: The fishery, *Journal of Political Economy*, **62**, 124–42.

Hilborn, R. (1976), Optimal exploitation of multiple stocks by a common fishery: a new methodology, *Journal of the Fisheries Research Board of Canada*, **33**(1), 1–5.

Huppert, D.D. (1981), Economic analysis for Northern Anchovy management, in *Economic Analysis for Fisheries Management Plans*, L.G. Anderson (ed.), Ann Arbor Science, Ann Arbor, Michigan.

Hutchinson, C.E. and Fischer, T.R. (1979), Stochastic control theory applied to fishery management, *Institute of Electrical and Electronic Engineers Transactions on Systems, Man, and Cybernetics*, SMC-**9**(5), 253–9.

Jaquette, L. (1972), A discrete-time population-control model, *Mathematical Biosciences*, **15**, 231–52.

Kennedy, J.O.S. and Watkins, J.W. (1986), Time-dependent quotas for the southern bluefin tuna fishery, *Marine Resource Economics*, **2**(4), 293–313.

Leslie, P.H. (1945), On the use of matrices in certain population mathematics, *Biometrika*, **35**, 183–212.

Levhari, D., Michener, R. and Mirman, L.J. (1981), Dynamic programming models of fishing: Competition, *American Economic Review*, **71**(4), 649–61.

Lewis, E.G. (1942), On the generation and growth of populations, *Sankhya*, **6**, 93–6.

Lewis, T.R. (1975), Optimal resource management under conditions of uncertainty: the case of an ocean fishery, PhD thesis, University of California, San Diego.

Lewis, T.R. (1977), Optimal fishery management under conditions of uncertainty, in *Economic Impacts of Extended Fisheries Jurisdiction*, L.G. Anderson, (ed.), Ann Arbor Science, Ann Arbor, Michigan.

Lord, G.E. (1973), Characterization of the optimum data acquisition and management of a salmon fishery as a stochastic dynamic program, *Fisheries Bulletin*, **71**(4), 1029–37.

Lovejoy, W. (1983), Policy bounds for Markov decision processes with applications to fisheries management, PhD dissertation, University of Delaware.

Lovejoy, W.S. (1984), Decision problems in marine fisheries, *Proceedings of 23rd Conference of Decision and Control*, Las Vegas, Nevada.

Mann, S.H. (1970), A mathematical theory for the harvest of natural animal populations when birth rates are dependent on total population size, *Mathematical Biosciences*, **7**, 97–110.

Mendelssohn, R. (1976), Optimization problems associated with the Leslie matrix, *American Naturalist*, **110**(973), 339–49.

Mendelssohn, R. (1978), Optimal harvesting strategies for stochastic single-species, multiage class models, *Mathematical Biosciences*, **41**, 159–74.

Mendelssohn, R. (1980), A systematic approach to determining mean-variance trade-offs when managing randomly varying populations, *Mathematical Biosciences*, **50**, 75–84.

Mendelssohn, R. (1982), Discount factors and risk aversion in managing random fish populations, *Canadian Journal of Fisheries and Aquatic Sciences*, **39**(9), 1252–7.

Meuriot, E. and Gates, J.M. (1983), Fishing allocations and optimal fees: A single- and multilevel programming analysis, *American Journal of Agricultural Economics*, **65**(4), 711–21.

Munro, G.R. (1979), The optimal management of transboundary renewable resources, *Canadian Journal of Economics*, **12**, 355–76.

Munro, G.R. (1981), The economics of fishing: an introduction, in *Economics and Resources Policy*, J.A. Butlin (ed.), Longman, London.

Munro, G.R. and Scott, A.D. (1984), The economics of fisheries management, Discussion Paper 84-09, Department of Economics, University of British Columbia, to appear in *Handbook of Natural Resource and Energy Economics*, A.V. Kneese and J.L. Sweeney (eds) North-Holland, Amsterdam.

Pindyck, R.S. (1984), Uncertainty in the theory of renewable resource markets, *Review of Economic Studies*, **51**(2), 289–303.

Reed, W.J. (1974), A stochastic model of the economic management of a renewable animal resource, *Mathematical Biosciences*, **22**, 313–37.

Reed, W.J. (1978), The steady-state of a stochastic harvesting model, *Mathematical Biosciences*, **41**, 273–307.

Reed, W.J. (1979), Optimal escapement levels in stochastic and deterministic models, *Journal of Environmental Economics and Management*, **6**, 350–63.

Ricker, W.E. (1954), Stock and recruitment, *Journal of the Fisheries Research Board of Canada*, **11**, 559–623.

Sancho, N.G.F. and Mitchell, C. (1975), Economic optimization in controlled fisheries, *Mathematical Biosciences*, **27**, 1–7.

Schaefer, M.B. (1957), Some considerations of population dynamics and economics in relation to the management of marine fisheries, *Journal of the Fisheries Research Board of Canada*, **14**, 669–81.

Smith, J.B. (1985), A discrete model of replenishable resource management under uncertainty, *Marine Resource Economics*, **1**(3), 283–308.

Smith, W.R. and Silvert, W. (1977), Dynamic programming techniques in fisheries management, *Fisheries Management Project.*, Volume 2, Institute for Environmental Studies, Dalhousie University, Halifax, Nova Scotia.

Spulber, D.F. (1982), Adaptive harvesting of a renewable resource and stable equilibrium, in *Essays in the Economics of Renewable Resources*, L.J. Mirman and D.F. Spulber, (eds.), North-Holland, Amsterdam.

Spulber, D.F. (1985), The multicohort fishery under uncertainty, *Marine Resource Economics*, **1**(3), 265–82.

United States Department of Commerce, NOAA (1978), Implementation of Northern Anchovy fishery management plan, *Federal Register*, **43**(141), Book 2, 31651–879.

Walters, C.J. (1975), Optimal harvest strategies for salmon in relation to environmental variability and uncertain production parameters, *Journal of the Fisheries Research Board of Canada*, **32**(10), 1777–84.

Walters, C.J. (1978), Some dynamic programming applications in fisheries management, in *Dynamic Programming and its Applications*, M.L. Puterman (ed.), Academic Press, New York.

Walters, C.J. and Hilborn, R. (1976), Adaptive control of fishing systems, *Journal of the Fisheries Research Board of Canada*, **33**(1), 145–59.

Waugh, G. (1984), *Fisheries Management: Theoretical Developments and Contemporary Applications*, Westview Press, Boulder, Colorado.

PART V

Conclusion

Chapter 11

THE SCOPE FOR DYNAMIC PROGRAMMING APPLIED TO RESOURCE MANAGEMENT

The basic procedures for solving different types of dynamic programming problem were explained in Part II. This was followed in Parts III and IV by an examination of how some of the problems encountered in agriculture and natural resource industries can be solved by dynamic programming. In this final chapter some reflections are presented on the usefulness of the dynamic programming approach, the extent to which the technique has been applied in agriculture and natural resource industries to date, and the scope for its use in the future.

11.1 Dynamic Programming as a Method of Conceptualizing Resource Problems

Many agricultural economists have noted the need for an operational approach for solving dynamic problems. Johnson and Rausser (1977, p. 164) comment that 'many agricultural economists for a number of years have been busily applying static neoclassical theory to intrinsically dynamic systems'. Heady (1981, p. 38) has suggested 'perhaps conventional optimizing theory was used more widely in recent decades because theory related to time and stochastic phenomena was not yet sufficiently operational'. Hanf and Schiefer (1983, p. 16) note that 'in most operational decision models the time dimension of managerial decisions is not considered adequately'. One of the aims of this book has been to draw together examples of how dynamic programming, since its advocacy by practitioners such as Throsby (1964) and Burt (1965) in the early 1960s, has already been applied to a wide variety of dynamic and stochastic resource problems. In so doing the image of dynamic programming as an arcane or esoteric technique may be rebutted, and its simplicity and practicality demonstrated.

A useful insight into the nature of a dynamic resource problem is obtained just by formulating the basic recursive equation, regardless of whether the problem is solved using dynamic programming. The process of identifying state variables, decision variables, stage return functions and state transformation functions helps to formalize the problem.

To summarize what is gained from the recursive equation, consider once more the equation for the general resource problem introduced in Chapter 1:

$$
\begin{aligned}
V_i\{x_i\} &= \max_{0 \le u_i \le \bar{u}_i} [a_i\{x_i, u_i\} + \alpha V_{i+1}\{x_i + g_i\{x_i\} - h_i\{x_i, u_i\}\}] \\
&= \max_{0 \le u_i \le \bar{u}_i} [z_i\{x_i, u_i\}] \\
&= [a_i\{x_i, u_i^*\} + \alpha V_{i+1}\{x_i + g_i\{x_i\} - h_i\{x_i, u_i^*\}\}] \qquad (i = n, \ldots, 1)
\end{aligned}
\tag{11.1}
$$

where 0 and $\bar{u}_i$ are lower and upper bounds on the control variable u_i, and u_i^* is the optimal value. The value of the resource stock at stage i is given the usual economic imputation of being the capitalized value of the stream of optimal stage returns. Thus (11.1) is consistent with

$$
V_i\{x_i\} = \sum_{j=1}^{n} \alpha^{j-i} a_j\{x_j, u_j^*\}
$$

It can be seen from (11.1) that at any stage i, u_i^* depends simultaneously on the impact of u_i on the current stage return and on the present value of the resulting resource stock at stage $i+1$.

The method of stage maximization is quite open. If the problem is solved numerically, the method may be a search procedure, or it may involve the use of mathematical programming. If the problem is solved analytically, calculus methods may be used. For example, if $z_i\{\cdot\}$ is strictly concave in u_i, the Kuhn–Tucker conditions for a constrained maximum may be used. If $z_i\{\cdot\}$ is strictly convex or linear in u_i, u_i^* equals either 0 or $\bar{u}_i$. The same arguments apply if x_{i+1} is a stochastic function of x_i and u_i, and the expectation operator precedes $V_{i+1}\{x_{i+1}\}$ so that $V_i\{x_i\}$ is the expected resource value at stage i. Examples of analytical results for stochastic problems are derived in Section 6.5.4 for fertilizer carryover, and in Section 10.3.2 for the fishery.

If all relevant functions are differentiable, some general results follow. Differentiating (11.1) with respect to x_i gives

$$\frac{dV_i}{dx_i} = \frac{\partial a_i}{\partial x_i} + \alpha \frac{dV_{i+1}}{dx_{i+1}} \left(1 + \frac{dg_i}{dx_i} - \frac{\partial h_i}{\partial x_i}\right) \qquad (i = 1, \ldots, n) \qquad (11.2)$$

Equation (11.2) can be rearranged to give (4.8) which shows the change in the shadow price of the resource, dV/dx, from stage i to stage $i+1$. For the special case where $z\{\cdot\}$ in (11.1) is strictly concave in u_i and the solution is an interior solution, some interesting results are obtained. In particular, the shadow price of the resource at stage i can be specified in terms of the stage i functions only. Setting the partial derivative of $z_i\{x_i, u_i\}$ with respect to u_i equal to zero gives

$$\alpha \frac{dV_{i+1}}{dx_{i+1}} = \frac{\partial a_i/\partial u_i}{\partial h_i/\partial u_i} \qquad (i = 1, \ldots, n) \qquad (11.3)$$

which on substitution in (11.2) gives

$$\frac{dV_i}{dx_i} = \frac{\partial a_i}{\partial x_i} + \frac{\partial a_i/\partial u_i}{\partial h_i/\partial u_i} \left(1 + \frac{dg_i}{dx_i} - \frac{\partial h_i}{\partial x_i}\right) \qquad (i = 1, \ldots, n) \qquad (11.4)$$

Depending on the type of resource problem and the nature of the stage return and state transformation functions, it may be possible to use (11.3) and (11.4) to deduce how x and u^* change from stage to stage, and how they are affected by the rate of discount.

11.2 Dynamic Programming as a Solution Technique

An important feature of a dynamic programming solution is that optimal decisions are specified for the entire range of possible states which may be reached at each stage. This feature is exploited when state transitions are subject to uncertainty. If the problem is stochastic, and the certainty-equivalence conditions described in Section 4.3.2 hold, then the solution for the equivalent deterministic problem is also optimal for the stochastic problem. The important point is that an optimal decision is known for whatever state is actually reached. Of course, optimal decisions in terms of states are also specified in stochastic solutions. The solutions are closed-loop or feedback policies. Jacobs (1967, p. 99) uses dynamic programming theory to show that the solution to any stochastic control problem must be a feedback policy.

Whilst it was explained in Chapter 4 how quantitative solutions could be obtained analytically for some problems, and an example of optimal fertilizer application was given in Chapter 6, solutions to most problems are determined numerically. In Part II a general purpose computer routine (GPDP) for solving dynamic programming problems numerically was introduced. It was used to obtain solutions to a variety of resource problems in Parts III and IV. This routine and others should help those who want to solve relatively simple problems but are daunted by the prospect of a lot of tedious calculation, or of having to start programming a solution algorithm from scratch. Because for many problems the same procedures must be performed in obtaining a solution, it makes sense to code these procedures so that they can be used for solving different problems. At the same time, because there is no standard type of dynamic programming problem, it is useful to be able to write problem-specific routines to generate the problem data. Routines for generating data for a range of resource problems have been presented in this book. They may be edited to generate problem data for other problems with a similar structure.

Inevitably the structures of some problems are such as to make it inefficient to use the GPDP routine without modification. Some problems of this type were encountered in the applications chapters. For example, in Chapter 7, a problem of feeding and replacing cattle was best formulated with nested maximization at each stage.

11.3 Applications to Date

Although many research applications of dynamic programming to agricultural and natural resource problems have been mentioned in this book, there are few reports of the commercial use of dynamic programming on a day-to-day basis. It is interesting to speculate whether dynamic programming will retain its status as essentially a research technique, or whether it will eventually come to be regarded as an operational technique, helping to determine actual resource management.

One reason for the lack of operational use of dynamic programming may be that it cannot be used to answer the practical questions faced by resource managers. Certainly this is the impression conveyed by some commentators. For example, Martin (1972), in an interesting exchange with Burt (1972), refers to Burt's 1971 paper on range management as 'another article in a continuing series of the possible but impractical uses of dynamic programming'. Day and Sparling (1977, p. 98), in reviewing the use of optimization models in agricultural and resource economics, make this comment about dynamic programming: 'After a first flash of excitement induced by the flexibility and potential of Bellman's approach as typified by Throsby (1964), applications have been limited to rather simple subsystems of total farm systems.' Barnard and Nix (1979, p. 428), in introducing dynamic programming in their book on farm planning and control, state: 'While fascinating intellectually, dynamic programming has little significance for practical farm planning at the present time and so will be dealt with relatively briefly.'

The notion that there are few uses for dynamic programming in practical resource management has to be seriously questioned.[1] Dixon and Howitt (1980) note that 'most empirical natural resource management problems are intertemporal optimization problems under uncertainty'. The same is true of most farm management problems. By definition, intertemporal optimization problems can only be solved by using dynamic optimization techniques. Basically, the techniques available for solving deterministic dynamic problems are: various forms of mathematical programming, such as linear, separable and quadratic programming, with matrices of dated activities; non-linear programming or gradient methods; and dynamic programming. Dynamic programming is a natural technique to choose for solving stochastic dynamic problems. However, the point to be made is not that dynamic programming is the most appropriate technique for solving all resource problems, merely that it can be used for solving many resource problems.

If there are plenty of potential applications of dynamic programming in agricultural and natural resource industries, why is it not used more as a management tool? One reason may be institutional in the case of natural resource industries. Land, forest and fishery resources are often national or social resources. Policy, framed at the government level, often defines the parameters within which private resource users can operate. In such a planning environment dynamic programming results cannot be

[1] The practical relevance of dynamic programming has been queried in other fields of application as well. Larson and Zadeh (1981, p. 986) as editors introduce a special Bellman issue of the Automatic Control transactions of the Institute of Electrical and Electronic Engineers. The issue bears witness to applications of dynamic programming to control problems in many fields (including pest management), but the editors note that 'Bellman's first paper on dynamic programming was turned down by the proceedings of the IRE (the forerunner of the IEEE) because the reviewers felt that his ideas had no demonstrable relevance to practical applications'.

used as an operational tool. However, they can be an important source of advice to government.

In the case of agriculture there do not appear to be institutional constraints of this type to the same extent. Prerequisites for applying dynamic programming which may be lacking are managers' understanding of the technique, relevant data, computer software and computing power. The prospects for an increase in the provision of these resources is considered in the final section.

11.4 Expected Developments

Computing power is becoming increasingly available on farms with the spread of microcomputers. Although computers are being used primarily for book-keeping and taxation purposes, many software packages have been developed to aid management decision-making in the areas of budgeting, control of farm inputs and predicting yields of crop and livestock products. An indication of the interest in the field is the publication of relevant newsletters and journals, such as *Farm Computer News* in Australia and the new international journal *Computers and Electronics in Agriculture*.

It is unrealistic to expect many farm managers to have the time and inclination to invest in any detailed understanding of dynamic programming. However, the continued development of dynamic programming software should make only a general understanding of the scope of dynamic programming necessary in order to apply it.

Perhaps the greatest problem is the lack of relevant data. However, as Burt (1972) has pointed out, data are not collected unless there is a perceived need for them, and the need is often not perceived until models designed to answer practical questions are developed. There is a symbiotic relationship between data available and models used. In practically all the applications of dynamic programming considered in this book there are areas of data shortfall. Such applications help to identify priorities in scientific research.

Some requirements are for data specific to the farm. By using dynamic programming within an adaptive control framework, it is possible to generate farm-specific data. Thus, if a farm-specific parameter in a dynamic programming model is initially unknown, it may be set at some value which is an average across farms. Bayesian or other methods may be used to periodically revise estimates of the parameter. An example application to estimating the yield potential of milking cows was studied in Chapter 7.

An important area of uncertainty in fisheries management which has still to be addressed in dynamic programming models concerns the size of fish stocks. In most applications to agricultural and natural resource industries the state variables are variables which can be measured to a satisfactory degree of accuracy. This cannot be said for applications to fisheries. However, at any point in time it should be possible to estimate a probability distribution of current fish stocks, perhaps dependent on previous harvests relative to previous fishing effort. It would be possible to formulate optimal policies taking account of uncertainty about current fish stocks, drawing on the ideas of Kalman filtering.

The power of dynamic programming as a technique for solving problems numerically is limited by the curse of dimensionality. Some methods of alleviating the restriction on the number of state variables were considered in Chapter 4. However, an important practical question to ask when faced with a problem with potentially many state variables is how much precision is lost by reducing the number of state variables. Can the problem be simplified so that optimal policies for the simplified problem are approximately optimal for the original problem? It was pointed out in Chapter 10 that Lovejoy (1983) has shown that the optimal policy for a highly simplified version of a large problem can bound quite closely the optimal policy for the large problem. Whether this result only holds for special examples or holds for many real-world problems is a question of considerable interest.

There are likely to be further developments in the use of other techniques in conjunction with dynamic programming. The power of dynamic programming to solve some problems can be increased by using a second optimizing technique as a subroutine. It is usually assumed in deterministic problems that there is only one path from a state node at stage i to another node at stage $i+1$. However, it may be that there are many possible transitions from one node to another. An optimizing technique may be used to find the best transition. Linear programming has been used in this way in a number of applications reviewed in previous chapters. For example, linear programming has been used for finding: the profit-maximizing cropping pattern which draws on a given volume of water; the least-cost ration which results in a given weight gain; the maximum annual net revenue from continued use or replacement of a tractor, and the optimal development of fishing fleets which leads to a given change in fish stocks.

Simulation is another technique which will continue to be used advantageously with dynamic programming. Its use is complementary. Models solved by dynamic programming can be formulated more flexibly than models solved by most other techniques. For example, models can be specified with non-linear and stochastic functions, integer variables and constraints on state and decision variables. However, the feasible number of state variables may be limiting, and there are separability conditions on the objective function. There are not even these restrictions on the structure of a simulation model. At the same time, it is not always straightforward to find optimal policies with a simulation model. Simulation has been used in two ways in conjunction with dynamic programming models. One use is to determine state transition probabilities for stochastic dynamic programming models. This has been done for some of the models reported in the summaries of applications, particularly where the state variable is population numbers as in pest and fisheries management. Another use is to evaluate the outcome of dynamic programming policies where the dynamic programming model has been simplified, for example by eliminating state variables, or by making the model deterministic instead of stochastic.

There are areas of application to agricultural and natural resource industries which have not been touched in this book. For example, dynamic programming can be usefully applied to problems in farm financial management. Entering into futures contracts, auction bidding and taking out flood and drought insurance can also be characterized as stochastic multi-stage decision processes. Another area is farm–firm growth under uncertainty. Already there have been applications to these areas and more can be expected. As regards applications to natural resources, the scope is well summarized by Denardo (1978, p. 394) in a closing address at an international conference on dynamic programming and its applications. Speculating on future applications, he singled out computer science, economic growth models and natural resources as promising areas for further developments. On natural resources he said:

> 'The management of natural resources, be they anchovies, elephants, salmon or trees, is a fascinating subject. One faces the need to measure, to control, to aggregate, to devise criteria, and to trade the present with the future. One must face the inherently stochastic nature of the situation. In short, one faces the issues that are the meat of dynamic programming.'

11.5 References

Barnard, C.S. and Nix, J.S. (1979), *Farm Planning and Control*, 2nd edition, Cambridge University Press, Cambridge.

Burt, O.R. (1965), Operations research techniques in farm management: potential contribution, *Journal of Farm Economics*, **47**(5), 1418–26.

Burt, O.R. (1972), More sophisticated tools for less important problems: the history of range improvement research: reply, *American Journal of Agricultural Economics*, **54**(1), 134–5.

Day, R.H. and Sparling, E. (1977), Optimization models in agricultural and resource economics, in *A Survey of Agricultural Economics Literature*, G.G. Judge, R.H. Day, S.R. Johnson, G.C. Rausser and L.R. Martin (eds), Volume 2, University of Minnesota Press, Minneapolis.

Denardo, E.V. (1978), Comments, in *Dynamic Programming and its Applications*, M.L. Puterman (ed.), Academic Press, New York.

Dixon, B.L. and Howitt, R.E. (1980), Resource production under uncertainty: a stochastic control approach to timber harvest scheduling, *American Journal of Agricultural Economics*, **62**(3), 499–507.

Hanf, C.H. and Schiefer, G.W. (eds) (1983), *Planning and Decision in Agribusiness: Principles and Experiences*, Elsevier Scientific, Amsterdam.

Heady, E.O. (1981), Micro-level accomplishments and challenges for the developed world, in *Rural Change: The Challenge for Agricultural Economists*, G. Johnson and A. Maunder (eds), Gower, Westmead, Farnborough, Hampshire.

Jacobs, O.L.R. (1967), *An Introduction to Dynamic Programming*, Chapman and Hall, London.

Johnson, S.R. and Rausser, G.C. (1977), Systems analysis and simulation: a survey of applications in agricultural and resource economics, in *A Survey of Agricultural Economics Literature*, G.G. Judge, R.H. Day, S.R. Johnson, G.C. Rausser and L.R. Martin (eds), Volume 2, University of Minnesota Press, Minneapolis.

Larson, R.E. and Zadeh, L.A. (1981), Foreword: Bellman Special Issue, *Institute of Electrical and Electronic Engineers Transactions on Automatic Control*, **26**(5), 985–6.

Lovejoy, W. (1983), Policy bounds for Markov decision processes with applications to fisheries management, PhD dissertation, University of Delaware.

Martin, W.E. (1972), More sophisticated tools for less important problems: the history of range improvement research: a comment, *American Journal of Agricultural Economics*, **54**(1), 132–4.

Throsby, C.D. (1964), Some dynamic programming models for farm management research, *Journal of Agricultural Economics*, **16**(1), 98–110.

Appendices

Appendix A1

CODING SHEETS FOR ENTERING DATA USING DPD

TABLE A1.1
Coding Sheets Required for Type of problem

Type of problem	*Sheet on page*	*Data blocks*
(1) Deterministic, stationary	310	A and B
	311	CD
	313	D[a]
(2) Deterministic, non-stationary	310	A and B
	311	CD
	313	D[a]
	314	ED
(3) Stochastic, stationary	310	A and B
	312	CS
	313	D[a]
(4) Stochastic, non-stationary	310	A and B
	312	CS
	313	D[a]
	315	ES

[a] Labels—optional

GENERAL PURPOSE DYNAMIC PROGRAMMING (GPDP)
Coding sheet for entering data BLOCKS A and B using DPD

Notes:
(a) Forms designed for problems with *up to 20 states* and *up to 5 decisions.*
(b) For each stage, number all possible states from 1 upwards.

(1) BLOCK A data—problem parameters

Problem name (up to 6 characters)	
Number of decision stages	
Rate of discount (per cent)	
Problem type—D (deterministic) or S (stochastic)	
Stationary problem?—Y (yes) or N (no)	
Labels available?—Y (yes) or N (no)	

(2) BLOCK B data—values V(I) of terminal state I

I, V(I)
1,
2,
3,
4,
5,
6,
7,
8,
9,
10,

I, V(I)
11,
12,
13,
14,
15,
16,
17,
18,
19,
20,

Final entry	0, 0

Next sheet: BLOCK CD data—if problem deterministic
BLOCK CS data—if problem stochastic

1/...

GENERAL PURPOSE DYNAMIC PROGRAMMING (GPDP)

Coding sheet for entering data BLOCK CD using DPD

Notes:
(a) Enter data for the *last* decision stage.
(b) DPD prompts for data entry line by line across decision columns.

(3) BLOCK CD data—next state J and stage return R(I, J)—stage number . . .

Decision 1	*Decision 2*	*Decision 3*	*Decision 4*	*Decision 5*
I, J, R(I, J)	*I, J, R(I, J)*	*I, J, R(I, J)*	*I, J, R(I, J)*	*I, J, R(I, J)*
1, ,	1, ,	1, ,	1, ,	1, ,
2, ,	2, ,	2, ,	2, ,	2, ,
3, ,	3, ,	3, ,	3, ,	3, ,
4, ,	4, ,	4, ,	4, ,	4, ,
5, ,	5, ,	5, ,	5, ,	5, ,
6, ,	6, ,	6, ,	6, ,	6, ,
7, ,	7, ,	7, ,	7, ,	7, ,
8, ,	8, ,	8, ,	8, ,	8, ,
9, ,	9, ,	9, ,	9, ,	9, ,
10, ,	10, ,	10, ,	10, ,	10, ,
11, ,	11, ,	11, ,	11, ,	11, ,
12, ,	12, ,	12, ,	12, ,	12, ,
13, ,	13, ,	13, ,	13, ,	13, ,
14, ,	14, ,	14, ,	14, ,	14, ,
15, ,	15, ,	15, ,	15, ,	15, ,
16, ,	16, ,	16, ,	16, ,	16, ,
17, ,	17, ,	17, ,	17, ,	17, ,
18, ,	18, ,	18, ,	18, ,	18, ,
19, ,	19, ,	19, ,	19, ,	19, ,
20, ,	20, ,	20, ,	20, ,	20, ,

Final entry for this stage 0, 0, 0

Next sheet: BLOCK D data—if labels available
BLOCK ED data—if no labels and problem is non-stationary—otherwise data entry complete

2/. . .

GENERAL PURPOSE DYNAMIC PROGRAMMING (GPDP)

Coding sheet for entering data BLOCK CS using DPD

Notes:
(a) Enter data for the *last* decision stage.
(b) DPD prompts for data entry line by line across columns.
(c) $P(J)$'s summed across each row must equal 1·000 exactly.

(3) BLOCK CS data—next state J with probability P(J) for decision D and associated expected stage return R(I, D)—stage number . . .

$I, D, R(I, D)$	$J, P(J)$	$J, P(J)$	$J, P(J)$	$J, P(J)$	$J, P(J)$
, 1,	,	,	,	,	,
, 2,	,	,	,	,	,
, 3,	,	,	,	,	,
, 4,	,	,	,	,	,
, 5,	,	,	,	,	,
, 1,	,	,	,	,	,
, 2,	,	,	,	,	,
, 3,	,	,	,	,	,
, 4,	,	,	,	,	,
, 5,	,	,	,	,	,
, 1,	,	,	,	,	,
, 2,	,	,	,	,	,
, 3,	,	,	,	,	,
, 4,	,	,	,	,	,
, 5,	,	,	,	,	,
, 1,	,	,	,	,	,
, 2,	,	,	,	,	,
, 3,	,	,	,	,	,
, 4,	,	,	,	,	,
, 5,	,	,	,	,	,

Enter any data for further states I on another BLOCK CS data sheet

Final entry for this stage | 0, 0, 0 |

Next sheet: BLOCK D data—if labels available
BLOCK ES data—if no labels and problem is non-stationary—otherwise data entry complete

GENERAL PURPOSE DYNAMIC PROGRAMMING (GPDP)

Coding sheet for entering data BLOCK D using DPD

Notes:

(a) Names and labels may be up to 6 characters.

(b) Labels for states and decisions must be the same for all stages.

(4) BLOCK D data—labels for each state I and decision D

State name	

Decision name	

I	*Label for state number I*
1	
2	
3	
4	
5	
6	
7	
8	
9	
10	
11	
12	
13	
14	
15	
16	
17	
18	
19	
20	

D	*Label for decision number D*
1	
2	
3	
4	
5	

Next sheet: BLOCK ED or ES data—if problem is non-stationary—otherwise data entry complete

3/. . .

GENERAL PURPOSE DYNAMIC PROGRAMMING (GPDP)

Coding sheet for entering data BLOCK ED using DPD

Notes:
(a) Data for remaining stages are entered in reverse temporal order.
(b) DPD prompts for data entry line by line across decision columns.

(5) BLOCK ED data—next state J and stage return R(I, J)—stage number . . .

Decision 1	*Decision 2*	*Decision 3*	*Decision 4*	*Decision 5*
I, J, R(I, J)	*I, J, R(I, J)*	*I, J, R(I, J)*	*I, J, R(I, J)*	*I, J, R(I, J)*
1, ,	1, ,	1, ,	1, ,	1, ,
2, ,	2, ,	2, ,	2, ,	2, ,
3, ,	3, ,	3, ,	3, ,	3, ,
4, ,	4, ,	4, ,	4, ,	4, ,
5, ,	5, ,	5, ,	5, ,	5, ,
6, ,	6, ,	6, ,	6, ,	6, ,
7, ,	7, ,	7, ,	7, ,	7, ,
8, ,	8, ,	8, ,	8, ,	8, ,
9, ,	9, ,	9, ,	9, ,	9, ,
10, ,	10, ,	10, ,	10, ,	10, ,
11, ,	11, ,	11, ,	11, ,	11, ,
12, ,	12, ,	12, ,	12, ,	12, ,
13, ,	13, ,	13, ,	13, ,	13, ,
14, ,	14, ,	14, ,	14, ,	14, ,
15, ,	15, ,	15, ,	15, ,	15, ,
16, ,	16, ,	16, ,	16, ,	16, ,
17, ,	17, ,	17, ,	17, ,	17, ,
18, ,	18, ,	18, ,	18, ,	18, ,
19, ,	19, ,	19, ,	19, ,	19, ,
20, ,	20, ,	20, ,	20, ,	20, ,

Final entry for this stage: 0, 0, 0

Next sheet: BLOCK ED data—if this sheet is not for stage number 1—otherwise data entry complete

. . ./. . .

GENERAL PURPOSE DYNAMIC PROGRAMMING (GPDP)

Coding sheet for entering data BLOCK ES using DPD

Notes:

(a) Data for remaining stages are entered in reverse temporal order.
(b) DPD prompts for data entry line by line across columns.
(c) $P(J)$'s summed across each row must equal 1·000 exactly.

(5) BLOCK ES data—next state J with probability P(J) for decision D and associated expected stage return R(I, D)—stage number . . .

$I, D, R(I, D)$	$J, P(J)$	$J, P(J)$	$J, P(J)$	$J, P(J)$	$J, P(J)$
, 1,	,	,	,	,	,
, 2,	,	,	,	,	,
, 3,	,	,	,	,	,
, 4,	,	,	,	,	,
, 5,	,	,	,	,	,
, 1,	,	,	,	,	,
, 2,	,	,	,	,	,
, 3,	,	,	,	,	,
, 4,	,	,	,	,	,
, 5,	,	,	,	,	,
, 1,	,	,	,	,	,
, 2,	,	,	,	,	,
, 3,	,	,	,	,	,
, 4,	,	,	,	,	,
, 5,	,	,	,	,	,
, 1,	,	,	,	,	,
, 2,	,	,	,	,	,
, 3,	,	,	,	,	,
, 4,	,	,	,	,	,
, 5,	,	,	,	,	,

Enter any data for further states *I* on another BLOCK ES data sheet

Final entry for this stage | 0, 0, 0 |

Next sheet: BLOCK ES data—if this sheet is not for stage number 1—otherwise data entry complete

. . ./. . .

PROGRAM LISTINGS

(For details of ordering these programs and other files on disk, please refer to the order form at the back of the book).

The GPDP programs have been tested using a VAX-11 BASIC V2.2 compiler and a BASIC interpreter for an IBM PC. Any modifications for running on other systems are likely to be minimal because the programs do not include any of the advanced features of some versions of BASIC.

A2.1 Listing of DPD

```
100 REM    ****************************************************
110 REM    DYNAMIC PROGRAMMING ROUTINE ---DPD---   MK4 JAN 1986
120 REM       COPYRIGHT 1986 BY  J KENNEDY
130 REM    SCHOOL OF ECONOMICS  LA TROBE UNIVERSITY  MELBOURNE
140 REM
150 REM    DPD WRITES PROBLEM DATA TO FILE
160 REM    ****************************************************
170 LET C$=","
180 LET D9=1
190 LET L$="-----------------------------------------------------------"
200 PRINT "PROBLEM NAME ";
210 INPUT N$
220 LET N$=N$+".DAT"
230 OPEN N$ FOR OUTPUT AS #1
240 PRINT
250 REM-----ENTER BLOCK A DATA - PROBLEM PARAMETERS-----
260 PRINT "NO. OF DECISION STAGES ";
270 INPUT N
280 PRINT #1, N
290 PRINT
300 PRINT "RATE OF DISCOUNT - PER CENT";
310 INPUT A2
320 PRINT #1, A2
330 PRINT
340 PRINT "PROBLEM TYPE - D(DETERMINISTIC) OR S(STOCHASTIC) ";
350 INPUT D$
360 PRINT #1, D$
370 PRINT
380 PRINT "ARE STATES,DECISIONS AND RETURNS THE SAME FOR ALL"
390 PRINT "STAGES - Y(YES) OR N(NO)";
```

```
400 INPUT Q$
410 PRINT #1, Q$
420 PRINT
430 PRINT "DO YOU HAVE LABELS FOR STATES AND DECISIONS -"
440 PRINT "(THE SAME FOR ALL STAGES) - Y(YES) OR N(NO) ";
450 INPUT Z$
460 PRINT #1, Z$
470 REM-----ENTER BOCK B DATA - VALUES OF TERMINAL STATES-----
480 PRINT L$
490 PRINT "VALUES OF TERMINAL STATES AT STAGE ";N+1;":"
500 PRINT "YOU WILL BE ASKED TO ENTER I,V(I) FOR ALL I"
510 PRINT "WHERE  I    = TERMINAL STATE NO."
520 PRINT "       V(I) = VALUE OF 'I'"
530 PRINT "ENTER '0,0' AFTER FINAL ENTRY"
540 PRINT L$
550 PRINT "ENTER I,V(I)";
560 INPUT I,F
570 PRINT #1, I;C$,F
580 IF I>0 THEN 550
590 PRINT
600 PRINT L$
610 IF D$="D" THEN 1010
620 REM-----ENTER BLOCKS CS AND ES DATA - (STOCHASTIC)-----
630 PRINT "RETURNS AND TRANSITION PROBABILITIES FOR EACH ";
640 PRINT "POSSIBLE DECISION:"
650 PRINT "YOU WILL BE ASKED TO ENTER I,D,R(I,D) FOR ALL I"
660 PRINT "WHERE  I    = CURRENT STATE NO."
670 PRINT "       D    = DECISION NO."
680 PRINT "    R(I,D) = STAGE RETURN"
690 PRINT "AFTER ENTERING I,D,R(I,D) YOU WILL BE ASKED TO ENTER"
700 PRINT "J,PROB(I,D,J) FOR ALL J"
710 PRINT "WHERE     J    = POSSIBLE STATE NO. REACHED AT NEXT STAGE"
720 PRINT "  PROB(I,D,J) = PROBABILITY OF REACHING J"
730 PRINT "ENTER '0,0,0' AFTER FINAL ENTRY FOR EACH STAGE"
740 PRINT L$
750 FOR T=N TO 1 STEP -1
760    PRINT #1, " T "," I  ";"D   ";"R   "," J   ";"P"
770    LET S=0
780    PRINT "FOR STAGE";T;", ENTER I,D,R(I,D)";
790    INPUT I,D,R
800    IF I=0 THEN 930
810    IF D<=D9 THEN 830
820    LET D9=D
830    LET M=I
840    PRINT "FOR I= ";I;", D= ";D;", ENTER J,PROB(I,D,J)";
850    INPUT J,P
860    PRINT #1, T;C$,I;C$;D;C$;R;C$,J;C$;P
870    LET S=S+P
880    IF S<= .9999 THEN 840
890    IF S<=1.0001 THEN 770
900    PRINT "PROBABILITIES DO NOT SUM TO ONE AT T= ";T;"  I= ";I
910    PRINT "EDIT ";N$;" AFTER ENTERING REMAINING DATA"
920    GO TO 770
930    PRINT #1, T;C$,I;C$;I;C$;I;C$,I;C$;I
940    PRINT
950    IF T<N THEN 990
960    IF Z$="N" THEN 980
970    GOSUB 1350
980    IF Q$="Y" THEN 1000
990 NEXT T
1000 GO TO 1560
1010 REM-----ENTER BLOCKS CD AND ED DATA - (DETEMINISTIC)-----
1020 PRINT "STAGE RETURNS FOR EACH POSSIBLE DECISION:"
1030 PRINT "YOU WILL BE ASKED TO ENTER I,J,R(I,J) FOR ALL I"
1040 PRINT "WHERE  I    = CURRENT STATE NO."
1050 PRINT "       J    = NEXT STATE NO. (DEPENDENT ON DECISION)"
```

```
1060 PRINT "     R(I,J) = STAGE RETURN"
1070 PRINT "ENTER '0,0,0' AFTER FINAL ENTRY FOR EACH STAGE"
1080 PRINT L$
1090 FOR T=N TO 1 STEP -1
1100    PRINT #1,  " T "," I  ";"D   ";"R   "," J   ";"P"
1110    LET P=1
1120    LET M=0
1130    PRINT "FOR STAGE";T;" ENTER I,J,R(I,J)";
1140    INPUT I,J,R
1150    IF I=0 THEN 1250
1160    IF I>M THEN 1210
1170    LET D=D+1
1180    IF D<=D9 THEN 1200
1190    LET D9=D
1200    GO TO 1230
1210    LET D=1
1220    LET M=I
1230    PRINT #1, T;C$,I;C$;D;C$;R;C$,J;C$;P
1240    GO TO 1130
1250    PRINT #1,T;C$,I;C$;I;C$;I;C$;I;C$;I
1260    PRINT
1270    IF T<N THEN 1320
1280    PRINT
1290    IF Z$="N" THEN 1310
1300    GOSUB 1350
1310    IF Q$="Y" THEN 1330
1320 NEXT T
1330 GO TO 1560
1340 REM-----ENTER BLOCK D DATA - LABELS-----
1350 REM---SUBROUTINE---LABELS---
1360 IF Z$="N" THEN 1560
1370 PRINT
1380 PRINT "STATE NAME ";
1390 INPUT X$
1400 PRINT #1, X$
1410 FOR I=1 TO M
1420    PRINT "LABEL FOR STATE NO. ";I;
1430    INPUT Y$
1440    PRINT #1, Y$
1450 NEXT I
1460 PRINT "DECISION NAME ";
1470 INPUT U$
1480 PRINT #1, U$;C$;D9
1490 FOR J=1 TO D9
1500    PRINT "LABEL FOR DECISION NO. ";J;
1510    INPUT V$
1520    PRINT #1, V$
1530 NEXT J
1540 PRINT
1550 RETURN
1560 PRINT
1570 PRINT  "*** DATA WRITTEN TO ";N$;" ***"
1580 CLOSE #1
1590 END
```

A2.2 Listing of FDP

```
100 REM    **************************************************
110 REM    DYNAMIC PROGRAMMING ROUTINE ---FDP---   MK4 JAN 1986
120 REM       COPYRIGHT 1986 BY  J KENNEDY
130 REM    SCHOOL OF ECONOMICS  LA TROBE UNIVERSITY  MELBOURNE
140 REM
150 REM    FDP SOLVES THE FINITE-STAGE PROBLEM
160 REM    **************************************************
```

```
170 PRINT "-----FINITE STAGE DP PROBLEM-----"
180 REM-----THE CURRENT MAXIMUM SIZE OF PROBLEM
190 REM-----IS GIVEN BY VALUES OF
200 REM-----S (MAXIMUM NO. OF STATES AT ANY STAGE)
210 REM-----G (MAXIMUM NO. OF DECISIONS AT ANY STATE)
220 REM-----C (MAXIMUM NO. OF POSSIBLE FOLLOWING STATES FROM ANY DECISION)
230 REM-----N (MAXIMUM NO. OF DECISION STAGES)
240 REM-----SHOWN IN LINES 310 TO 350 AS
250 REM-----LINE 310   A(S,N+1),B(S,N+1),D(S,N+1)
260 REM-----LINE 320   F(S,N+1),Y$(S)
270 REM-----LINE 330   I(S*G*C),J(S*G*C),P(S*G*C)
280 REM-----LINE 340   R(S,G),E(G),V$(G)
290 REM-----LINE 350   Q(N+1),U(N+1),V(N+1),W(N+1)
300 REM-----FOR LARGER PROBLEMS,CHANGE LINES 310 TO 350
310 DIM A( 20 , 11 ),B( 20 , 11 ),D( 20 , 11 )
320 DIM F( 20 , 11 ),Y$( 20 )
330 DIM I( 2000 ),J( 2000 ),P( 2000 )
340 DIM R( 20 , 10 ),E( 10 ),V$( 10 )
350 DIM Q( 11 ),U( 11 ),V( 11 ),W( 11 )
360 LET L$="------------------------------------------------------------"
370 PRINT "PROBLEM NAME";
380 INPUT M$
390 LET N$=M$+".DAT"
400 REM----------INPUT DATA FROM FILE----------
410 OPEN N$ FOR INPUT AS #1
420 REM-----READ BLOCK A DATA - PROBLEM PARAMETERS-----
430 INPUT #1, N
440 LET N1=N+1
450 REM----------READ PROBLEM PARAMETERS----------
460 INPUT #1, A2
470 LET D1=1/(1+A2/100)
480 INPUT #1, D$
490 INPUT #1, Q$
500 INPUT #1, Z$
510 REM----------PRINT PROBLEM PARAMETERS----------
520 PRINT
530 PRINT
540 PRINT L$
550 IF D$="S" THEN 580
560 PRINT "      DETERMINISTIC DYNAMIC PROGRAMMING SOLUTION"
570 GO TO 590
580 PRINT "        STOCHASTIC DYNAMIC PROGRAMMING SOLUTION"
590 PRINT "                   PROBLEM ---";M$;"---"
600 PRINT
610 PRINT "PROBLEM PARAMETERS:"
620 PRINT "NO. OF DECISION STAGES        = ";N
630 PRINT "RATE OF DISCOUNT (PER CENT)   = ";A2
640 PRINT L$
650 LET C1=.9999
660 LET C0=1-C1
670 REM-----READ BLOCK B DATA - VALUES OF TERMINAL STATES-----
680 INPUT #1,I2,F2
690 IF I2=0 THEN 720
700 LET F(I2,1)=F2
710 GO TO 680
720 REM----------BACKWARD RECURSION STARTS----------
730 REM-----READ BLOCKS C AND E DATA-----
740 LET N2=2
750 INPUT #1, T$
760 LET K=1
770 INPUT #1, T2,I(K),D2,R2,J(K),P(K)
780 IF I(K)=0 THEN 830
790 IF P(K)<=C0 THEN 770
800 LET R(I(K),D2)=R2
810 LET K=K+1
820 GO TO 770
```

```
830 IF N2>2 THEN 940
840 IF Z$="N" THEN 940
850 REM-----READ BLOCK D DATA - LABELS-----
860 INPUT #1, X$
870 FOR I9=1 TO I(K-1)
880    INPUT #1, Y$(I9)
890 NEXT I9
900 INPUT #1, U$,D9
910 FOR J9=1 TO D9
920    INPUT #1, V$(J9)
930 NEXT J9
940 REM-----ASSIGN RETURNS FOR ALL DECISIONS-----
950 LET K=0
960 LET L=0
970 LET S=0
980 LET L=L+1
990 LET E(L)=R(I(K+1),L)
1000 LET K=K+1
1010 LET E(L)=E(L)+D1*P(K)*F(J(K),N2-1)
1020 LET S=S+P(K)
1030 IF S<C1 THEN 1000
1040 IF I(K+1)=0 THEN 1060
1050 IF P(K+1)=0 THEN 1000
1060 LET I2=I(K)
1070 IF I(K)=I(K+1) THEN 970
1080 REM-----FIND OPTIMAL DECISION-----
1090 LET K9=1
1100 IF L=1 THEN 1150
1110 FOR L9=2 TO L
1120    IF E(L9)<=E(K9) THEN 1140
1130    LET K9=L9
1140 NEXT L9
1150 LET D(I2,N2)=K9
1160 LET A(I2,N2)=R(I2,K9)
1170 LET B(I2,N2)=J(K-L+K9)
1180 LET F(I2,N2)=E(K9)
1190 IF I(K+1)>0 THEN 960
1200 IF D$="S" THEN 1620
1210 IF N2=N1 THEN 1250
1220 LET N2=N2+1
1230 IF Q$="Y" THEN 950
1240 GO TO 750
1250 REM----------PRINT RESULTS FOR DETERMINISTIC PROBLEM----------
1260 PRINT
1270 PRINT "OPTIMAL STATE SEQUENCE FOR ALL INITIAL STATES"
1280 PRINT L$
1290 PRINT TAB(2*N1-2); "STAGE";TAB(9+4*N1);"VALUE"
1300 FOR T=1 TO N1
1310    PRINT T;TAB(1+4*T);
1320 NEXT T
1330 PRINT
1340 PRINT L$
1350 FOR I2=1 TO I(K)
1360    GOSUB 1780
1370    FOR T=1 TO N1
1380       PRINT Q(T);TAB(1+4*T);
1390    NEXT T
1400    PRINT TAB(8+4*N1);W(1)
1410 NEXT I2
1420 PRINT L$
1430 PRINT
1440 PRINT
1450 PRINT "DETAILS OF OPTIMAL PATH"
1460 PRINT "ENTER INITIAL STATE NO. (OR '0' TO FINISH) ";
1470 INPUT I2
1480 IF I2=0 THEN 1900
```

```
1490 PRINT
1500 PRINT L$
1510 PRINT "STAGE    STATE       DECISION       DISCOUNTED       VALUE"
1520 PRINT " NO.   -----------   -----------    STAGE RETURN"
1530 PRINT "        NO. ";X$;TAB(21);"NO. ";U$
1540 PRINT L$
1550 GOSUB 1780
1560 FOR T=1 TO N1
1570    PRINT TAB(1);T;TAB(7);Q(T);TAB(11);Y$(Q(T));TAB(21);U(T);
1580    PRINT TAB(25);V$(U(T));TAB(36);V(T);TAB(51);W(T)
1590 NEXT T
1600 PRINT L$
1610 GO TO 1430
1620 REM----------PRINT RESULTS FOR STOCHASTIC PROBLEM----------
1630 LET T=N+2-N2
1640 PRINT
1650 PRINT "STOCHASTIC SOLUTION FOR STAGE ";T
1660 PRINT L$
1670 PRINT "   STATE        DECISION       STAGE            EXPECTED"
1680 PRINT "-----------     -----------";TAB(31);"RETURN           VALUE"
1690 PRINT " NO. ";X$;TAB(16);"NO. ";U$
1700 PRINT L$
1710 FOR I2=1 TO I(K)
1720    PRINT TAB(1);I2;TAB(5);Y$(I2);TAB(16);D(I2,N2);TAB(20);V$(D(I2,N2));
1730    PRINT TAB(31);A(I2,N2);TAB(46);F(I2,N2)
1740 NEXT I2
1750 PRINT L$
1760 IF N2=N1 THEN 1900
1770 GO TO 1220
1780 REM---SUBROUTINE---FORWARD TRACKING---
1790 LET Q(1)=I2
1800 FOR N2=N1 TO 2 STEP -1
1810    LET T=N+2-N2
1820    LET U(T)=D(Q(T),N2)
1830    LET V(T)=A(Q(T),N2)*D1^(T-1)
1840    LET W(T)=F(Q(T),N2)
1850    LET Q(T+1)=B(Q(T),N2)
1860 NEXT N2
1870 LET W(N1)=F(Q(N1),1)
1880 LET V(N1)=W(N1)*D1^(T-1)
1890 RETURN
1900 END
```

A2.3 Listing of IDP

```
100 REM   ****************************************************
110 REM   DYNAMIC PROGRAMMING ROUTINE ---IDP---   MK4 JAN 1986
120 REM   COPYRIGHT 1986 BY  J KENNEDY
130 REM   SCHOOL OF ECONOMICS  LA TROBE UNIVERSITY  MELBOURNE
140 REM
150 REM   IDP SOLVES THE INFINITE-STAGE PROBLEM
160 REM   ****************************************************
170 PRINT "-----INFINITE STAGE DP PROBLEM-----"
180 REM-----THE CURRENT SIZE OF PROBLEM
190 REM-----IS GIVEN BY VALUES OF
200 REM-----S (MAXIMUM NO. OF STATES AT ANY STAGE)
210 REM-----G (MAXIMUM NO. OF DECISIONS AT ANY STATE)
220 REM-----C (MAXIMUM NO. OF POSSIBLE FOLLOWING STATES FROM ANY STATE)
230 REM-----SHOWN IN LINES 300 TO 340 AS
240 REM-----LINE 300   G(S,S+1),P1(S,S)
250 REM-----LINE 310   R(S,G),Z(S,G),E(G),V$(G)
260 REM-----LINE 320   D(S),Y(S),H(S)
270 REM-----LINE 330   U(S),Y$(S)
```

```
280 REM-----LINE 340   I(S*G*C),J(S*G*C),P(S*G*C)
290 REM-----FOR LARGER PROBLEMS CHANGE LINES 300 TO 340
300 DIM G( 20 , 21 ),P1( 20 , 20 )
310 DIM R( 20 , 10 ),Z( 20 , 10 ),E( 10 ),V$( 10 )
320 DIM D( 20 ),Y( 20 ),H( 20 )
330 DIM U( 20 ),Y$( 20 )
340 DIM I( 2000 ),J( 2000 ),P( 2000 )
350 LET L$="-------------------------------------------------------"
360 LET L2$="-----------"
370 LET L3$="----"
380 PRINT "PROBLEM NAME ";
390 INPUT M$
400 LET N$=M$+".DAT"
410 REM----------INPUT DATA FROM FILE----------
420 OPEN N$ FOR INPUT AS #1
430 REM-----READ BLOCK A DATA - PROBLEM PARAMETERS-----
440 INPUT #1, N3
450 INPUT #1, A2
460 LET D1=1/(1+A2/100)
470 INPUT #1, D$
480 INPUT #1, Q$
490 IF Q$="Y" THEN 530
500 PRINT "NO INFINITE-STAGE SOLUTION FOR NON-STATIONARY"
510 PRINT "PROBLEM. RUN TERMINATED."
520 GO TO 3810
530 INPUT #1, Z$
540 REM----------PRINT PROBLEM PARAMETERS----------
550 PRINT
560 PRINT
570 PRINT L$
580 IF D$="S" THEN 610
590 PRINT "        DETERMINISTIC DYNAMIC PROGRAMMING SOLUTION"
600 GO TO 620
610 PRINT "         STOCHASTIC DYNAMIC PROGRAMMING SOLUTION"
620 PRINT "                  PROBLEM ---";M$;"---"
630 PRINT
640 PRINT "PROBLEM PARAMETERS:"
650 PRINT "INFINITE NO. OF STAGES"
660 PRINT "RATE OF DISCOUNT (PER CENT)  = ";A2
670 PRINT L$
680 LET C1=.9999
690 LET C0=1-C1
700 REM-----READ BLOCK B DATA - VALUES OF TERMINAL STATES-----
710 INPUT #1, I2,B2
720 IF I2=0 THEN 740
730 GO TO 710
740 REM-----READ BLOCK C DATA-----
750 LET Z3=1
760 LET K=1
770 INPUT #1, T$
780 INPUT #1, T2,I(K),D2,R2,J(K),P(K)
790 IF I(K)=0 THEN 880
800 IF P(K)<=C0 THEN 780
810 LET Z2=I(K)-D2
820 IF Z2=Z3 THEN 840
830 LET Z(I(K),D2)=K
840 LET R(I(K),D2)=R2
850 LET Z3=Z2
860 LET K=K+1
870 GO TO 780
880 LET M1=I(K-1)
890 IF Z$="N" THEN 990
900 REM-----READ BLOCK D DATA - LABELS-----
910 INPUT #1, X$
920 FOR I9=1 TO M1
930    INPUT #1, Y$(I9)
```

```
940 NEXT I9
950 INPUT #1, U$,D9
960 FOR J9=1 TO D9
970    INPUT #1, V$(J9)
980 NEXT J9
990 REM----------INITIAL GUESS AT D VECTOR----------
1000 FOR I9=1 TO M1
1010    LET Y(I9)=0
1020    LET D(I9)=1
1030    FOR J9=1 TO M1
1040       LET P1(I9,J9)=0
1050    NEXT J9
1060 NEXT I9
1070 LET C=0
1080 REM----------SET UP G MATRIX GIVEN D VECTOR----------
1090 FOR I2=1 TO M1
1100    FOR J2=1 TO M1
1110       LET G(I2,J2)=0
1120    NEXT J2
1130    LET G(I2,I2)=1
1140 NEXT I2
1150 LET C=C+1
1160 FOR I2=1 TO M1
1170    LET S=0
1180    LET K2=Z(I2,D(I2))
1190    LET G(I2,J(K2))=G(I2,J(K2))-D1*P(K2)
1200    IF G(I2,J(K2))<>0 THEN 1210
1210    LET S=S+P(K2)
1220    IF S>=C1 THEN 1250
1230    LET K2=K2+1
1240    GO TO 1190
1250    LET H(I2)=R(I2,D(I2))
1260 NEXT I2
1270 IF D1<1 THEN 1320
1280 REM----------MATRIX SET FOR NO DISCOUNTING----------
1290 FOR I2=1 TO M1
1300    LET G(I2,1)=1
1310 NEXT I2
1320 GOSUB 3080
1330 IF S9=0 THEN 1370
1340    PRINT "NO SOLUTION WITH THIS ALGORITHM"
1350    PRINT "  TRY DISCOUNTING ?"
1360    GO TO 3810
1370 IF D1<1 THEN 1400
1380 LET V=U(1)
1390 LET U(1)=0
1400 REM----------FIND NEW OPTIMAL D VECTOR----------
1410 LET K=0
1420 FOR I2=1 TO M1
1430    LET L=0
1440    LET S=0
1450    LET L=L+1
1460    LET E(L)=R(I(K+1),L)
1470    LET K=K+1
1480    LET E(L)=E(L)+D1*P(K)*U(J(K))
1490    LET S=S+P(K)
1500    IF S<C1 THEN 1470
1510    LET I2=I(K)
1520    IF I(K)=I(K+1) THEN 1440
1530    LET K9=1
1540    IF L=1 THEN 1590
1550    FOR L9=2 TO L
1560       IF E(L9)<=E(K9) THEN 1580
1570       LET K9=L9
1580    NEXT L9
1590    LET Y(I2)=K9
```

```
1600     IF I2>1 THEN 1620
1610     LET U1=E(K9)
1620 NEXT I2
1630 REM----------TEST FOR CHANGE IN D VECTOR----------
1640 FOR I2=1 TO M1
1650     IF D(I2)<>Y(I2) THEN 1680
1660 NEXT I2
1670 GO TO 1740
1680 FOR I9=1 TO M1
1690     LET D(I9)=Y(I9)
1700 NEXT I9
1710 IF C<15 THEN 1080
1720 PRINT "SOLUTION NOT FOUND IN ";C;" ITERATIONS"
1730 GO TO 1780
1740 REM----------PRINT RESULTS----------
1750 LET P$="?"
1760 PRINT
1770 PRINT "NO. OF ITERATIONS TO FIND SOLUTION = ";C
1780 PRINT
1790 PRINT
1800 IF D1<1 THEN 1850
1810 LET R$="RELATIVE"
1820 GOSUB 2780
1830 PRINT "      OPTIMAL AVERAGE STAGE RETURN = ";V
1840 GO TO 1870
1850 LET R$="PRESENT"
1860 GOSUB 2780
1870 IF D$="S" THEN 2280
1880 REM---FOR DETERMINISTIC PROBLEM
1890 REM   PRINT OPTIMAL STAGE CYCLE---
1900 LET M2=M1+1
1910 PRINT
1920 PRINT "OPTIMAL STAGE CYCLE"
1930 PRINT "ENTER INITIAL STATE NO. (OR '0' TO FINISH) ";
1940 INPUT I3
1950 IF I3=0 THEN 3810
1960 FOR N=0 TO M2
1970     LET P(N)=0
1980 NEXT N
1990 LET I8=1
2000 PRINT L2$;L2$;L3$;L2$
2010 PRINT "  STAGE        STATE         DECISION"
2020 PRINT "   NO.      ";L2$;"     ";L2$
2030 PRINT "            ";" NO. ";X$;TAB(27);"NO. ";U$
2040 PRINT L2$;L2$;L3$;L2$
2050 LET P(I3)=P(I3)+I8
2060 IF I8=1 THEN 2080
2070 PRINT TAB(4);"1";TAB(12);I3;TAB(16);Y$(I3);TAB(27);D(I3);TAB(3
2080 LET Q1=I3
2090 FOR N=2 TO M2
2100     LET Q2=J(Z(Q1,D(Q1)))
2110     LET P(Q2)=P(Q2)+I8
2120     IF P(Q2)<>2 THEN 2150
2130     LET I8=2
2140     GO TO 2050
2150     IF I8=1 THEN 2220
2160     IF P(Q2)<>4 THEN 2180
2170     PRINT
2180     PRINT TAB(3);N;TAB(12);Q2;TAB(16);Y$(Q2);TAB(27);D(Q2);
2190     PRINT TAB(31);V$(D(Q2))
2200     IF P(Q2)<>6 THEN 2220
2210     GO TO 2240
2220     LET Q1=Q2
2230 NEXT N
2240 PRINT L2$;L2$;L3$;L2$
2250 GO TO 1910
```

```
2260 REM---FOR STOCHASTIC PROBLEM
2270 REM   PRINT LONG-RUN PROBABILITIES---
2280 PRINT
2290 PRINT "REQUIRE LONG-RUN STATE PROBABILITIES - Y(YES) OR N(NO)";
2300 INPUT Q2$
2310 IF Q2$="N" THEN 3810
2320 REM---FOR OPTIMAL POLICY
2330 REM   DETERMINE PROBABILITY MATRIX---
2340 PRINT
2350 FOR I2=1 TO M1
2360    LET S=0
2370    LET K2=Z(I2,D(I2))
2380    LET P1(I2,J(K2))=P1(I2,J(K2))+P(K2)
2390    LET S=S+P(K2)
2400    IF S>=C1 THEN 2430
2410    LET K2=K2+1
2420    GO TO 2380
2430    IF P1(I2,I2)>=C1 THEN 2650
2440 NEXT I2
2450 IF S9=1 THEN 2670
2460 FOR I2=1 TO M1
2470    LET P1(I2,I2)=P1(I2,I2)-1
2480    FOR J2=1 TO M1
2490       LET G(I2,J2)=-P1(J2,I2)
2500    NEXT J2
2510    LET G(1,I2)=1
2520    LET H(I2)=0
2530 NEXT I2
2540 LET H(1)=1
2550 GOSUB 3080
2560 IF S9=0 THEN 2630
2570 FOR I2=1 TO M1
2580    FOR J2=1 TO M1
2590       LET P1(I2,J2)=0
2600    NEXT J2
2610 NEXT I2
2620 GO TO 2350
2630 GOSUB 2970
2640 GO TO 3810
2650 PRINT "STATE NO.";I2;"(";Y$(I2);") IS AN ABSORBING STATE"
2660 REM---IF LONG-RUN PROBABILITIES DEPEND ON INITIAL STATE---
2670 GOSUB 3590
2680 PRINT
2690 PRINT "ENTER INITIAL STATE FOR STATE PROBABILITIES"
2700 PRINT "AFTER 32 DECISION STAGES (OR '0' TO FINISH)";
2710 INPUT N2
2720 IF N2=0 THEN 3810
2730 FOR I2=1 TO M1
2740    LET U(I2)=G(N2,I2)
2750 NEXT I2
2760 GOSUB 2970
2770 GO TO 2680
2780 REM---SUBROUTINE---PRINT RESULTS---
2790 PRINT L$
2800 PRINT "  CURRENT       DECISION          NEXT";TAB(46);R$
2810 PRINT "   STATE";TAB(31);"STATE              VALUE"
2820 PRINT "-----------   -----------   -----------"
2830 PRINT " NO. ";X$;TAB(15);"NO. ";U$;TAB(29);"NO. ";X$
2840 PRINT L$
2850 FOR I2=1 TO M1
2860    IF D$="D" THEN 2900
2870    PRINT TAB(1);I2;TAB(5);Y$(I2);TAB(15);D(I2);TAB(19);V$(D(I2));TAB(30);
2880    PRINT P$;TAB(46);U(I2)
2890    GO TO 2930
2900    LET Q2=J(Z(I2,D(I2)))
2910    PRINT TAB(1);I2;TAB(5);Y$(I2);TAB(15);D(I2);TAB(19);V$(D(I2));TAB(29);
```

```
2920    PRINT Q2;TAB(33);Y$(Q2);TAB(46);U(I2)
2930 NEXT I2
2940 PRINT L$
2950 PRINT
2960 RETURN
2970 REM---SUBROUTINE---PRINT LONG-RUN PROBABILITIES---
2980 PRINT L2$;L2$;L2$
2990 PRINT "   STATE        PROBABILITY"
3000 PRINT L2$
3010 PRINT " NO. ";X$
3020 PRINT L2$;L2$;L2$
3030 FOR I2=1 TO M1
3040    PRINT TAB(1);I2;TAB(5);Y$(I2);TAB(17);U(I2)
3050 NEXT I2
3060 PRINT L2$;L2$;L2$
3070 RETURN
3080 REM---SUBROUTINE---SIMULTANEOUS EQUATIONS---
3090 REM  FIND U WHICH SOLVES  G*U=H
3100 REM  WHERE G IS M1*M1, U IS M1, AND H IS M1
3110 REM  BY GAUSSIAN ELIMINATION
3120 REM-------------------------------------------
3130 REM
3140 LET S9=0
3150 REM-----AUGMENT G MATRIX-----
3160 LET M2=M1+1
3170 FOR J2=1 TO M1
3180    LET G(J2,M2)=H(J2)
3190 NEXT J2
3200 REM
3210 FOR I2=1 TO M1
3220 REM-----PARTIAL PIVOTING-----
3230    LET I3=I2+1
3240    LET G4=G(I2,I2)
3250    FOR I4=I3 TO M1
3260       IF G(I4,I2)<=G4 THEN 3290
3270       LET G4=G(I4,I2)
3280       LET L4=I4
3290    NEXT I4
3300    IF G4<1E-06 THEN 3560
3310    IF G4=G(I2,I2) THEN 3370
3320    FOR J2=I2 TO M2
3330       LET H(J2)=G(I2,J2)
3340       LET G(I2,J2)=G(L4,J2)
3350       LET G(L4,J2)=H(J2)
3360    NEXT J2
3370 REM-----GUASSIAN ELIMINATION-----
3380    FOR I4=I3 TO M1
3390       LET M9=G(I4,I2)/G(I2,I2)
3400       FOR J2=I2 TO M2
3410          LET G(I4,J2)=G(I4,J2)-G(I2,J2)*M9
3420       NEXT J2
3430    NEXT I4
3440 NEXT I2
3450 REM-----BACK SUBSTITUTION-----
3460 FOR I2=M1 TO 1 STEP -1
3470    LET S=G(I2,M2)
3480    IF I2=M1 THEN 3530
3490    LET J3=I2+1
3500    FOR J4=J3 TO M1
3510       LET S=S-G(I2,J4)*U(J4)
3520    NEXT J4
3530    LET U(I2)=S/G(I2,I2)
3540 NEXT I2
3550 GO TO 3580
3560 REM---MATRIX SINGULAR---
3570 LET S9=1
3580 RETURN
3590 REM---SUBROUTINE---LONG-RUN PROBABILITIES
```

```
3600 REM                     DEPENDING ON INITIAL STATE---
3610 LET I5=0
3620 FOR I2=1 TO M1
3630    FOR J2=1 TO M1
3640       LET S=0
3650       FOR K2=1 TO M1
3660          LET S=S+P1(I2,K2)*P1(K2,J2)
3670       NEXT K2
3680       LET G(I2,J2)=S
3690    NEXT J2
3700 NEXT I2
3710 LET I5=I5+1
3720 IF I5=5 THEN 3790
3730 FOR I2=1 TO M1
3740    FOR J2=1 TO M1
3750       LET P1(I2,J2)=G(I2,J2)
3760    NEXT J2
3770 NEXT I2
3780 GO TO 3620
3790 RETURN
3800 REM-----------------------------------------------------
3810 END
```

A2.4 Listing of DIM

```
100 REM   *****************************************************
110 REM   DYNAMIC PROGRAMMING ROUTINE ---DIM---   MK4 JAN 1986
120 REM   COPYRIGHT 1986 BY J KENNEDY
130 REM   SCHOOL OF ECONOMICS  LA TROBE UNIVERSITY  MELBOURNE
140 REM
150 REM   DIM WRITES DIMENSION STATEMENTS TO FILE
160 REM   *****************************************************
170 OPEN "DIMFDP.BAS" FOR OUTPUT AS #1
180 OPEN "DIMIDP.BAS" FOR OUTPUT AS #2
190 PRINT "THIS PROGRAM WRITES FILES WHICH CAN BE USED TO ADJUST THE"
200 PRINT "SIZES OF ARRAYS IN FDP AND IDP TO THE SIZE OF YOUR PROBLEM"
210 PRINT
220 PRINT "MAXIMUM NO. OF STATES AT ANY STAGE";
230 INPUT S
240 PRINT "MAXIMUM NO. OF DECISIONS AT ANY STATE";
250 INPUT G
260 PRINT "MAXIMUM NO. OF POSSIBLE FOLLOWING STATES FROM ANY DECISION";
270 INPUT C
280 PRINT "MAXIMUM NO. OF DECISION STAGES";
290 INPUT N
300 LET C$=","
310 LET N1=N+1
320 LET S2=S*G*C
330 REM----------REDIMENSIONING FOR FDP (#1)----------
340 PRINT #1, "310 DIM A(";S;C$;N1;"),B(";S;C$;N1;"),D(";S;C$;N1;")"
350 PRINT #1, "320 DIM F(";S;C$;N1;"),Y$(";S;")"
360 PRINT #1, "330 DIM I(";S2;"),J(";S2;"),P(";S2;")"
370 PRINT #1, "340 DIM R(";S;C$;G;"),E(";G;"),V$(";G;")"
380 PRINT #1, "350 DIM Q(";N1;"),U(";N1;"),V(";N1;"),W(";N1;")"
390 REM----------REDIMENSIONING FOR IDP (#2)----------
400 PRINT #2, "300 DIM G(";S;C$;S+1;"),P1(";S;C$;S;")"
410 PRINT #2, "310 DIM R(";S;C$;G;"),Z(";S;C$;G;"),E(";G;")";
420 PRINT #2, ",V$(";G;")"
430 PRINT #2, "320 DIM D(";S;"),Y(";S;"),H(";S;")"
440 PRINT #2, "330 DIM U(";S;"),Y$(";S;")"
450 PRINT #2, "340 DIM I(";S2;"),J(";S2;"),P(";S2;")"
460 CLOSE #1,2
470 PRINT
480 PRINT "REDIMENSIONED ARRAY STATEMENTS ARE ON FILES ";
490 PRINT "DIMFDP.BAS AND DIMIDP.BAS "
500 END
```

AUTHOR INDEX

Numbers in italic type indicate those pages on which references are given in full

SUBJECT INDEX

The GPDP programs and other files compiled by the author as described in this book are available on disk for IBM-PC, most CP/M formats, and many other formats.

Disks will be sent airmail, price £28 including postage and local taxes where applicable. Disk contents:

GPDP routines: DPD.BAS, FDP.BAS, IDP.BAS, DIM.BAS
Example programs for generating data files:
IRDF.BAS, IRSF.BAS, IRSI.BAS, STORE.BAS, BEEF.BAS, POL.BAS, FOREST.BAS, TUNA.BAS
Example data files: ROUTE.DAT, CISF.DAT, CIDI.DAT.

Programs are copyright of the author; single back-up copies may be made by users. Orders should be sent directly to the distributors:

Elsevier Applied Science Publishers
Crown House
Linton Road
Barking
Essex IG11 8JU
UK

Owing to the large number of possible formats, it is not possible to supply disks to bookstores; all orders must be sent direct to the address above.

It is essential that you state full details of the disk size and format required when ordering. Please note that the disks are not accompanied by manuals; users should refer to *Dynamic Programming: Applications to Agriculture and Natural Resources* for details of program use.

To: Elsevier Applied Science Publishers

Please supply .... disk(s) of GPDP and other files for *Dynamic Programming: Applications to Agriculture and Natural Resources* by John O. S. Kennedy.

Disk size and format required: ....

Payment can be made in the following manner at £28 per disk.

☐ Cheque enclosed for .... ☐ Please send invoice

☐ Charge my credit card

Mastercard/Access ☐ Visa/Barclaycard ☐ American Express ☐

Card No. .... Issuing Bank No. ....

Signature .... Expiry date ....

PLEASE PRINT

Name .... Position ....

Organisation ....

Address ....

....

Post/Zip Code .... Country ....